Ole Valente Mortensen
Editor

Monoamine Transporters

Springer

Editor
Ole Valente Mortensen
Department of Pharmacology and Physiology
Drexel University College of Medicine
Philadelphia, PA, USA

ISSN 2190-5215 ISSN 2190-5223 (electronic)
Advances in Neurobiology
ISBN 978-3-031-96363-6 ISBN 978-3-031-96364-3 (eBook)
https://doi.org/10.1007/978-3-031-96364-3

This Springer imprint is published by the registered company Springer Nature Switzerland AG
The registered company address is: Gewerbestrasse 11, 6330 Cham, Switzerland

Contents

About the Editor

Ole Valente Mortensen, PhD, is Associate Professor in the Department of Pharmacology & Physiology at Drexel University College of Medicine, Philadelphia, Pennsylvania, USA. He is also Director of the Pharmacology & Physiology Graduate Program. He has a long-standing research interest in how monoamine transporters are involved in modulating monoamine signaling and associated behaviors. He obtained an MS degree from Tokyo Institute of Technology in Japan and a PhD degree from Aarhus University in Denmark. He continued his postdoctoral training under the guidance of Dr. Susan Amara at the Vollum Institute in Portland, Oregon. He was a research assistant professor at the University of Pittsburgh School of Medicine before he joined Drexel University in 2009.

About the Editor

Pharmacological Mechanisms of SLC6 Neurotransmitter Transporter Inhibition

Dipanjana Bandyopadhyay, Smruti Ranjan Nayak, and Aravind Penmatsa

Contents

Abstract Solute carrier 6 family members comprise of neurotransmitter sodium symporters that are the transporters primarily involved in the reuptake of released neurotransmitters from the synaptic space. The family includes structurally related transporters involved in biogenic amine and amino acid neurotransmitter uptake and is a well-known therapeutic target for several classes of inhibitors for the treatment of ailments ranging from depression, pain, addiction, seizures and anxiety. Inhibition of NSS transporters can work either through competitive inhibition or through allosteric inhibition at diverse sites that target multiple conformational states of the transporters. This chapter explores the diverse inhibition strategies observed with numerous inhibitors that target this class of transporters and potential for improvements in inhibition strategies.

Keywords Solute carrier 6 (SLC6) · Neurotransmitter sodium symporters (NSS) · Reuptake inhibitors · Competitive inhibition · Allosteric inhibitors

Dipanjana Bandyopadhyay and Smruti Ranjan Nayak contributed equally with all other contributors.

D. Bandyopadhyay · S. R. Nayak · A. Penmatsa (✉)
Molecular Biophysics Unit, Indian Institute of Science, Bangalore, India
e-mail: penmatsa@iisc.ac.in

O. V. Mortensen (ed.), *Monoamine Transporters*, Advances in Neurobiology 46,
https://doi.org/10.1007/978-3-031-96364-3_1

1 Introduction

Neural communication, the fundamental currency of information exchange in the brain, depends on a well-choreographed quantal release of neurotransmitters into the synaptic space. Chemical neurotransmission involves an action potential triggering the release of neurotransmitters. Upon binding to their cognate metabotropic or ionotropic receptors, these neurotransmitters initiate a response, which can be either excitatory or inhibitory [1]. To terminate the signalling process and prevent prolonged and aberrant stimulation, neurotransmitters have to be cleared from the synaptic cleft. This is facilitated by neurotransmitter transporters present in the presynaptic neuron and glial cells [2] (Fig. 1).

Most neurotransmitters are either amino acids like L-glutamate and glycine or amino acid derivatives like monoamines and GABA. Aromatic amino acids L-tyrosine and L-tryptophan are the precursors for the biogenic amine neurotransmitters, whereas glutamate is the precursor for GABA, the major inhibitory neurotransmitter. Neurotransmitters in the cytosol are transported into the synaptic vesicles by vesicular neurotransmitter transporters (SLC17, 18 and 32), and are subsequently released in the synaptic cleft by vesicle exocytosis, thus establishing the neurotransmitter cycle [3].

Secondary active transporters play vital roles in the spatiotemporal control of neurotransmitters released in the synaptic space and their uptake into synaptic vesicles. Released neurotransmitters are transported back into neurons through the activity of sodium and chloride-coupled neurotransmitter sodium symporters (NSS) that are localized in the presynaptic regions of the neurons and also in surrounding

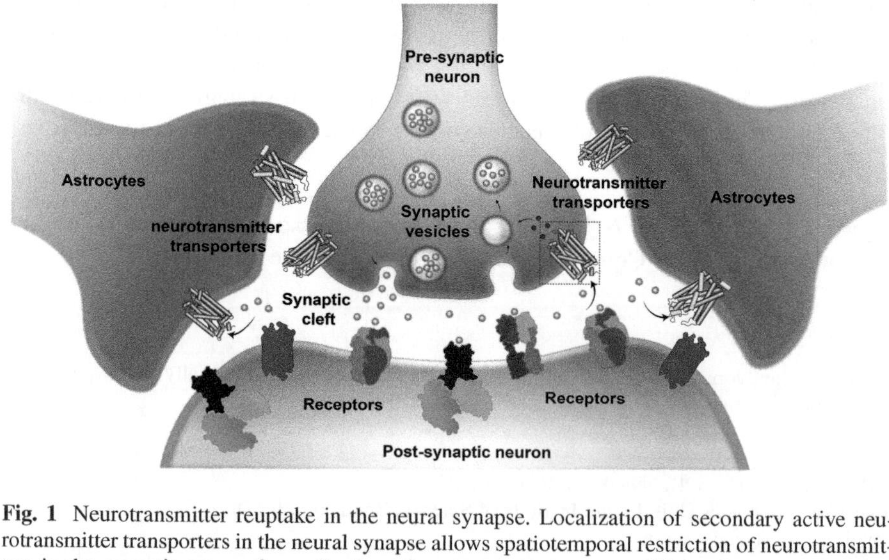

Fig. 1 Neurotransmitter reuptake in the neural synapse. Localization of secondary active neurotransmitter transporters in the neural synapse allows spatiotemporal restriction of neurotransmitters in the synaptic space after activation of postsynaptic receptors. Glial cells surrounding the synapse also express NSS members to facilitate uptake and recycling of neurotransmitters

glial cells [3, 4] (Fig. 1). The ionic electrochemical gradient provided by Na^+ and Cl^- ions across the membrane bilayer provides the necessary driving force to move the neurotransmitters against their concentration gradient. Upon uptake, the neurotransmitters are loaded into synaptic vesicles through the activity of vesicular neurotransmitter transporters that are proton-coupled antiporters. Together, the two transport systems are vital to establish the neurotransmitter cycle during chemical neurotransmission.

Despite the discovery and cloning of these transport systems in the early 1990s, the earliest insights into their structural organization were derived from bacterial homologues like LeuT that transport amino acids [5]. Also, pharmacological insights have come through drug discovery analyses done extensively prior to the elucidation of NSS structures [6]. The recent explosion of structural information of multiple SLC6 members in complex with multiple inhibitors and cognate neurotransmitters merits a holistic update of the latest information that is available to understand the pharmacology of neurotransmitter uptake inhibition. This chapter is being written with this primary focus to provide a summary of the latest advances in SLC6 transporters and the insights derived from structural studies that reveal underlying pharmacology.

2 SLC6 Family of Transporters

Solute carrier (SLC) superfamily of transporters are an indispensable part of every cell and are required for the survival of both prokaryotic and eukaryotic life forms. SLC transporters are categorized into 55 families of which SLC6 family is one of the largest, constituting about 20 genes encoding for a group of similar proteins [4, 7, 8]. Most of these neurotransmitter transporters belong to the solute carrier 6 (SLC6) family of transporters, also known as neurotransmitter sodium symporters (NSS), since they cotransport Na^+ ions to drive the movement of substrates against their chemical gradient. Glutamate transporters (EAATs) are, however, SLC1 transporters with a trimeric organization, and in this chapter, we will restrict ourselves to pharmacological and structural aspects of SLC6 transporters. These transporters are secondary active transporters, which couple uphill transport of neurotransmitters with downhill transport of Na^+ and Cl^- ions leading to accumulation of neurotransmitters in the presynaptic neuron [9]. The ability to transport is governed by a combination of concentration gradient and membrane potential as described in the following equation [10].

$$\mathrm{nS}\frac{\mathrm{RT}}{\mathrm{zF}}\ln\frac{Sin}{Sout} = \mathrm{nI}\frac{\mathrm{RT}}{\mathrm{zF}}\ln\frac{Iout}{Iin} - \Delta\psi\left(\mathrm{nSzS} + \mathrm{nIzI}\right) \tag{1}$$

nS represents the number of substrate molecules and zS represents charge on the substrate. Similarly, nI and zI represent the number of co-transported ions and charge on each ion, respectively. *I*out and *I*in represent molar concentrations of ions outside and inside the cell; correspondingly *S*in and *S*out represent concentrations of substrate inside and outside the cell, respectively.

Mammalian SLC6 transporters can be further subdivided into subfamilies based on the substrates they transport which are amino acid neurotransmitters (γ-amino butyric acid (GABA), glycine, proline and taurine), monoamine neurotransmitters (dopamine, serotonin, norepinephrine and epinephrine) and osmolytes (betaine and creatine) [2]. These transporters are predominantly expressed in the central nervous system (CNS), where they play a major role in neurotransmitter homeostasis. Few members are also found in other tissues, where they are involved in amino acid transport, nutrient absorption, maintenance of osmotic levels and spermatogenesis [4, 11, 12]. In neuronal systems, these transporters are almost exclusively expressed in their respective neural circuits and are primarily located in perisynaptic regions at the ends of axons. They are also distributed in other parts of axons, dendrites and glial cells [13, 14]. The behaviour of these transporters follows Michaelis-Menten kinetics, with K_M values falling within the lower micromolar range and maximal turnover rates ranging from 1 to 20 substrate molecules per second [4]. The transporters use electrochemical gradient of Na^+ and Cl^- ions to drive substrate transport, where co-transport of Na^+ ions is essential for the transport process but the role of Cl^- is not properly understood.

The monoamine transporters (MATs) among SLC6 members are widely distributed throughout the brain and are implicated in various physiological and behavioural functions. Most of the MATs are expressed in the CNS, although NET and SERT are also expressed in different peripheral sites (Table 1). NSS transporters regulate synaptic levels of neurotransmitters through uptake, thereby influencing synaptic signalling. Additionally, they facilitate recycling of neurotransmitters, which can be packaged into synaptic vesicles by vesicular transporters [15]. Given their major role in neurotransmitter homeostasis, NSS transporters are targeted by many drugs, including psychostimulants, antidepressants and antiepileptics. Monoamine transporters are also targeted by commonly abused drugs like cocaine, amphetamine/methamphetamine and 3,4-methylenedioxymethamphetamine (MDMA). Inhibition of these transporters can elevate neurotransmitter concentrations in the synaptic cleft, leading to restored neurotransmission, with potential therapeutic benefits. We will delve into specifics of SLC6 neurotransmitter transporters, their transport mechanism, their physiology and pharmacology in the following section.

2.1 Structural Organization of SLC6 Transporters

Extensive biochemical and mutagenesis studies on SLC6 transporters have provided information about their topology and sites, which were important for ligand interaction, however it was not until the structure of LeuT, a bacterial homologue of SLC6 transporters from thermophilic bacterium *Aquifex aeolicus*, was solved at 1.65 Å, providing a wealth of information about the structure, the organization of

Table 1 List of SLC6 neurotransmitter transporters

Transporter	Substrate	Isoforms/localization	Inhibitors
Dopamine transporter (SLC6A3)	Dopamine (DA), norepinephrine	Isoforms: None Location: Dopaminergic neurons throughout the brain	Cocaine, amphetamine,
Serotonin transporter (SLC6A4)	Serotonin (5HT)	Isoforms: None Location: Neurons in brain, thalamus, limbic cortex, hypothalamus, retina, enterochromaffin cells of intestine and blood platelets	Fluoxetine, paroxetine, citalopram, ibogaine, vilazodone
Norepinephrine transporter (SLC6A2)	Norepinephrine (NE), dopamine	Isoforms: None Location: Noradrenergic neurons in hypothalamus, thalamus, amygdala, cerebellum and spinal cord. Also present in chromaffin cells	Nisoxetine, duloxetine, milnacipran, nortriptyline
GABA transporter (SLC6A1)	γ-Aminobutyric acid (GABA), β-alanine	GAT1: GABAergic neurons from cerebral cortex, hippocampus, cerebellum, thalamus GAT2: hepatocytes, proximal tubules in renal cortex GAT3: brain glial cells, olfactory bulb, retina, brainstem, diencephalon, hippocampus and cortex BGT1: kidneys, liver, cerebellum, cerebral cortex	Tiagabine, NO711, SKF89976A
Glycine transporter (SLC6A5)	Glycine	GlyT1: astrocytes, hypothalamus, thalamus, diencephalon, retina, olfactory bulb GlyT2: brainstem, spinal cord, cerebellum	Bitopertin analogues, SSR504734, PF03463275

helices and residues involved in interaction with the substrate [5]. It was seen that the transporter had 12 transmembrane helices, and the first 10 helices (TM1-TM10) had a striking internal structure repeat among them whereby TM1-TM5 and TM6-TM10 are related by a pseudo-twofold axis located in the plane of the membrane. TMs 1 and 6 are discontinuous helices and have a disordered linker in the middle of the transmembrane segment where they harbour the primary substrate- and ion-binding sites (Fig. 2a).

LeuT has low overall sequence identity (20–25%) with its mammalian counterparts, but the overall architecture has many similarities. LeuT emerged as an invaluable template for the creation of computational models prior to attaining the structures of the neurotransmitter transporters [16, 17]. The binding of two sodium ions at site 1 and site 2 is observed to stabilize the outward-open state in both LeuT and other NSS members as sodium-bound substrate-free structures have been solved for both LeuT and dDAT in outward-open state [18]. The binding of the substrate

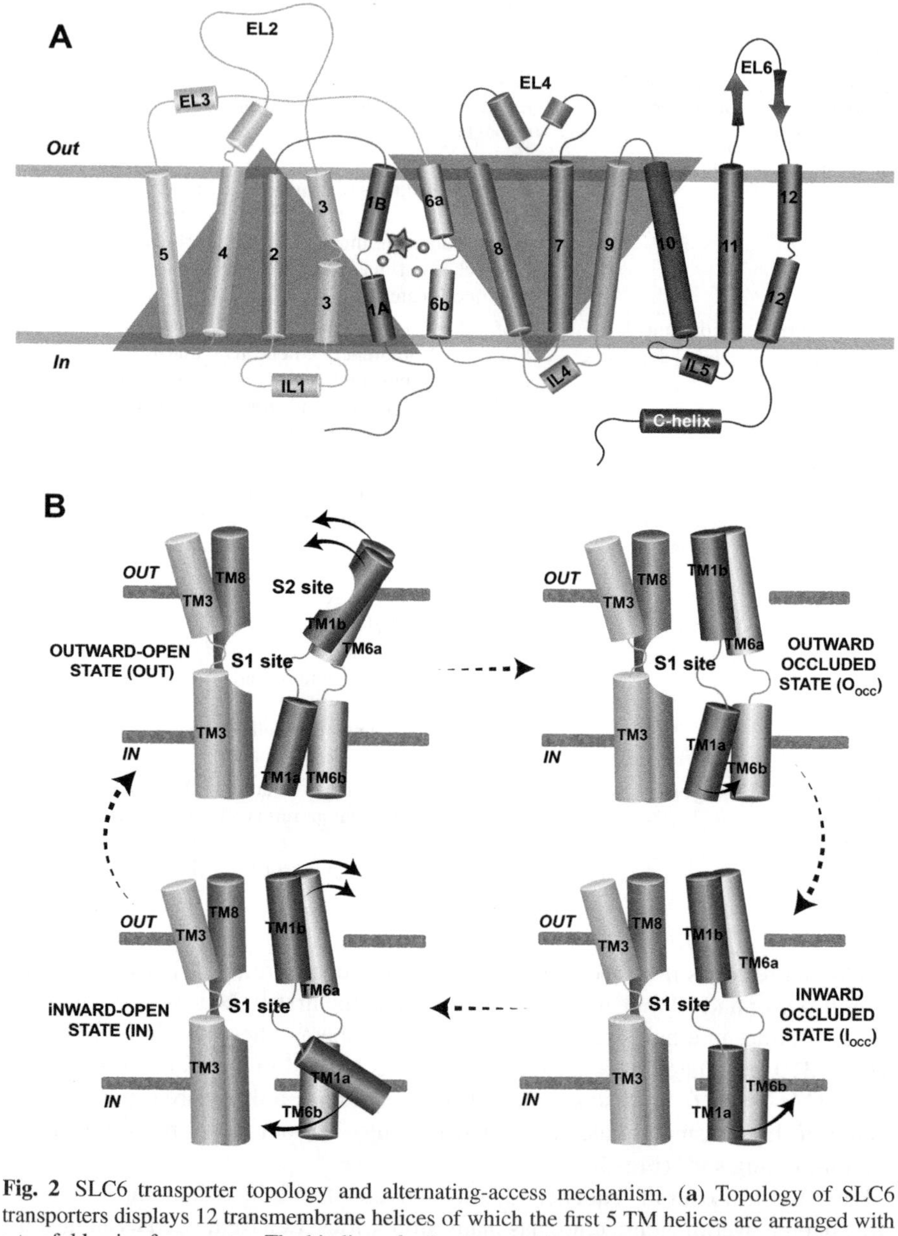

Fig. 2 SLC6 transporter topology and alternating-access mechanism. (**a**) Topology of SLC6 transporters displays 12 transmembrane helices of which the first 5 TM helices are arranged with a twofold axis of symmetry. The binding of substrates and ions happens proximal to two discontinuous gating helices made up of TMs 1 and 6 that allow rocking-bundle movement of the transporters in this class of solute carriers. (**b**) Alternating-access mechanism through rocking-bundle motion of gating helices. Helices TM1a, 1b and 6a, 6b are involved in gating of transporters through multiple conformations of outward-open, occluded and inward-open states to facilitate uptake of the neurotransmitter through the membrane. S1 site represents primary binding site for substrates and competitive inhibitors. S2 site represents non-competitive binding site for allosteric inhibitors in the extracellular vestibule

induces the formation of the occluded state with the substrate occluded from solvent access from both extracellular and the intracellular solvent. The ion gradients induce the release of sodium bound at site 2 followed by the weakening of the intracellular gate and opening of TM1a that creates solvent access to the substrate at the cytosol (Figs. 2b and 3) [19, 20]. Unlike SLC6 members, LeuT does not retain a chloride-binding site and is capable of transporting multiple α-amino acids [21]. SLC6 members are generally more specific to cognate neurotransmitters and tend to have a narrow substrate uptake ability.

Reuptake of neurotransmitters can be blocked at various stages depending on the pharmacophore being used. The transport block happens through competitive inhibition of the inhibitor occupying the substrate-binding site and blocking the movement of extra- and intracellular gates. The primary binding site called S1 site is situated approximately halfway through the transporter and is responsible for binding the substrates, sodium and chloride ions as well as competitive inhibitors. This site comprises mainly of residues from helical breaks of TM1 and TM2 and residues from scaffolding helices TM3 and TM8. The site S1 can be divided into three subsites: subsite A, subsite B and subsite C depending on their preference for interaction with various components of substrate and inhibitors (Fig. 4) (described in detail later). From the SLC6 structures, an additional substrate and inhibitor binding site was observed, which is conserved among both bacterial homologues and NSS

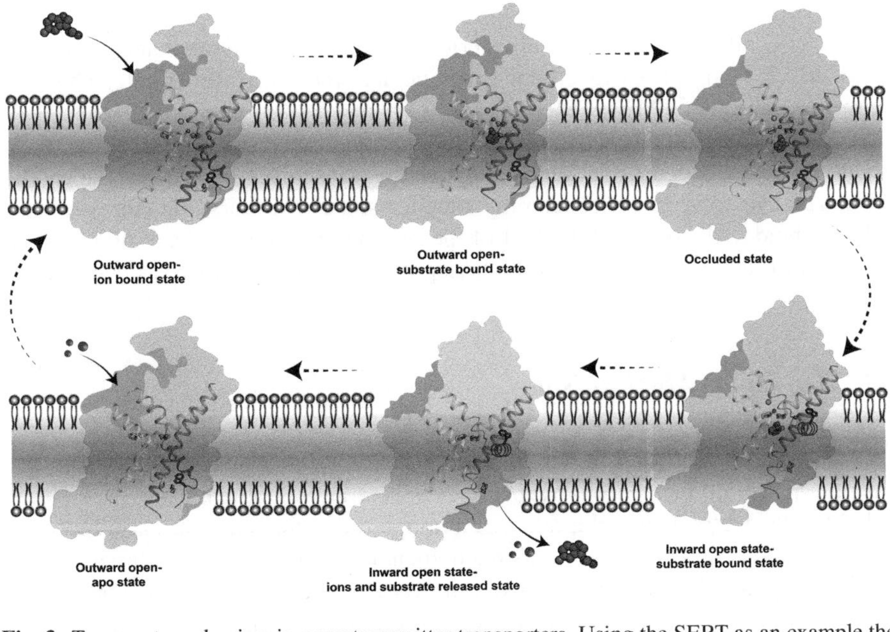

Fig. 3 Transport mechanism in neurotransmitter transporters. Using the SERT as an example the transport cycle of a neurotransmitter is displayed in multiple conformational states beginning from an outward-open state, followed by substrate-bound occluded state, followed by release of ions and substrate in the cytosol facing state and recovery of the outward-open state

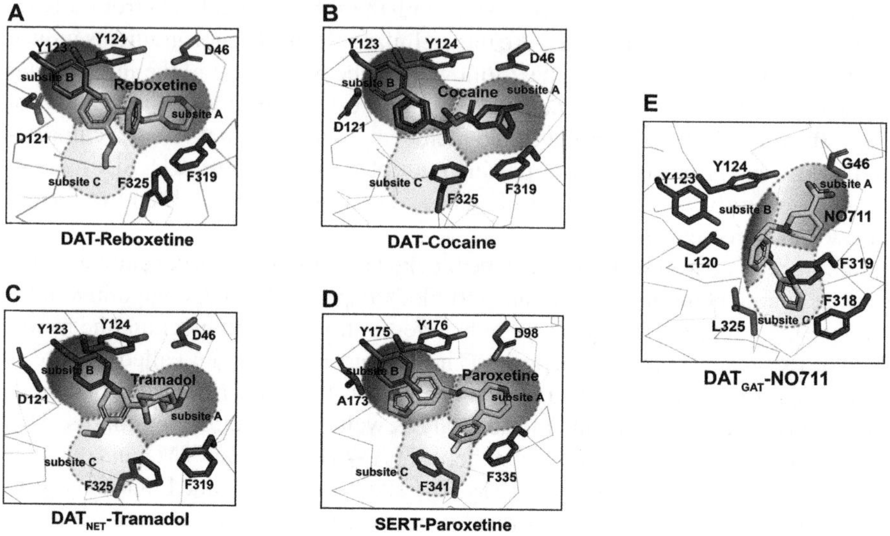

Fig. 4 Subsite architecture among neurotransmitter transporters. Subsite representation in biogenic amine and GABA transporters. Subsites A, B and C represent regions in the primary binding site (S1). Residues surrounding the binding pocket are represented for NET, DAT and SERT inhibitors. In the GAT1 structure, the subsite B is occluded by side chains and the subsite C shifts into the TM6 linker to form subsite C′

transporters called the S2 site or allosteric site. This site is situated above the primary site S1, along the extracellular vestibule near the extracellular gate. This site comprises mainly of TM1b, TM6a, TM10, TM3 and a sharp turn of EL4 loop. The S2 site is also the target site for multiple non-competitive inhibitors, where they bind to prevent any conformational change in the transporter to translocate the substrate. The cytosolic gates are also the sites for some inhibitors that can permeate cells and bind to the intracellular half of the transporter in the TM1a region leading to a block of the reorganization of the transporter back to the outward-open state.

2.2 *Transport Mechanism*

The transport of neurotransmitters through the SLC6 transporters occurs from the synaptic space into the neurons against the concentration gradient using the sodium and chloride electrochemical gradient. Although these molecules are also reported to work in the reverse direction, this phenomenon is much less characterized compared to reuptake. The stoichiometry of this process is generally agreed to be around $2Na^+$ and $1Cl^-$ ions for each molecule of neurotransmitter that moves into the cell [22]. NSS members undergo a myriad of conformational transitions in the process of neurotransmitter uptake. Since Na^+ and Cl^- are available in excess in the extracellular medium, the transporter initially binds Na^+ and Cl^- and is stabilized in an outward-open conformation where the substrate-binding site has solvent accessibility towards the synaptic space. The NSS members function through the "rocking-bundle" mode of alternating access where two discontinuous helices move to control the extra- and intracellular gates of the transporter [23]. The role is played by TM1 and TM6 wherein TM1b and TM6a control the extracellular gate and TM1a and TM6b line the intracellular gate. There is an asymmetry in the movement of these helices at both levels (Fig. 2a, b). In the extracellular gate, TM1b and TM6a move in synchrony by around 10–15° to open and close the gate; however, at the intracellular level, the movement is restricted to TM1a that swings out by 40–45° to allow solvent access to the cytosolic region of the substrate-binding site (Fig. 2b). Structural information of NSS organization, ligand and ion-binding and structure–function relationships were unknown until the structure of LeuT, an amino acid transporter from *Aquifex aeolicus*, was first solved in 2005 by Gouaux and colleagues [5]. Since LeuT shares the same fold as that of NSS members, it became a useful tool for generating homology models of other NSS members and opened the possibility for computational and biochemical studies [17, 20]. Further structures of LeuT with competitive inhibitor tryptophan and antidepressants were solved which provided indirect insights about ligand binding in NSS transporters [24–26]. Inferences regarding substrate and inhibitor binding in DAT could be made through experiments based on LeuT structures, although LeuT is an amino acid transporter and several aspects like the mode of substrate and inhibitor recognition differed in LeuT from NSS transporters. Some antidepressants, which were competitive inhibitors of NSS transporters, display a non-competitive mode of binding to LeuT [25]. Besides this, there were unanswered questions regarding the regions of NSS unrelated to LeuT differing both in sequence and folding pattern, determinants of substrate selectivity in NSS and finally the role of lipids and cholesterol molecule in NSS structure–function, which are absent in bacterial systems. Conformational transitions were studied primarily among LeuT and MhsT. The latter is a hydrophobic amino acid transporter from *Bacillus halodurans* whose structure was elucidated in the inward-occluded state. Together with LeuT, the structures were determined in outward-open (O_{out}), outward-occluded (O_{occ}), inward-occluded (In_{occ}) and inward-open (O_{in}) states demonstrating the multiple steps in the transport cycle [27] (Fig. 3). Numerous structures of neurotransmitter transporters that were elucidated in the recent past demonstrate a conserved transport mechanism similar to bacterial amino acid transporters like LeuT. The transporter gates tend to form through both electrostatic and H-bond networks at the extracellular and intracellular halves. The interactions are primarily between the gating helices TM1 and 6 and the scaffold helices TMs 3, 8 and 10. The occlusion of the substrate-binding site at the extracellular end is mediated by hydrophobic residues like F335 (hSERT) and F319 (dDAT) located in TM 6a. Above this residue, a network of H-bonds form between gating helices and scaffold helices to close the extracellular gate upon substrate binding. Substrate binding occurs through interactions of the biogenic amine with the binding site aspartate (D46, dDAT) and (D98, hSERT), which induces the occlusion of the binding pocket to close the extracellular gate (Fig. 4a–d). In GABA and glycine transporters, the aspartate is replaced by glycine to accommodate the carboxylate of the neurotransmitter which also participates in Na^+ ion coordination at site 1 (Fig. 4e). Substrate binding is followed by the weakening of the sodium bound at sodium site 2 whose release is induced by the unwinding of TM5 followed by the movement of TM1a as a trapdoor to release the substrate. The floor of the substrate-binding pocket is lined by another set of aromatic residues Y95 (hSERT), F43 (dDAT) and Y60 (rGAT1) that open to provide solvent access upon movement of the TM1a (Fig. 3). Incidentally a cholesterol molecule was found in the dDAT and hSERT structures, interacting with TM1a that could allosterically influence the opening of the intracellular gate in some SLC6 members. The release of the neurotransmitter into the cytosol has to facilitate a reorganization of the transporter back to the outward-open state which is facilitated by the binding of K^+ to hSERT that allows the transporter to reorganize. However, this state is yet to be structurally characterized to understand how the reorganization induces the transporter to shift from cytosol-facing state to the outward-open state. Each of these varied conformational states could be a potential target of inhibitors where binding of inhibitors stabilizes the state to an extent where transition between conformational states is not possible leading to block of neurotransmitter uptake. In the following sections, we explore the currently known strategies of inhibition and the exploration of novel inhibitors through modern strategies.

2.3 *Pharmacological Significance of Neurotransmitter Transport Inhibition*

Since the NSS transporters mediate a critical step of neurotransmission by participating in neurotransmitter reuptake from the synaptic space, inhibition of these molecules tends to have neurophysiological effects that can yield therapeutic outcomes. Diverse natural and synthetic compounds have been used as inhibitors or substrates, modulating NSS function. These transporters are intricately involved in many neurological disorders and psychostimulant abuse, and so a deeper understanding of these transporters is essential for new therapeutic interventions. Consequently, SLC6 members have the distinction of being the most targeted members in the entire solute carrier protein family [28]. A few instances of these are discussed in this section to highlight this aspect (Fig. 5).

Psychostimulants like cocaine, a natural alkaloid, and its analogues inhibit DAT, NET and SERT non-selectively. It leads to increase in dopamine levels in the reward pathways, by binding to the dopamine-binding site of DAT, thereby blocking reuptake of dopamine from synaptic cleft and reinforcing the drug's effects on the postsynaptic dopaminergic receptors in the reward pathways of the brain [29]. Several analogues of cocaine have been developed like methylphenidate, benztropine, WIN-35,428 (β-CFT), RTI-55 and others that have similar pharmacological effects

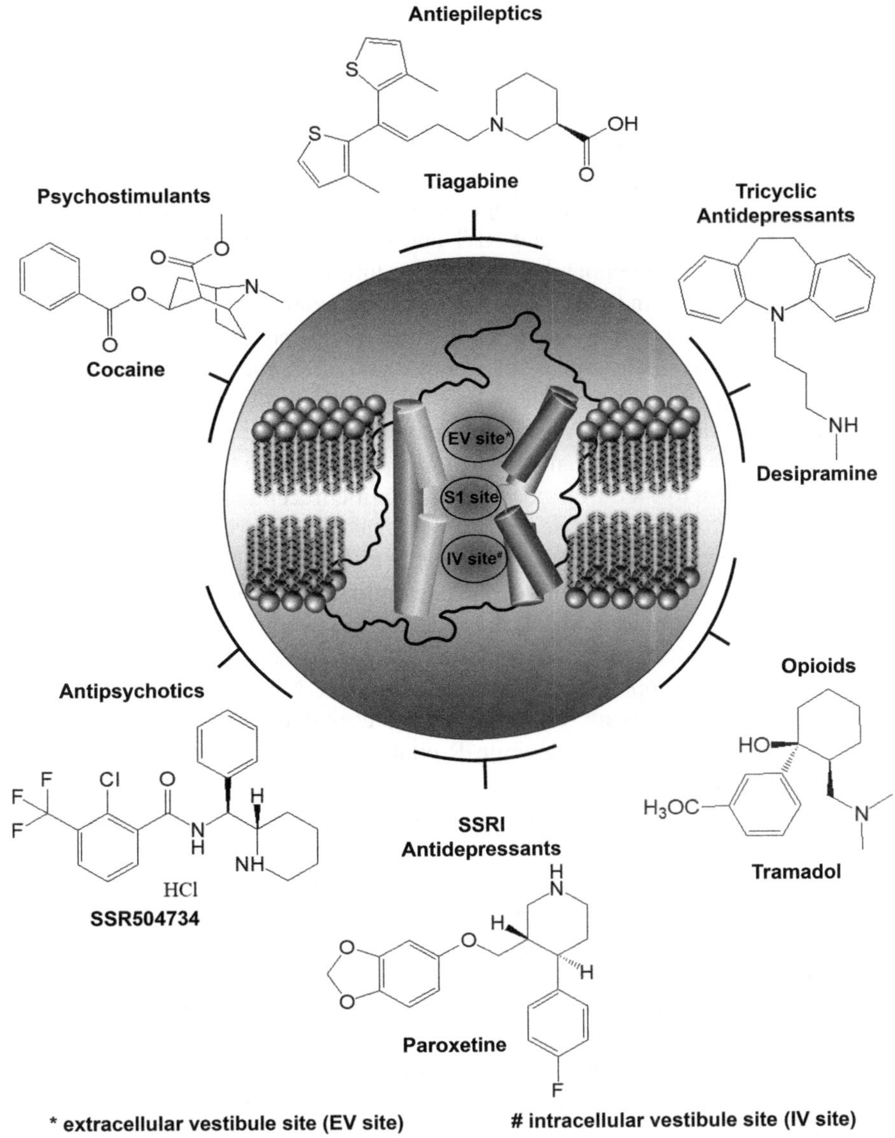

Fig. 5 Sites of inhibition and diversity of transport inhibitors. Modes of inhibition observed in NSS transporters, with competitive inhibitors interacting with the primary site (S1) and non-competitive allosteric inhibitors interacting at the extracellular vestibule (EV) or the intracellular vestibule (IV) site. Inhibitor types are provided as chemical structures around the transporter to indicate the range of inhibitors that target SLC6 transporters

like that of cocaine. Compounds like WIN-35,428 and RTI-55, however, have reduced therapeutic use over time because of their similar psychostimulant effects to that of cocaine leading to addiction and abuse. The role of DAT is implicated in many neurological disorders like Parkinson's disease, addiction, schizophrenia, attention deficit hyperactivity disorder (ADHD) making it an important molecule for pharmacological studies.

Most major antidepressants are biogenic amine transport inhibitors that mediate their action by inhibiting SERT and NET and in doing so prevent the reuptake of serotonin (5-HT) and norepinephrine (NE) leading to increased concentration of these neurotransmitters in the synaptic cleft. Despite extensive research into finding alternatives from conventional antidepressants, the attempts have largely been unsuccessful and biogenic amine uptake blockers still remain the most popular pharmacophores to treat depression making this a multi-billion dollar drug industry [30]. While the first-generation MAT inhibitors like tricyclic antidepressants (clomipramine, imipramine, desipramine and amitriptyline) block both SERT and NET, they are also known to act on various neurotransmitter receptors which often lead to adverse side effects [31]. Selective serotonin reuptake inhibitors (SSRIs) are a group of newer antidepressant medications that include fluoxetine (Prozac), citalopram (celexa), paroxetine and are an improvement over first-generation antidepressants in terms of selectively inhibiting SERT, thus generating fewer side effects. Citalopram, in particular, is a potent and selective SERT inhibitor that does not bind to other transporters like the dopamine transporter (DAT) or NET. When citalopram is resolved into its enantiomers, S-citalopram (or escitalopram) is found to be the active, high-efficacy enantiomer, while R-citalopram has weaker affinity for the SERT-binding site [32].

The continued development of newer antidepressants include combined inhibition of norepinephrine reuptake with the established selective serotonin reuptake inhibitors (SSRIs). Medications like duloxetine and venlafaxine, classified as serotonin–norepinephrine reuptake inhibitors (SNRIs), have the unique capability of increasing both serotonin and norepinephrine levels. SNRIs have also been repurposed to treat chronic pain conditions. Norepinephrine and serotonin are released in the dorsal horn of the spinal cord as a response to the activation of descending pain pathways. Norepinephrine release activates inhibitory postsynaptic receptors like α2-adrenergic receptors that reduce stimulation of the afferent pain pathways leading to alleviation of pain [33]. This property is harnessed by SNRIs like duloxetine and milnacipran to supresses pain in conditions like fibromyalgia and neuropathic pain. Tramadol, a weak opioid, blocks norepinephrine uptake. Its dual ability to act as an opioid and norepinephrine uptake blocker has allowed tramadol to act as a powerful antinociceptive agent in postoperative pain and restless leg syndrome [34].

GABA reuptake from inhibitory synapses is also a major pharmacological target in the context of anxiety and epilepsy [35]. Inhibitory inputs from GABA are vital to balance the excitatory inputs from glutamatergic neurons. Consequently a loss of GABA release in neural circuits due to injury or shock could lead to a lowered inhibitory neurotransmitter release causing an imbalance with excitatory inputs. This unbalanced load of excitatory stimulations results in seizures and epilepsy

[36]. Inhibition of GABA reuptake therefore is a strategy to enhance GABAergic neurotransmission in the treatment of certain forms of epilepsy and anxiety. Glycine which is another inhibitory neurotransmitter is vital for the activation of NMDA receptors [37]. Glycine transporters are considered to be a therapeutic target to indirectly modulate NMDA receptor activity and are considered a potential therapeutic target for treating schizophrenia [38]. Sarcosine-based inhibitors have been known to block glycine transport in isoform1 although newer and more effective inhibitors interact with alternate binding sites. As is evident in this section, the inhibition of NSS members is of crucial pharmacological importance. It is therefore vital to understand the basis of these mechanisms in greater detail to facilitate enhancing the specificity of the inhibitors to their individual transporters.

3 Competitive Inhibition of Biogenic Amine Uptake

A majority of inhibitors of monoamine uptake are observed to display competitive inhibition of neurotransmitter transport. This was also bolstered by the identification of antidepressants and psychostimulant binding in the substrate binding pocket of the *Drosophila melanogaster* dopamine transporter [39]. Unlike the neurotransmitters that are relatively small and occupy a fraction of the NSS primary binding site, the S1/primary binding site in NSS transporters is large, flexible and is capable of binding neurotransmitters, antidepressants, psychostimulants and antiepileptics of varying sizes, while still maintaining specificity for the type of molecules it can bind or transport. To understand the substrate, inhibitor interactions and associated plasticity within the binding pocket with better clarity, the S1 site is further dissected into three subsites A, B and C based on the region and the mode of interactions they have with the bound substrate or inhibitor [40, 41] (Fig. 4). Subsite A is predominantly polar consisting of mainly polar residues and free amine and carboxy groups of unwound regions of TM1 and TM6. Subsite A generally accommodates the polar groups of substrates and inhibitors, including carbonyls in case of amino acid transporters and amine groups in case of monoamine transporters. The residue recognizing the amine group of biogenic amines is usually an aspartate (D46-dDAT, D98-hSERT, D75-hNET) that interacts with the bound substrate. In the case of amino acid transporters like GATs and GlyTs that have a carboxylate group in the substrate, the residue is substituted to a glycine. The carboxylate group of these substrates coordinates sodium ion at site 1 and interacts with a conserved tyrosine hydroxyl (Y124-dDAT, Y152-hNET, Y176-hSERT, Y140-rGAT1 and Y142-hGlyT1) that is part of the extracellular gate (Fig. 4). Subsite B consists of residues from TM3 and TM8 containing nonpolar residues forming hydrophobic interactions with hydrophobic rings of substrates and inhibitors. Some polar interactions are also observed in subsite B, in case of monoamines and antidepressant bound structures, where hydroxyl groups of monoamines and polar groups in the drugs' aromatic rings are accommodated. Subsite C comprises of unwound helical regions of TM6 and displays plasticity to accommodate substrates and inhibitors of varying sizes. In case of the GABA transporter, the subsite B is substantially reduced due to the presence of multiple residues that occlude the access to this pocket. In GATs, the subsite C is displaced (referred to as subsite C′) and is accommodated within the disordered linker of TM6 and accommodates the amine group of substrate GABA and aromatic ring of antiepileptic drugs (Fig. 4e) [42, 43]. The shape of the S1 site varies among different groups of transporters displaying a trilobed subsite architecture in case of monoamine transporters, whereas in GABA transporters it displays a bean-shaped architecture [42]. The mode of competitive inhibition, however, needs further evaluation in a GABA transporter isoform.

3.1 Dopamine Transporter Inhibition

DAT is localized mainly in DA neurons and is involved in active reuptake of DA from synaptic cleft. Termination of dopaminergic neurotransmission is either brought about by enzymatic degradation, commonly mediated by monoamine oxidases (MAO) or catechol-O-methyl transferase (COMT), or through uptake of dopamine from synaptic cleft by DAT [44]. The plasma membrane DAT utilizes the ionic gradient across the membrane and facilitates the transport of dopamine. Dopamine is further transported into the synaptic vesicles by VMAT and is sequestered in the vesicles, till it is again released into the synapse by fusion of the vesicles with the plasma membrane.

Initial structural insights into DA reuptake came from dDAT that primarily functions as a dopamine transporter (Fig. 6). Cocaine and its derivatives are competitive inhibitors of DAT, and they compete for the binding pocket of DAT and lock it in an outward-open conformation, resulting in prolonged accumulation of DA in the synaptic cleft [16, 45]. Amphetamines are DAT substrates, which increase the synaptic DA levels by promoting efflux of DA from synaptic vesicles into cytoplasm and further into the synaptic cleft via DAT [45, 46].

In 2013, Gouaux and colleagues reported the structure of *Drosophila melanogaster* dopamine transporter (dDAT) in complex with the tricyclic antidepressant nortriptyline [39]. The *Drosophila melanogaster* dopamine transporter (dDAT) was chosen for this study which displayed better stability compared to mammalian orthologues. Although dDAT shows preference to transport dopamine over norepinephrine, it is more sensitive towards antidepressants than amphetamines and cocaine when compared to mammalian DATs [47]. Despite that, the affinity of dDAT for dopamine, its analogues and hDAT inhibitors was enough to capture their densities in crystal structures. Multiple structures of dDAT were solved, with the substrate dopamine, a substrate analogue 3,4-dichlorophenethylamine (3,4-DCP), the psychostimulant substrates D-amphetamine and methamphetamine, cocaine and its analogues and antidepressants like nortriptyline, reboxetine and nisoxetine [45, 48].

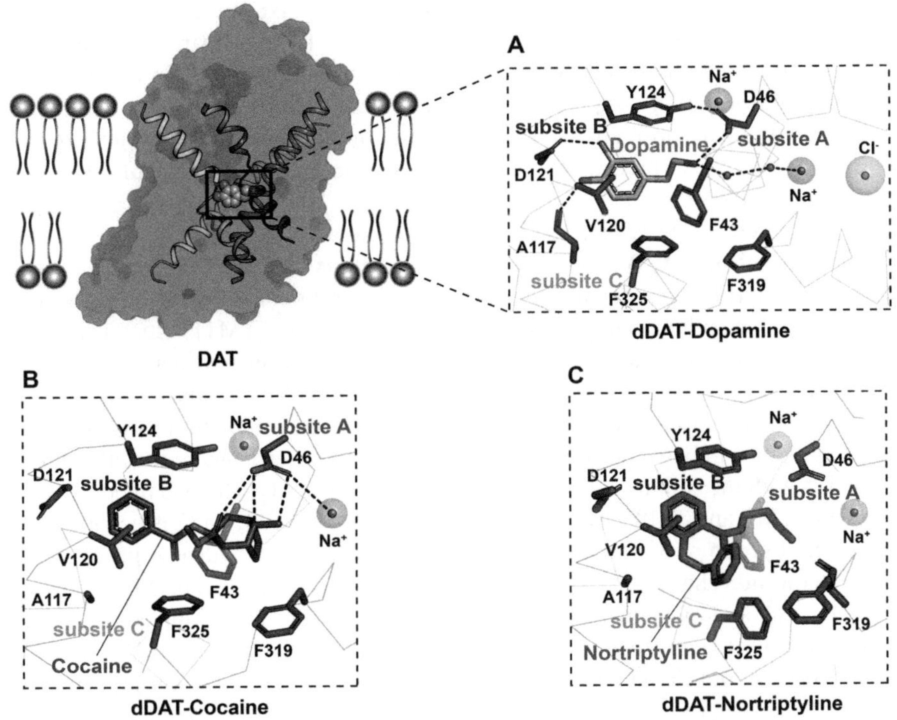

Fig. 6 Substrate and competitive inhibition of dopamine transporter. Dopamine-bound dDAT (PDB id. 4XP1) demonstrating interactions of the neurotransmitter in the primary site. Plasticity in dDAT-binding pocket is demonstrated by changes to the side chain positions to accommodate the smaller inhibitor cocaine (PDB id. 4XP4) in comparison to a large tricyclic antidepressant nortriptyline (PDB id. 4M48)

The structure of a thermostabilized construct of dDAT was determined in complex with a tricyclic antidepressant, nortriptyline, capturing the transporter in an outward-open conformation, where the antidepressant was bound to the S1 site and is solvent accessible from the extracellular vestibule. The DAT outward-open structure looks like an hourglass and displays a LeuT-like fold with 12 TM helices in which TM1–5 and 6–10 are related by pseudo twofold symmetry. The presence of helical breaks in TM1 and TM6 allows the backbone polypeptide to participate in ion coordination and substrate interaction along with residues from TM3 and TM8 in the S1 site. Just like the tryptophan-bound LeuT structure, nortriptyline locks DAT in an outward-open state by preventing the closure of extracellular gates to transition into an inward-open state and acting as a "foot-in-door mechanism" [39]. The DAT-nortriptyline complex just like the engineered LeuT that binds antidepressants (Leu-BAT) [41] inhibits substrate binding to the transporter through competitive inhibition by binding to the S1 substrate-binding site (Fig. 6c). Subsequent structures of a functional construct of dDAT with substrate dopamine, psychostimulants D-amphetamine and methamphetamine and competitive inhibitor cocaine displayed an outward-open conformation just like the DAT-nortriptyline and LeuT-tryptophan complex, whereas the substrate analogue 3,4-dichlorophenethylamine (DCP) captured the structure of DAT in an outward-occluded conformation [45]. Taken together, these studies found that DAT can accommodate many different sizes of substrates and inhibitors, by reorienting the residues in the helical breaks of TM6 along with the aromatic chains of F319 and F325 (Fig. 6a–c).

As discussed in the earlier sections, the subsite architecture of S1 site can be divided into subsites A, B and C where each subsite is dedicated to a specific kind of interaction (Fig. 4). In subsite A the amine group of substrates dopamine, D-amphetamine and methamphetamine interacts with D46 (TM1), which is known to be an important residue for substrate selectivity and transport activity through computational and mutagenesis study. The amine group of dopamine also indirectly interacts with a sodium ion through a network of H-bond interaction including two water molecules in between (Fig. 6a). This differs from LeuT where the interaction of leucine with the sodium ion is direct. The subsite B cavity which is lined by residues from TM3 and TM8 including A117, V120, D121, Y124, S422 is occupied by the catechol group of dopamine in which the meta- and para-hydroxyl group interacts with the side chain carbonyl of D121 and main chain carbonyl of A117, respectively. In case of amphetamines, due to lack of hydroxyl groups in amphetamine, the binding affinity to subsite B is weaker. Indeed, in case of mammalian DAT, where the subsite B is less polar due to substitution of D121 and S426 by glycine and methionine, respectively, the binding affinity for amphetamines is higher. This suggests subsite B plays a critical role in substrate interactions owing to the differences in substitutions on the catechol group. The subsite C which contains non-helical regions from TM6 displays plasticity to accommodate substrate and drugs of various sizes. Although subsite C residues are not involved in direct interactions with the substrate dopamine, F325 interacts with the phenyl group of amphetamines through edge-to-face aromatic interaction. Apart from this, the residue F319 serves as a gating residue and occludes the binding pocket of DAT upon substrate binding by closing solvent access to extracellular vestibule. So, the subsite C plays a major role in substrate translocation by transitioning the outward open structure of DAT to an occluded conformation.

In case of DAT inhibitors, D46 (TM1) of subsite A interacts with the amine groups of DCP antidepressants and cocaine (Fig. 6b). Mutation of this aspartate substantially reduces transport activity and cocaine-binding affinity [49]. The amine group of nortriptyline forms a hydrogen bond with the main chain carbonyl of F43 and a cation-π interaction with the aromatic side chain of F43. S421 (TM8) simultaneously forms hydrogen bond with the main chain carbonyl of F43 and sodium ion at site 2. This indirectly connects the amine group of nortriptyline and sodium at site 2 through residues F43 and S421 (Fig. 6c). In the subsite B, the aromatic rings of the drugs are involved in hydrophobic interactions with V120, Y124 and A117. Negative charge in subsite B results in a reduced binding affinity of dDAT for cocaine compared to hDAT. Mutations D121G and S421M to mimic hDAT, which made the subsite B more hydrophobic, displayed higher binding affinity for cocaine and its analogue β-CFT. It was observed for halide-substituted cocaine analogues including β-CFT and RTI-55 that the architecture of subsite B did not differ much from the architecture in the cocaine-bound structure. Subsite C displays the highest plasticity among all structures to make room for drugs of varying sizes. F325 forms edge-to-face interactions with the aromatic rings of the inhibitors; its position is different in dopamine-, DCP-, cocaine- and nortriptyline-bound structures to accommodate them all. F319 position is conserved among all structures except for the DCP bound structure, where F319 rotates by about 100° to occlude the binding pocket and prevent solvent access from extracellular vestibule, like that of the LeuT-leucine structure in occluded state (Fig. 6b, c) [5, 45].

Together, the substrate- and inhibitor-bound structures of DAT suggest the mechanism of ligand selectivity and accommodation in the S1-binding pocket. Substrate/inhibitor is recognized at the binding pocket through the interaction of its amine group with carboxylate of D46 or carbonyl of F43 and F319, followed by stabilization of the aromatic group through hydrophobic and/or polar interactions in subsite B. To accommodate ligands of different sizes, the F325 rotates inward in case of substrate dopamine, amphetamine and dopamine analogue DCP, and in case of inhibitors like nortriptyline, cocaine and its analogues, F325 moves out to create room for the extra aromatic rings. Finally, F319 is responsible for rotating and occluding the binding pocket upon ligand binding. Substrates interact at the opposite ends of S1 site, i.e. subsite A and subsite B, whereas the inhibitors utilize the plasticity of subsite C to accommodate their extra aromatic rings and lock the transporter in outward-open state like a wedge in a door.

3.2 *Norepinephrine Transporter-Inhibition Specificity*

Norepinephrine transporter (NET) is present on the plasma membrane of noradrenergic neurons and is involved in active reuptake of norepinephrine from the synapse. Since norepinephrine regulates various behavioural and physiological aspects like alertness, depression, pain, cognition and heart rate, NET is a potential candidate for many drug targets. Malfunctioning of NET can lead to various neurological disorders like insomnia, chronic pain, depression and post-traumatic stress disorder (PTSD) [50–52].

Like the other members of SLC6 family, NET utilizes the Na^+ gradient across the plasma membrane to actively transport norepinephrine inside presynaptic neuronal cells. Binding of 2 Na^+ and 1 Cl^- ions to NET is essential for the substrate to bind and subsequently move across the membrane. The large extracellular loop between TM3 and TM4 has N-glycosylation sites, which are important for protein stability, surface expression, transport activity but are not essential for ligand recognition. The biogenic amine transporters closely share their architecture, mechanistic properties and even pharmacology to some extent [39, 53–55]. This makes the questions regarding substrate selectivity and ligand recognition in NET highly relevant. Although an experimental structure for NET was resolved very recently, structures of *Drosophila melanogaster* dopamine transporter (dDAT) and a dDAT construct with NET like mutations called $dDAT_{NET}$ in complex with norepinephrine and norepinephrine-reuptake inhibitors (SNRIs) have improved our understanding of mode of recognition and selectivity in NETs.

For a long time, the structural insights about hNET were limited to dDAT [39, 48]. Surprisingly while dDAT transports dopamine more efficiently, it is inhibited better by drugs that are generally NET specific, making it a very useful model to study norepinephrine uptake inhibition [47]. dDAT, for instance, binds NRIs with a nanomolar affinity, and the structures of dDAT bound to highly specific norepinephrine-reuptake inhibitors (SNRIs), nisoxetine, duloxetine, reboxetine and tramadol were resolved earlier (Fig. 7) [48, 53]. These inhibitors display high specificity to NET compared to DAT or SERT, and reboxetine is a known drug for treatment of panic disorder and ADHD. These inhibitors contain discontinuous aromatic groups branching from central carbon which differs from TCAs where the aromatic

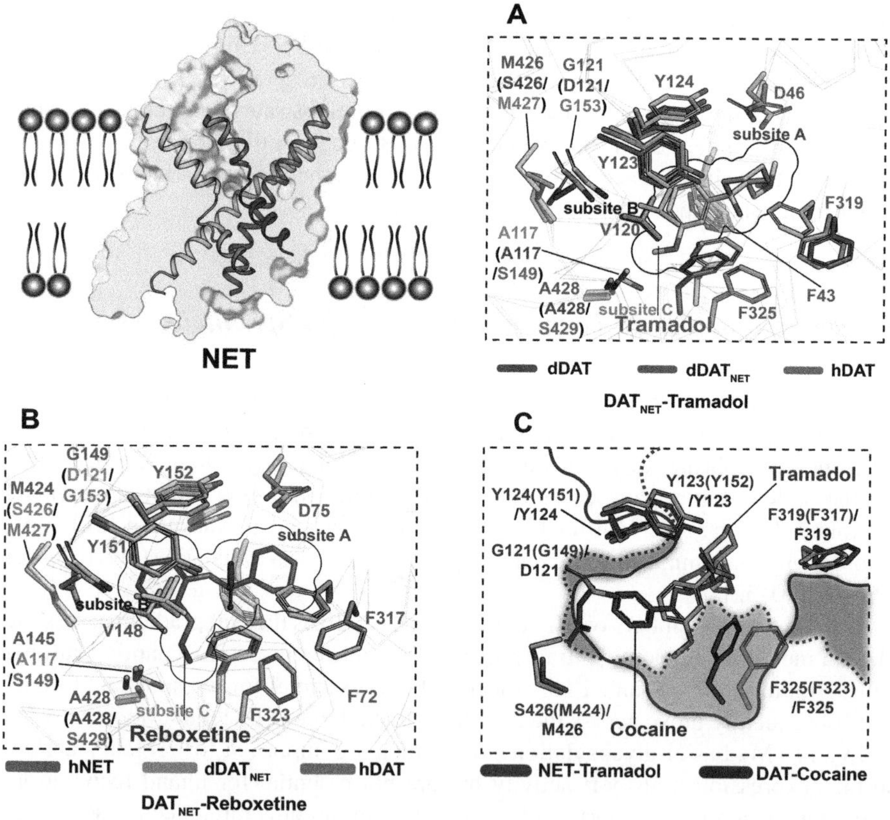

Fig. 7 NET-specific competitive inhibition. Reboxetine- and tramadol-bound structures (PDB ids. 4XNX, 6M47) of dDAT interact in regions of subsite C which differ between DAT and NET, thereby rendering specificity to the interactions

rings are fused to form a tricyclic framework. Nisoxetine, reboxetine and tramadol bind to the central binding pocket through their amine groups, in subsite A with the main chain carbonyl of F43 and possibly with the carboxylate of D46 which is at a distance of 3.6 Å (Fig. 7a, b). Subsite B, which is known to harbour the aromatic ring of NSS ligands, comprises of V120, Y123 and Y124 in TM3, and S422 in TM8 accommodates the discontinuous aromatic rings of both nisoxetine and reboxetine. Mutation of subsite B residues D121G and S426M to mimic mammalian NETs makes this site more hydrophobic and enhances nisoxetine-binding affinity [45]. Interestingly, the two drugs have methoxy and ethoxy groups on one of the aromatic groups that face subsite C. In hNET, the residues that surround this region are A145 (TM3) and A426 (TM8) which happen to be S149 and S429 in human DAT. The presence of polar side chains in this region creates clashes with the aromatic substitutions on NRIs and lowers the affinity towards hDAT and NRIs.

This aspect was displayed by making the equivalent substitution of A117S and A426S, and combination of both in dDAT to observe that affinity towards the transporter for duloxetine and tramadol is significantly lowered. A study that involved complexes between duloxetine and tramadol with dDAT utilized the dDAT construct with mutations D121G and S426M, which made the subsite B more hydrophobic to mimic the hNET-binding pocket (Fig. 7a). Post changes to the subsite, the modified construct $dDAT_{NET}$ displayed a binding affinity for nisoxetine closer to hNET [53]. The structure of $dDAT_{NET}$ construct was solved in a substrate-free, norepinephrine-bound state and bound to NET-specific inhibitors S-duloxetine, milnacipran, and a synthetic opioid, tramadol. Initially it was a common perception and observed through computational studies that norepinephrine and dopamine, being monoamines and sharing similar backbone, should bind to the S1-binding pocket in a similar manner [56, 57]. However, it was observed that although the subsite architecture was similar in both cases and the amine group interactions were similar, the pose of the catechol group of norepinephrine was more towards subsite C in the outward-open state of dDAT in contrast to dopamine, which orients towards subsite B. Also, the observed edge-to-face aromatic interaction of F325 with catechol group of dopamine was absent in case of norepinephrine binding (Fig. 7c). However, in the recent structural study of hNET in the occluded state bound to the norepinephrine, the neurotransmitter is bound in a similar fashion to that of dopamine. There is a possibility that the neurotransmitter binds in the primary binding site in a certain orientation and shifts to an alternate state once the transporter changes to an occluded state [55]. Drugs like nisoxetine and amitriptyline bound to hNET seemed to retain a very similar pose in comparison to the dDAT complexes with the same drug types [58].

There are close mechanistic similarities observed in antidepressant binding to hNET and dDAT including amine group interaction with side chain of D46 and main chain carbonyls of F43, hydrophobic interactions of aromatic rings with subsite B and edge-to-face interaction with F325 in subsite C. In the case of tramadol, interactions with subsite B are minimal, but the drug retains a high specificity to dDAT. Cocaine, on the other hand, does not interact with subsite C and entirely is bound in the subsite A and subsite B regions of the binding pocket. This has allowed the hypothesis that subsite B is primarily required for enhancing binding affinity and subsite C facilitates specific interactions between the closely related NET and DAT (Fig. 7c) [45, 53].

3.3 Serotonin Transporter

Serotonin transporter (SERT) has a very similar architecture to DAT and NET. SERT is involved in reuptake of 5-hydroxytryptamine (5-HT) and is dependent on Na^+ and Cl^- ions suggesting a coupling of these ions with 5-HT transport. Subsequent studies on SERT showed that 5-HT was co-transported alongside one Na^+ and one Cl^- ion, while a potassium ion (K^+) was transported in the opposite direction to facilitate transporter reorientation during the reuptake cycle [59]. Studies carried out by Rudnick and Nelson showed that equilibrium accumulation of 5-HT is electrically neutral and is directly facilitated by K^+ antiport but that presence of K^+ ion was not an absolute requirement for the transport process [60, 61]. This led to the speculation that the K^+-binding site can also be occupied by H^+ ions. In physiological conditions, this site is likely to be occupied by K^+ ions, but in low concentrations of K^+, the site can be occupied by a proton mediating the transport process [60].

Structural and functional studies have shown that upon binding with sodium ion, the transporter attains an outward facing conformation while interaction with the substrate and chloride ions, in the absence of sodium, results in an inward-open conformation (Fig. 8). According to the present model, first the transporter binds to Na^+ followed by 5-HT and Cl^-. Serotonin binding leads to a conformational change in the transporter resulting in an inward-facing conformation, finally releasing the neurotransmitter and ions into the cytoplasm. K^+ binds to the inward-facing conformation of the transporter, reorienting it for another cycle. Structural studies have shown that helices comprising of TM1, TM2, TM6 and TM7 undergo conformational changes facilitating the transport of serotonin (Fig. 3) [62]. During this process, a plethora of molecular interactions are made and broken according to a specific gating mechanism. The gating residues are present in intracellular and extracellular vestibule of the transporter. In the extracellular vestibule, there are two gates: one is due to a salt bridge interaction between TM1b and TM10 and another involving an interaction between aromatic residues present in TM3 and TM6a which forms a hydrophobic lid shielding the bound substrate from the extracellular side. Conversely, the gating of the intracellular vestibule is mediated by a comprehensive network of residues spanning parts of TM5, TM6, TM8, TM9 and the N-terminus.

The structure of SERT was solved in complex with citalopram and paroxetine by Gouaux and colleagues in 2016 [54]. These inhibitors locked the transporter in an outward-open conformation, by forcing the TM1 and TM6 to adopt a pose incapable of transitioning to the occluded conformation. Just like other NSS members, the amine group of the inhibitors localize at subsite A and interact with the conserved aspartate D98 (equivalent to D46 of dDAT). Aromatic rings of the inhibitors occupy

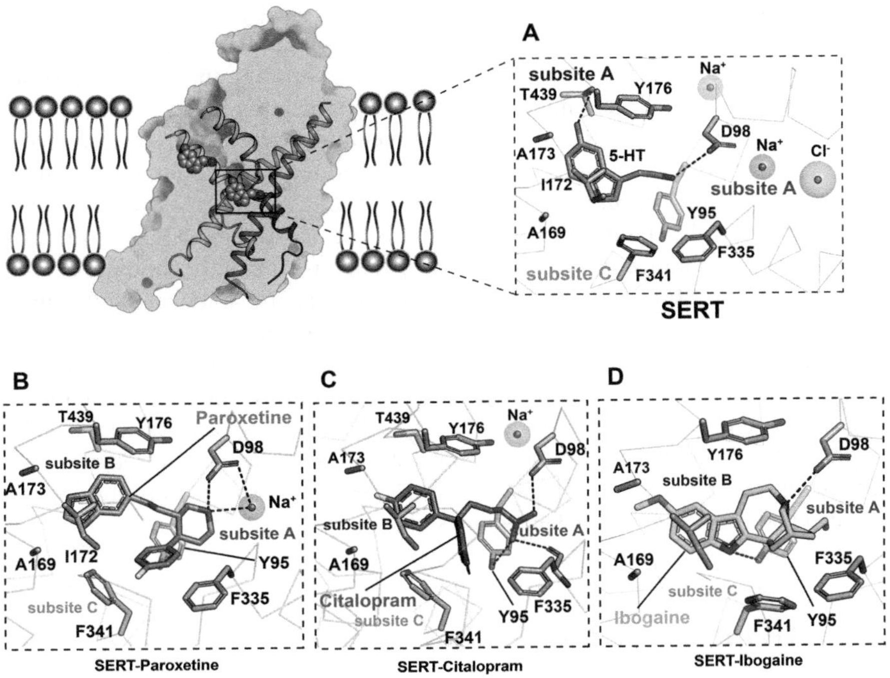

Fig. 8 Substrate and competitive inhibition of SERT. (**a**) 5HT interacts with the S1 site of SERT wherein D98 interacts with the amine group of 5HT. The 5-hydroxy group interacts with side chain hydroxyl of T439 of SERT (PDB id. 7LIA). (**b**, **c**) Panels display the interactions of high-affinity inhibitors paroxetine and citalopram that also act as competitive inhibitors (PDB ids. 5I6X, 5I75). (**d**) Ibogaine is a non-competitive inhibitor although it interacts with the S1 site (PDB id. 6DZY) and stabilizes the inward-open state of the transporter

subsite B and some parts of subsite C, where major hydrophobic interactions occur with Y176, I172 and F341. Mutations of these residues severely affect the ligand-binding affinity as well as substrate transport activity [63, 64]. F335 in subsite C is involved in edge-to-face aromatic interactions with the inhibitors (Fig. 8b, c).

Another study by the same group, captured SERT in three different states: outward-open, occluded, and inward-open state, each bound to the inhibitor ibogaine [62]. The outward-open and occluded structures of SERT were obtained in the presence of NaCl which is known to stabilize the outward-open and occluded states. In these two states, ibogaine interacts at the S1 binding site with the amine group interacting with D98 in subsite A while the tricyclic ring fitting into subsite B lined by aromatic rings of Y95 and Y176, just like the SERT-citalopram structure. F341 interacts with the indole ring of ibogaine at subsite C. The inward-open state was captured by replacing NaCl with KCl. Upon comparison of localization of ibogaine in the inward-open state, with respect to other states, it was observed that ibogaine is pushed towards TM1a and TM8, towards the cytoplasmic-permeation pathway. The F335 position in subsite C is different in each state, continuously rotating to accommodate the drug in the pocket and finally covering it like a lid blocking its exit from the extracellular side. This suggests F335 may play a major role in gating the substrate translocation process, controlling the exit of substrate from central binding pocket into the cytoplasmic-permeation pathway (Fig. 8b–d). The structure of porcine SERT determined in complex with cocaine also revealed a similar binding pose to cocaine bound to dDAT reported earlier within the primary binding site primarily occupying subsites A and B [65].

Compared to other inhibitors like citalopram and paroxetine, ibogaine is a relatively weaker inhibitor and displays a non-competitive mode of inhibition, even though it binds to SERT in the active site. It is speculated that ibogaine binding to SERT in the inward-open state might be the reason for the non-competitive mode of inhibition as serotonin cannot compete for a transporter locked in an inward-open state. The SERT structures determined in diverse conformational and inhibitor-bound states unveil the transport mechanism of NSS with substantial clarity. The presence of the rocking-bundle movements in SERT is mediated through movements in different helices and loops as the molecule shuttles from outward-open to occluded and further into an inward-open state.

3.4 GABA Reuptake Inhibition

The GABA transporter was first cloned and reconstituted by Kanner and colleagues in 1990 [66]. The transporter is similar to other SLC6 members but has a completely distinct pharmacological profile wherein the drugs that block GABA uptake have structurally distinct features as compared to drugs that block biogenic amine transporters. For instance, the subsite A in the GABA transporter lacks an aspartate to accommodate the carboxylate of GABA, which also coordinates Na^+-binding at site 1. The carboxylate also interacts with Y140 at the entrance of the primary binding site and is located in TM3 (Fig. 9a). This is a critical residue whose substitution results in loss of substrate recognition [67]. Substrates of the GABA transporter span a wide array of compounds like GABA, β-alanine, nipecotic acid, guvacine and betaine. The transporter is also blocked by a wide array of inhibitors that structurally include a carboxylate carrying group linked to an aromatic group that competitively block GABA uptake. Traditional inhibitors of GABA reuptake include drugs like tiagabine, NO711 and SKF89976a that largely resemble the architecture mentioned earlier. Tiagabine is used to treat partial seizures and is the only medically prescribed inhibitor of GABA reuptake. However, recent structures bound to tiagabine have revealed that tiagabine binds towards the cytosolic face of the transporter and not within the primary binding site [68]. The interactions with tiagabine are therefore dealt as part of the allosteric inhibitors of GABA uptake. An alternate study done in the authors' lab approached this problem of understanding GAT inhibition through alternate means. Given the structural similarity between NSS transporters, a conversion of the dDAT primary binding site was performed to resemble the primary binding site of the GAT1. The engineered binding site was capable of

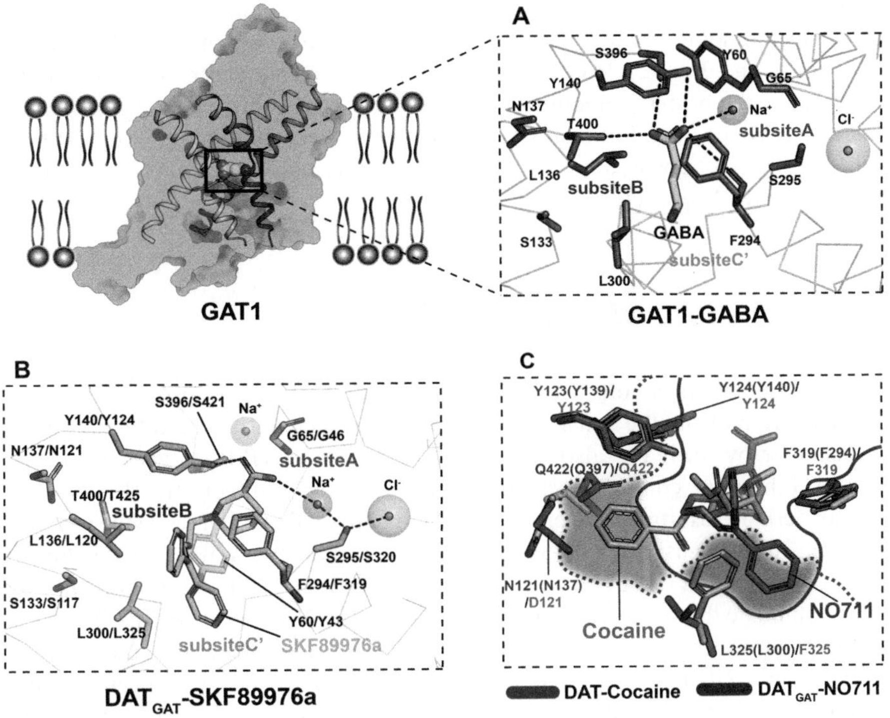

Fig. 9 Altered subsite architecture of GABA transporter. (**a**) GABA-bound inward-open state of the ratGAT1 (PDB id. 8GNK) displays interactions of GABA in the S1 site. (**b**) An engineered dDAT whose binding pocket is modified to resemble GAT1 binds GAT1-specific inhibitors with an altered binding site architecture. Structure of SKF89976a bound to the $dDAT_{GAT}$ is displayed (PDB id. 7WLW). (**c**) The changes in the binding pocket in dDAT vs $dDAT_{GAT}$ are displayed using cocaine and NO711 (PDB id. 7WGT)

accommodating two of the well-known GAT1 inhibitors NO711 and SKF89976a within the primary binding site with an altered subsite binding configuration (Fig. 9b, c). The guvacine or nipecotic acid groups of the two inhibitors interact at the subsite A with H-bond interactions with Y140 and simultaneously coordinating the bound sodium ion. A linker that connects the branched aromatic groups with the nipecotic acid or guvacine groups orients the drug away from interacting with the subsite B which is occluded through side chains of surrounding residues and cannot interact with the aromatic groups of inhibitor due to the residue substitutions. The region around the TM 6 linker that constitutes subsite C is also subtly shifted in the structure to accommodate the aromatic groups of the drugs. An important substitution in GAT1 is the presence of a leucine in the position of phenylalanine at position 300. This substitution allows the displacement of the leucine side chain during drug binding in a way so as to accommodate the aromatic group of NO711 and SKF89976a within the disordered region of the TM6 linker instead of having the groups interact adjacent to the linker as observed in biogenic amine transporters. This effectively changes the trilobed binding pocket observed in biogenic amine transporters to a bean-shaped binding pocket to accommodate competitive inhibitors of GAT1 function (Fig. 4e and 9c) [42]. Both NO711 and SKF89976a are observed to bind in the same pose in the primary binding site in this study done using an engineered transporter (Fig. 9b, c). An additional site for SKF89976a was observed in the vestibule of the engineered transporter, but this will also be discussed as part of the allosteric inhibition in the later section.

3.5 Glycine Reuptake Block

Glycine transporters among SLC6 members are involved in specific uptake of released glycine from the synapses. Glycine is considered to be an inhibitory neurotransmitter by itself but also acts as a co-activator of NMDA receptors in the synapse. Inhibitory control of glycine transporters has a direct effect on NMDA receptor activity resulting in altering long-term potentiation. Glycine is the smallest neurotransmitter transported by SLC6 members, and the two well-studied isoforms of GlyT (GlyT1 and GlyT2) have specific modifications within the binding pocket to accommodate the small size of the substrate. The F325 (dDAT) and L300 (GAT1) that are positioned in proximity to the substrate in the TM6 linker are substituted to a tryptophan in GlyTs (W322). This heavily constrains the size of the primary binding pocket and prevents larger neurotransmitters to interact (Fig. 10a–c). GlyT1 allows substrates like N-methyl and N-ethyl glycine, whereas GlyT2 is more specific due to the presence of an S479 residue that is substituted to a glycine in GlyT1, allowing greater substrate interactions. The ability of GlyTs to influence glycinergic neurotransmission and also NMDA receptor activity make them attractive targets for therapeutics for the treatment of pain, schizophrenia and epilepsy. Inhibitors derived from sarcosine, ORG-24598 and ALX5407 block GlyT1 and prevent glycine uptake. However, sarcosine derivatives tend to be toxic due to their irreversible interactions with GlyT1. Benzylpiperazine derivatives like bitopertin and iclepertin are known to be less toxic and are being explored for the treatment of schizophrenia. Bitopertin analogue cmpd1 was used to obtain a structure of GlyT1 in the inward-open state. However, it only partially occupies the primary binding site and mainly interacts with the inward-open state of GlyT1. A recent study on the GlyT1 structure demonstrates the binding of PF-034636725 within the vestibule and S1 site of GlyT1 in an outward-open state [69]. The inhibitor acts as a competitive inhibitor to glycine reuptake but can only partially occupy the binding pocket due to the limited size of the GlyT1 primary binding site. The rest of the molecule extends to the extracellular vestibule where the inhibitor wedges into the transporter in an outward-open state (Fig. 10b, c). The sarcosine-based derivative binds to the inward-open state of the transporter equivalent to bitopertin. Another GlyT1-specific inhibitor SSR504734 binds to the extracellular vestibule similar to citalopram in hSERT and SKF89976a in $dDAT_{GAT}$.

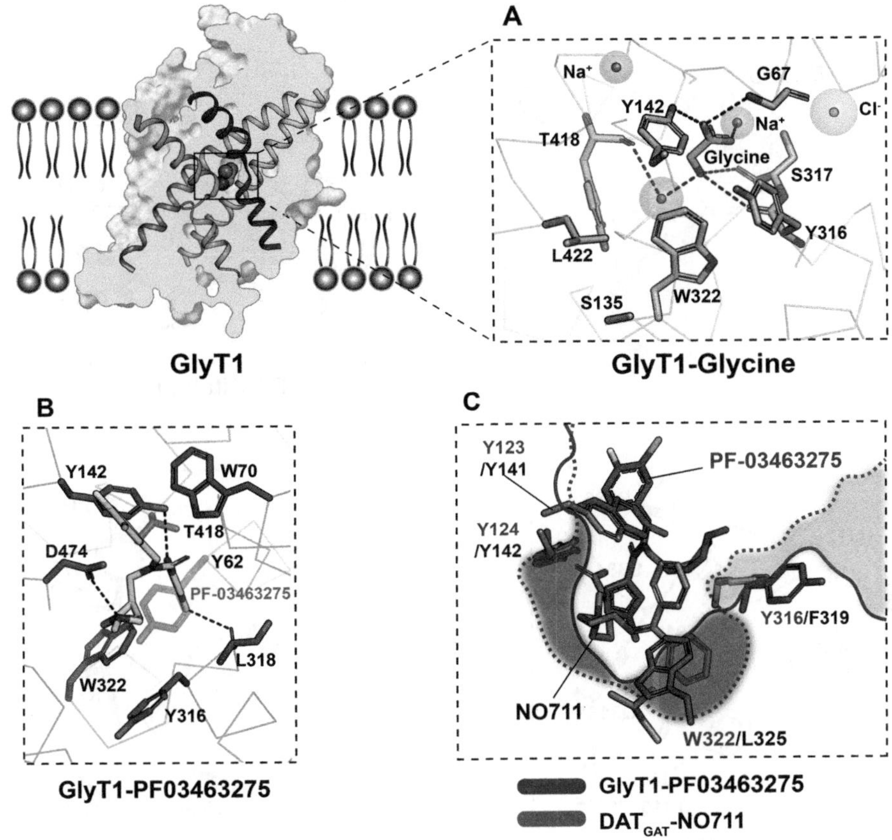

Fig. 10 Primary binding site in GlyT. (**a**) Glycine-bound structure of GlyT1 demonstrates the interactions of glycine within the primary S1 site of GlyT1 (PDB id. 8WFI). (**b**) Competitive inhibitor PF03463275 binds in the small S1 site, but the bulk of the drug interacts outside the S1 site due to the restriction of space in the binding pocket. (**c**) Comparison of the S1 sites of $dDAT_{GAT}$ and GlyT1 demonstrates the role of W322 in restricting GlyT's binding pocket

4 Allosteric Block of Neurotransmitter Transport

4.1 Biogenic Amine Transporters

Allosteric block of neurotransmitter uptake is an alternate strategy to competitive inhibition that is gaining prominence recently [70]. A majority of the known non-competitive blockers that block neurotransmitter uptake allosterically target the serotonin and dopamine transporters. The non-competitive inhibition of neurotransmitter uptake can happen at distinct sites around the transporters (Figs. 11 and 12). The obvious advantages to this type of inhibition lie in the greater specificity that allosteric inhibitors can have and also lesser toxicity due to lower interactions with

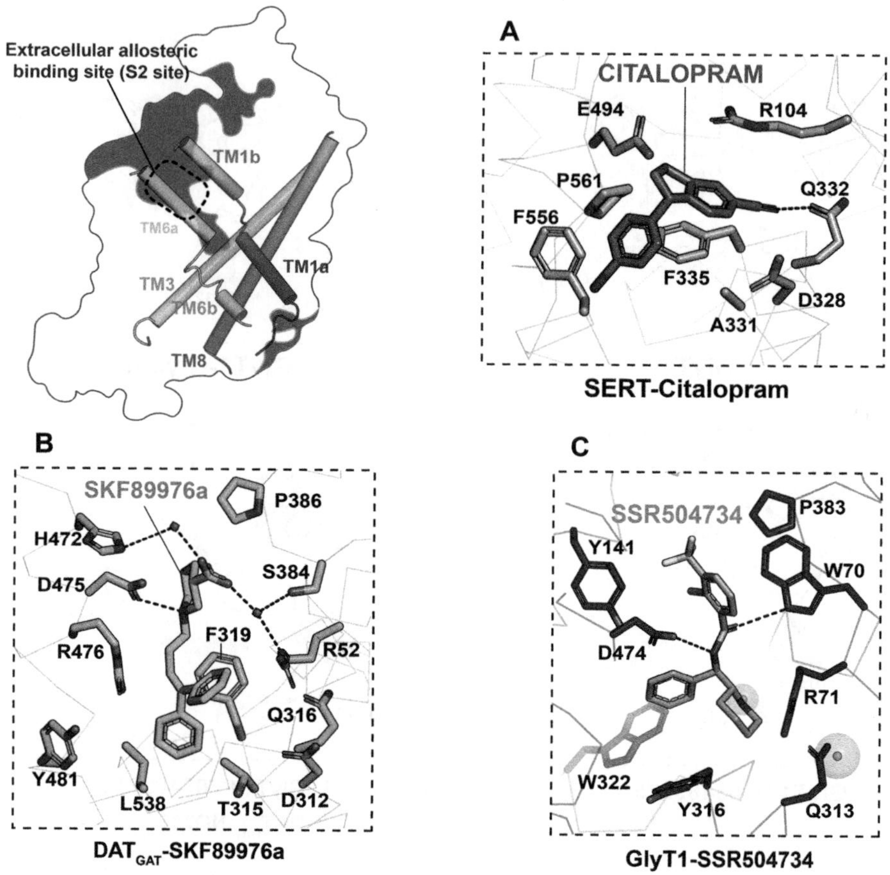

Fig. 11 Allosteric inhibition through interactions in the extracellular vestibule. (**a**–**c**) Structures of SLC6 transporters SERT, DAT_{GAT} and GlyT1 bound to citalopram (PDB id. 5I75), SKF899876a (PDB id. 7WLW) and PH (PDB id. 8WFK), respectively, provide details of interactions of non-competitive inhibitors bound to SLC6 members

alternate receptors that could be activated by the same compound. These factors make allosteric inhibitors/modulators desired compounds for drug discovery. It is therefore interesting to explore these mechanisms in the NSS family transporters that have traditionally been targets of competitive and nonspecific inhibitors leading to substantial and long-lasting side effects during treatment regimens. Presence of an alternate substrate-binding site in the outward-open conformation of the serotonin transporter was previously explored. Similarly some inhibitors like citalopram displayed the propensity to block the unbinding of the bound radiolabelled inhibitor suggesting that the drug can bind to an alternate site and prevent the dissociation of the drug bound in the primary site. In subsequent structural studies, citalopram was observed to bind to the extracellular vestibule and prevent the dissociation of the inhibitor from the bound primary site. Early studies had indicated that serotonin can

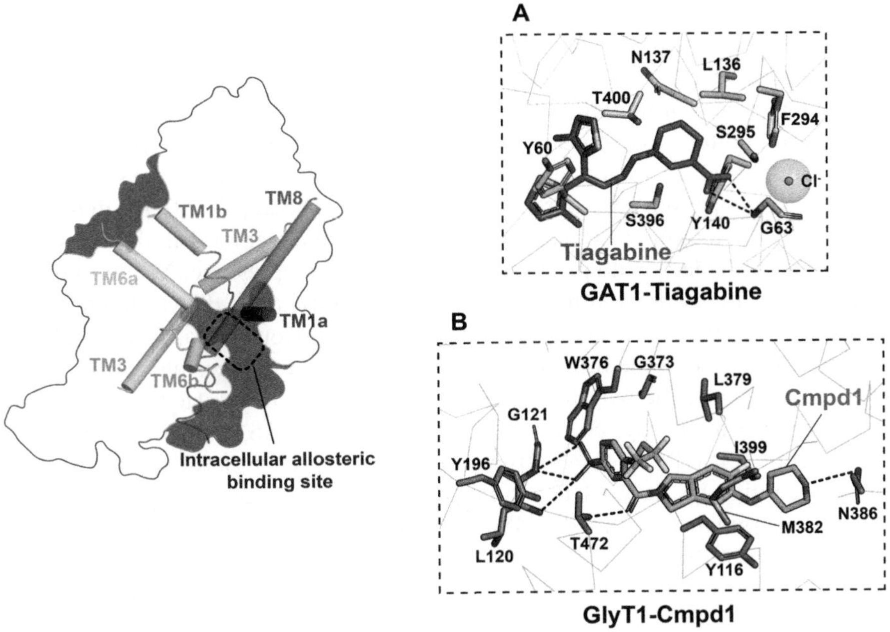

Fig. 12 Allosteric interactions at the intracellular vestibule: (**a**) Tiagabine-bound GAT1 (PDB id. 7Y7Z) and (**b**) bitopertin analogue (cmpd1) bound to GlyT1 (PDB id. 6ZBV) interacting in the cytosolic region of the transporters to wedge TM1a and the transporter in an inward-open state

prolong the dissociation rates of bound radiolabelled imipramine indicating that serotonin can bind to an alternate site and prevent unbinding of an inhibitor [71]. S-citalopram was also observed to have this allosteric effect, and mutations in the extracellular region of TM10 and TM12 affected the allosteric site but not the binding at the primary site [54, 72]. Several years after these initial findings, the structure of serotonin transporter revealed the presence of a citalopram-binding site in the extracellular vestibule in addition to the primary site, corroborating the biochemical evidence observed in earlier studies. The molecule was bound at the entry point of the extracellular vestibule in the vicinity of the gating and scaffold helices TMs 1,6, 10, 11 alongside the extracellular loops 4 and 6. The binding of citalopram disrupts the extracellular gate interactions between R104 (TM1b) and E494 (TM10) and likely blocks the closure of this gate during the transport cycle (Fig. 11a). More recently, specific allosteric inhibitors that exploit this site have been reported including vilazodone and Lu AF60097. Interestingly, these compounds are specific to the allosteric site and do not interact with the orthosteric site [73]. Vilazodone is a linear molecule that wedges deep in the extracellular vestibule and overlaps with the citalopram-binding site [74].

An interesting compound that also acts as a non-competitive 5-HT blocker is ibogaine, an alkaloid derived from *Tabernanthe iboga* [75]. The drug was discovered to have a non-competitive block of serotonin transport, but the cryoEM structures reveal that the molecule stabilizes the inward-open state of hSERT but still binds to the primary binding site making it an atypical non-competitive blocker of neurotransmitter transport [62].

The dopamine transporter in the recent years has also been the subject of the allosteric inhibition. The dopamine transporter has a unique extracellular zinc-binding site that allosterically modulates the activity [76]. Although the structure of a mammalian dopamine transporter was recently elucidated [77, 78], the dDAT has provided inputs into the pharmacology of inhibitors interacting with the dopamine transporter particularly with psychostimulants and antidepressants. While a bulk of the inhibitors were observed to be competitive inhibitors, the dDAT structure displayed the presence of a cholesterol molecule bound on the surface of TM1a. Presence of a lipid on the primary intracellular gate led to plausible role of lipid-mediated transporter activity regulation among NSS members. However, a bound cholesterol was not always consistently observed in the NSS structures at this location although hSERT and dDAT structures did display cholesterol at this location. Non-competitive blockers of dopamine transport were recently reported by the Mortensen group that identified KM822 as a molecule interacting with the extracellular vestibule of the dopamine transporter [79]. Using cysteine accessibility measurement, these compounds were observed to bind around TM1b (proximal to W84C) at a position similar to the citalopram interaction site of SERT. It is also specific for the outward-open conformation of the DAT and does not interact with the inward-open conformation.

4.2 Allosteric Inhibition of GABA and Glycine Transporters

Although the inhibitor tiagabine biochemically is a competitive inhibitor, the structure revealed it to be bound towards the cytosolic gate of GAT1 with the nipecotic acid moiety interacting at the primary binding site (Fig. 12a) [68]. The inhibitor at higher concentrations seems to have a non-competitive component that possibly allows it to permeate through the transporter and bind towards the cytosolic gate. Interestingly an engineered dopamine transporter made to resemble GAT1 did not display this behaviour, and the tiagabine analogues NO711 and SKF89976a were observed to be bound to the primary binding site [42]. It is also of interest to note that SKF89976a displayed an additional bound site in the extracellular vestibule of dDAT (Fig. 11b). It is suggested that SKF89976a blocks leak currents in GAT1 in a concentration-dependent fashion and SKF89976a binding to the $dDAT_{GAT}$ vestibule would be very interesting if a similar site is found in GAT1 [42, 80]. These sites are yet to be observed in the so far reported GAT1 structures, and future structures can unravel some of these aspects in future structural studies.

Glycine transporters: The binding of cmpd1, a bitopertin analogue, was found to occur in the inward open state of GlyT1 (Fig. 12b) [81]. The binding site overlaps with the binding site of tiagabine in GAT1. Although these compounds are competitive inhibitors, the interactions are observed to happen in a way that prevents the closure of the internal gate of the transporters. The GlyTs are also modulated by lipids allosterically. N-lauryl sarcosine is an allosteric modulator of GlyT2 function [82]. A sarcosine analogue, ALX 5407, was observed to bind to the cytosolic-facing state of the transporter similar to the bitopertin analogue [69]. The sarcosine group interacts by overlapping with the glycine-binding site, and the rest of the drug extends through the solvent accessible intracellular vestibule to lock the transporter in a cytosol facing state. Interestingly, a non-sarcosine-based inhibitor SSR504734 binds to the outward-open conformation of the hGlyT1 with the inhibitor wedged in the extracellular vestibule surrounded by gating helices TM1, TM6, El4 and the scaffold helix 3 [69]. SSR504734-binding site residues differ between GlyT1 and GlyT2, and therefore the inhibitor retains specificity to GlyT1. Unlike PF-034636725 that partially interacts with the S1 glycine-binding site, SSR504734 remains closer to the mouth of the vestibule in the outward-open states akin to citalopram in hSERT (Fig. 11c) [69].

5 Future of Drug Discovery Against NSS Transporters

The presence of numerous NSS structures in diverse conformational states and bound to cognate substrates and inhibitors is a very useful resource to perform drug discovery. One area that can benefit from this information is the design of conformational selective drugs. Incidentally the alkaloid ibogaine which was observed to stabilize the inward open conformation does so by interacting at the primary binding site [62]. Such information was harnessed in a recent study on hSERT, which highlighted the ability to generate novel high-affinity inhibitors against the cytosol-facing state of the transporters through extensive docking of a large-scale chemical database compound library ZINC. The best hits in the docked molecules were shortlisted followed by screening the affinity to the transporter followed by structure elucidation through cryoEM leading to an inward-open state SERT structure bound to a high affinity inhibitor [75]. These results are very encouraging for applying similar screening tools to other SLC6 members to obtain specific and potent inhibitors for clinical studies.

Allosteric inhibition has been a desired pharmacological strategy to block neurotransmitter uptake. The recent inroads into the discovery of several non-competitive inhibitors of neurotransmitter uptake have facilitated the use and development of this strategy to target select inhibitors. For instance, vilazodone is a non-competitive inhibitor of 5HT uptake that binds to the extracellular vestibule of hSERT and prevents the unbinding of the substrate from the primary binding site [74]. Similar inhibitors have been observed for human DAT and GlyT making this an exciting area of drug development among SLC6 members.

The development of AI-based tools for protein prediction could also be applied for drug discovery. Predictions of integral membrane transporter models are done in a highly accurate fashion by Alphafold2 [83]. This creates a potential for combining drug discovery with model prediction platforms to improve the predictive accuracies of inhibitors that could interact with NSS members. Besides small molecule inhibitors that target NSS transporters, novel antibodies could also be generated against the transporters to modulate their function. Directed evolution-based methods and technology to screen naïve libraries have dramatically enhanced the ability to develop binders against NSS members [84]. Consequently, antibody tools can be developed to obtain the desired activity against SLC6 members in non-neuronal targets, as more often than not, antibodies do not cross the blood–brain barrier. The availability of antibodies can also aid in imaging the transporters within neuronal tissues to facilitate the understanding of the clustering, trafficking and dynamics of the SLC6 members in synaptic membranes.

Although many human transporters can serve as druggable targets for physiological disorders, the vital roles that particularly the SLC6 members play in neurophysiology will retain them as a leading pharmacological target among all transporters [28]. These efforts could yield some very novel pharmacophores in the near future driven through structural biology of neurotransmitter transporters.

Acknowledgements DB and SRN are PhD students funded by GATE fellowships at IISc. AP is a senior fellow of the DBT-Wellcome Trust India Alliance, and research on neurotransmitter transporters is funded by India Alliance (IA/S/22/1/506242) and DST-Core Research Grant (CRG/2022/002301) in the Penmatsa lab.

Author Contributions The manuscript was authored by all the three authors. Figures were made by DB and SRN.

Competing Interests The authors declare no competing interests.

References

1. Jessell TM, Kandel ER (1993) Synaptic transmission: a bidirectional and self-modifiable form of cell-cell communication. Cell 72(Suppl):1–30
2. Amara SG, Kuhar MJ (1993) Neurotransmitter transporters: recent progress. Annu Rev Neurosci 16:73–93
3. Blakely RD, Edwards RH (2012) Vesicular and plasma membrane transporters for neurotransmitters. Cold Spring Harb Perspect Biol 4:a005595
4. Kristensen AS, Andersen J, Jorgensen TN, Sorensen L, Eriksen J, Loland CJ, Stromgaard K, Gether U (2011) SLC6 neurotransmitter transporters: structure, function, and regulation. Pharmacol Rev 63:585–640
5. Yamashita A, Singh SK, Kawate T, Jin Y, Gouaux E (2005) Crystal structure of a bacterial homologue of Na+/Cl--dependent neurotransmitter transporters. Nature 437:215–223
6. Iversen L (2006) Neurotransmitter transporters and their impact on the development of psychopharmacology. Br J Pharmacol 147(Suppl 1):S82–S88
7. Hediger MA, Romero MF, Peng JB, Rolfs A, Takanaga H, Bruford EA (2004) The ABCs of solute carriers: physiological, pathological and therapeutic implications of human membrane transport proteinsIntroduction. Pflugers Arch 447:465–468
8. Fredriksson R, Nordstrom KJ, Stephansson O, Hagglund MG, Schioth HB (2008) The solute carrier (SLC) complement of the human genome: phylogenetic classification reveals four major families. FEBS Lett 582:3811–3816

9. Masson J, Sagne C, Hamon M, El Mestikawy S (1999) Neurotransmitter transporters in the central nervous system. Pharmacol Rev 51:439–464
10. Rudnick G, Sandtner W (2019) Serotonin transport in the 21st century. J Gen Physiol 151:1248–1264
11. Broer S, Palacin M (2011) The role of amino acid transporters in inherited and acquired diseases. Biochem J 436:193–211
12. Chatterjee N, Rollins J, Mahowald AP, Bazinet C (2011) Neurotransmitter Transporter-Like: a male germline-specific SLC6 transporter required for Drosophila spermiogenesis. PLoS One 6:e16275
13. Ramamoorthy S, Shippenberg TS, Jayanthi LD (2011) Regulation of monoamine transporters: role of transporter phosphorylation. Pharmacol Ther 129:220–238
14. Jursky F, Tamura S, Tamura A, Mandiyan S, Nelson H, Nelson N (1994) Structure, function and brain localization of neurotransmitter transporters. J Exp Biol 196:283–295
15. Chaudhry FA, Edwards RH, Fonnum F (2008) Vesicular neurotransmitter transporters as targets for endogenous and exogenous toxic substances. Annu Rev Pharmacol Toxicol 48:277–301
16. Beuming T, Kniazeff J, Bergmann ML, Shi L, Gracia L, Raniszewska K, Newman AH, Javitch JA, Weinstein H, Gether U, Loland CJ (2008) The binding sites for cocaine and dopamine in the dopamine transporter overlap. Nat Neurosci 11:780–789
17. Penmatsa A, Gouaux E (2014) How LeuT shapes our understanding of the mechanisms of sodium-coupled neurotransmitter transporters. J Physiol 592:863–869
18. Claxton DP, Quick M, Shi L, de Carvalho FD, Weinstein H, Javitch JA, McHaourab HS (2010) Ion/substrate-dependent conformational dynamics of a bacterial homolog of neurotransmitter:sodium symporters. Nat Struct Mol Biol 17:822–829
19. Focht D, Neumann C, Lyons J, Eguskiza Bilbao A, Blunck R, Malinauskaite L, Schwarz IO, Javitch JA, Quick M, Nissen P (2021) A non-helical region in transmembrane helix 6 of hydrophobic amino acid transporter MhsT mediates substrate recognition. EMBO J 40:e105164
20. Krishnamurthy H, Gouaux E (2012) X-ray structures of LeuT in substrate-free outward-open and apo inward-open states. Nature 481:469–474
21. Kantcheva AK, Quick M, Shi L, Winther AM, Stolzenberg S, Weinstein H, Javitch JA, Nissen P (2013) Chloride binding site of neurotransmitter sodium symporters. Proc Natl Acad Sci USA 110:8489–8494
22. Rudnick G (1998) Ion-coupled neurotransmitter transport: thermodynamic vs. kinetic determinations of stoichiometry. Methods Enzymol 296:233–247
23. Drew D, Boudker O (2016) Shared molecular mechanisms of membrane transporters. Annu Rev Biochem 85:543–572
24. Singh SK, Piscitelli CL, Yamashita A, Gouaux E (2008) A competitive inhibitor traps LeuT in an open-to-out conformation. Science 322:1655–1661
25. Singh SK, Yamashita A, Gouaux E (2007) Antidepressant binding site in a bacterial homologue of neurotransmitter transporters. Nature 448:952–956
26. Zhou Z, Zhen J, Karpowich NK, Goetz RM, Law CJ, Reith ME, Wang DN (2007) LeuT-desipramine structure reveals how antidepressants block neurotransmitter reuptake. Science 317:1390–1393
27. Malinauskaite L, Quick M, Reinhard L, Lyons JA, Yano H, Javitch JA, Nissen P (2014) A mechanism for intracellular release of Na+ by neurotransmitter/sodium symporters. Nat Struct Mol Biol 21:1006–1012
28. Cesar-Razquin A, Snijder B, Frappier-Brinton T, Isserlin R, Gyimesi G, Bai X, Reithmeier RA, Hepworth D, Hediger MA, Edwards AM, Superti-Furga G (2015) A call for systematic research on solute carriers. Cell 162:478–487
29. Nestler EJ (2005) The neurobiology of cocaine addiction. Sci Pract Perspect 3:4–10
30. Berton O, Nestler EJ (2006) New approaches to antidepressant drug discovery: beyond monoamines. Nat Rev Neurosci 7:137–151
31. Gillman PK (2007) Tricyclic antidepressant pharmacology and therapeutic drug interactions updated. Br J Pharmacol 151:737–748

32. Plenge P, Gether U, Rasmussen SG (2007) Allosteric effects of R- and S-citalopram on the human 5-HT transporter: evidence for distinct high- and low-affinity binding sites. Eur J Pharmacol 567:1–9
33. Pertovaara A (2006) Noradrenergic pain modulation. Prog Neurobiol 80:53–83
34. Maxwell SR, Bateman DN (2007) Choice of opioid analgesics in postoperative care. Lancet 369:2000
35. Treiman DM (2001) GABAergic mechanisms in epilepsy. Epilepsia 42(Suppl 3):8–12
36. Madsen KK, Clausen RP, Larsson OM, Krogsgaard-Larsen P, Schousboe A, White HS (2009) Synaptic and extrasynaptic GABA transporters as targets for anti-epileptic drugs. J Neurochem 109(Suppl 1):139–144
37. Cummings KA, Popescu GK (2015) Glycine-dependent activation of NMDA receptors. J Gen Physiol 145:513–527
38. Mohler H, Boison D, Singer P, Feldon J, Pauly-Evers M, Yee BK (2011) Glycine transporter 1 as a potential therapeutic target for schizophrenia-related symptoms: evidence from genetically modified mouse models and pharmacological inhibition. Biochem Pharmacol 81:1065–1077
39. Penmatsa A, Wang KH, Gouaux E (2013) X-ray structure of dopamine transporter elucidates antidepressant mechanism. Nature 503:85–90
40. Sorensen L, Andersen J, Thomsen M, Hansen SM, Zhao X, Sandelin A, Stromgaard K, Kristensen AS (2012) Interaction of antidepressants with the serotonin and norepinephrine transporters: mutational studies of the S1 substrate binding pocket. J Biol Chem 287:43694–43707
41. Wang H, Goehring A, Wang KH, Penmatsa A, Ressler R, Gouaux E (2013) Structural basis for action by diverse antidepressants on biogenic amine transporters. Nature 503:141–145
42. Joseph D, Nayak SR, Penmatsa A (2022) Structural insights into GABA transport inhibition using an engineered neurotransmitter transporter. EMBO J 41:e110735
43. Nayak SR, Joseph D, Hofner G, Dakua A, Athreya A, Wanner KT, Kanner BI, Penmatsa A (2023) Cryo-EM structure of GABA transporter 1 reveals substrate recognition and transport mechanism. Nat Struct Mol Biol 30:1023–1032
44. Meiser J, Weindl D, Hiller K (2013) Complexity of dopamine metabolism. Cell Commun Signal 11:34
45. Wang KH, Penmatsa A, Gouaux E (2015) Neurotransmitter and psychostimulant recognition by the dopamine transporter. Nature 521:322–327
46. Khoshbouei H, Wang H, Lechleiter JD, Javitch JA, Galli A, Amphetamine-induced dopamine efflux. (2003) A voltage-sensitive and intracellular Na+-dependent mechanism. J Biol Chem 278:12070–12077
47. Porzgen P, Park SK, Hirsh J, Sonders MS, Amara SG (2001) The antidepressant-sensitive dopamine transporter in Drosophila melanogaster: a primordial carrier for catecholamines. Mol Pharmacol 59:83–95
48. Penmatsa A, Wang KH, Gouaux E (2015) X-ray structures of Drosophila dopamine transporter in complex with nisoxetine and reboxetine. Nat Struct Mol Biol 22:506–508
49. Kitayama S, Shimada S, Xu H, Markham L, Donovan DM, Uhl GR (1992) Dopamine transporter site-directed mutations differentially alter substrate transport and cocaine binding. Proc Natl Acad Sci USA 89:7782–7785
50. Marks DM, Shah MJ, Patkar AA, Masand PS, Park GY, Pae CU (2009) Serotonin-norepinephrine reuptake inhibitors for pain control: premise and promise. Curr Neuropharmacol 7:331–336
51. Pietrzak RH, Gallezot JD, Ding YS, Henry S, Potenza MN, Southwick SM, Krystal JH, Carson RE, Neumeister A (2013) Association of posttraumatic stress disorder with reduced in vivo norepinephrine transporter availability in the locus coeruleus. JAMA Psychiatry 70:1199–1205
52. Mitchell HA, Weinshenker D (2010) Good night and good luck: norepinephrine in sleep pharmacology. Biochem Pharmacol 79:801–809
53. Pidathala S, Mallela AK, Joseph D, Penmatsa A (2021) Structural basis of norepinephrine recognition and transport inhibition in neurotransmitter transporters. Nat Commun 12:2199

54. Coleman JA, Green EM, Gouaux E (2016) X-ray structures and mechanism of the human serotonin transporter. Nature 532:334–339
55. Zhang H, Yin YL, Dai A, Zhang T, Zhang C, Wu C, Hu W, He X, Pan B, Jin S, Yuan Q, Wang MW, Yang D, Xu HE, Jiang Y (2024) Dimerization and antidepressant recognition at noradrenaline transporter. Nature 630:247–254
56. Koldso H, Christiansen AB, Sinning S, Schiott B (2013) Comparative modeling of the human monoamine transporters: similarities in substrate binding. ACS Chem Neurosci 4:295–309
57. Schlessinger A, Geier E, Fan H, Irwin JJ, Shoichet BK, Giacomini KM, Sali A (2011) Structure-based discovery of prescription drugs that interact with the norepinephrine transporter, NET. Proc Natl Acad Sci USA 108:15810–15815
58. Binda F, Dipace C, Bowton E, Robertson SD, Lute BJ, Fog JU, Zhang M, Sen N, Colbran RJ, Gnegy ME, Gether U, Javitch JA, Erreger K, Galli A (2008) Syntaxin 1A interaction with the dopamine transporter promotes amphetamine-induced dopamine efflux. Mol Pharmacol 74:1101–1108
59. Rudnick G, Nelson PJ (1978) Platelet 5-hydroxytryptamine transport, an electroneutral mechanism coupled to potassium. Biochemistry 17:4739–4742
60. Hellsberg E, Boytsov D, Chen Q, Niello M, Freissmuth M, Rudnick G, Zhang YW, Sandtner W, Forrest LR (2024) Identification of the potassium-binding site in serotonin transporter. Proc Natl Acad Sci USA 121:e2319384121
61. Rudnick G, Nelson PJ (1978) Reconstitution of 5-hydroxytryptamine transport from cholate-disrupted platelet plasma membrane vesicles. Biochemistry 17:5300–5303
62. Coleman JA, Yang D, Zhao Z, Wen PC, Yoshioka C, Tajkhorshid E, Gouaux E (2019) Serotonin transporter-ibogaine complexes illuminate mechanisms of inhibition and transport. Nature 569:141–145
63. Henry LK, Field JR, Adkins EM, Parnas ML, Vaughan RA, Zou MF, Newman AH, Blakely RD (2006) Tyr-95 and Ile-172 in transmembrane segments 1 and 3 of human serotonin transporters interact to establish high affinity recognition of antidepressants. J Biol Chem 281:2012–2023
64. Andersen J, Olsen L, Hansen KB, Taboureau O, Jorgensen FS, Jorgensen AM, Bang-Andersen B, Egebjerg J, Stromgaard K, Kristensen AS (2010) Mutational mapping and modeling of the binding site for (S)-citalopram in the human serotonin transporter. J Biol Chem 285:2051–2063
65. Yang D, Zhao Z, Tajkhorshid E, Gouaux E (2023) Structures and membrane interactions of native serotonin transporter in complexes with psychostimulants. Proc Natl Acad Sci USA 120:e2304602120
66. Guastella J, Nelson N, Nelson H, Czyzyk L, Keynan S, Miedel MC, Davidson N, Lester HA, Kanner BI (1990) Cloning and expression of a rat brain GABA transporter. Science 249:1303–1306
67. Bismuth Y, Kavanaugh MP, Kanner BI (1997) Tyrosine 140 of the gamma-aminobutyric acid transporter GAT-1 plays a critical role in neurotransmitter recognition. J Biol Chem 272:16096–16102
68. Motiwala Z, Aduri NG, Shaye H, Han GW, Lam JH, Katritch V, Cherezov V, Gati C (2022) Structural basis of GABA reuptake inhibition. Nature 606:820–826
69. Wei Y, Li R, Meng Y, Hu T, Zhao J, Gao Y, Bai Q, Li N, Zhao Y (2024) Transport mechanism and pharmacology of the human GlyT1. Cell 187:1719–1732 e1714
70. Niello M, Gradisch R, Loland CJ, Stockner T, Sitte HH (2020) Allosteric modulation of neurotransmitter transporters as a therapeutic strategy. Trends Pharmacol Sci 41:446–463
71. Wennogle LP, Meyerson LR (1985) Serotonin uptake inhibitors differentially modulate high affinity imipramine dissociation in human platelet membranes. Life Sci 36:1541–1550
72. Zhong H, Sanchez C, Caron MG (2012) Consideration of allosterism and interacting proteins in the physiological functions of the serotonin transporter. Biochem Pharmacol 83:435–442
73. Plenge P, Abramyan AM, Sorensen G, Mork A, Weikop P, Gether U, Bang-Andersen B, Shi L, Loland CJ (2020) The mechanism of a high-affinity allosteric inhibitor of the serotonin transporter. Nat Commun 11:1491

74. Plenge P, Yang D, Salomon K, Laursen L, Kalenderoglou IE, Newman AH, Gouaux E, Coleman JA, Loland CJ (2021) The antidepressant drug vilazodone is an allosteric inhibitor of the serotonin transporter. Nat Commun 12:5063
75. Singh I, Seth A, Billesbolle CB, Braz J, Rodriguiz RM, Roy K, Bekele B, Craik V, Huang XP, Boytsov D, Pogorelov VM, Lak P, O'Donnell H, Sandtner W, Irwin JJ, Roth BL, Basbaum AI, Wetsel WC, Manglik A, Shoichet BK, Rudnick G (2023) Structure-based discovery of conformationally selective inhibitors of the serotonin transporter. Cell 186:2160–2175 e2117
76. Norregaard L, Frederiksen D, Nielsen EO, Gether U (1998) Delineation of an endogenous zinc-binding site in the human dopamine transporter. EMBO J 17:4266–4273
77. Srivastava DK, Navratna V, Tosh DK, Chinn A, Sk MF, Tajkhorshid E, Jacobson KA, Gouaux E (2024) Structure of the human dopamine transporter and mechanisms of inhibition. Nature 632:672–677
78. Nielsen JC, Salomon K, Kalenderoglou IE, Bargmeyer S, Pape T, Shahsavar A, Loland CJ (2024) Structure of the human dopamine transporter in complex with cocaine. Nature 632:678–685
79. Aggarwal S, Liu X, Rice C, Menell P, Clark PJ, Paparoidamis N, Xiao YC, Salvino JM, Fontana ACK, Espana RA, Kortagere S, Mortensen OV (2019) Identification of a novel allosteric modulator of the Human Dopamine Transporter. ACS Chem Neurosci 10:3718–3730
80. Krause S, Schwarz W (2005) Identification and selective inhibition of the channel mode of the neuronal GABA transporter 1. Mol Pharmacol 68:1728–1735
81. Shahsavar A, Stohler P, Bourenkov G, Zimmermann I, Siegrist M, Guba W, Pinard E, Sinning S, Seeger MA, Schneider TR, Dawson RJP, Nissen P (2021) Structural insights into the inhibition of glycine reuptake. Nature 591:677–681
82. Mostyn SN, Wilson KA, Schumann-Gillett A, Frangos ZJ, Shimmon S, Rawling T, Ryan RM, O'Mara ML, Vandenberg RJ (2019) Identification of an allosteric binding site on the human glycine transporter, GlyT2, for bioactive lipid analgesics. elife 8:e47150
83. Del Alamo D, Sala D, McHaourab HS, Meiler J (2022) Sampling alternative conformational states of transporters and receptors with AlphaFold2. elife 11:e75751
84. McMahon C, Baier AS, Pascolutti R, Wegrecki M, Zheng S, Ong JX, Erlandson SC, Hilger D, Rasmussen SGF, Ring AM, Manglik A, Kruse AC (2018) Yeast surface display platform for rapid discovery of conformationally selective nanobodies. Nat Struct Mol Biol 25:289–296

Substrate Binding and Conformational Dynamics of the Monoamine Transporters

Jeppe C. Nielsen and Claus J. Loland

Contents

Abstract The monoamine transporters move substrates across the plasma membrane by an alternating-access mechanism, in which a central substrate-binding site is alternately exposed to either the extracellular milieu or the cytoplasm at any given time. This process is driven by co-transport of sodium ions along the inwardly directed sodium gradient. Alternating access to the central substrate-binding site is facilitated by a stepwise series of changes to the transporter conformation, referred to as the transport cycle. The focus of this chapter is to discuss the conformational dynamics of the monoamine transporters that are orchestrated by the binding of substrate and ions, as part of the transport cycle. Firstly, we describe the substrate-binding event, and how it is fine-tuned to induce the conformational flexibility needed to initiate transport. Secondly, we discuss how sodium fuels the substrate transport as well as how it is aided by potassium and chloride. We also provide a mechanistic description of the cooperativity of the two sodium-binding sites and how they couple allosterically to the intracellular gating mechanism. Thirdly, we go over the amino acid residues of the intra- and extracellular gates and how they affect the transporter conformation.

J. C. Nielsen · C. J. Loland (✉)
Department of Neuroscience, University of Copenhagen, Copenhagen, Denmark
e-mail: jeppe.nielsen@sund.ku.dk; cllo@sund.ku.dk

O. V. Mortensen (ed.), *Monoamine Transporters*, Advances in Neurobiology 46,
https://doi.org/10.1007/978-3-031-96364-3_2

Keywords Substrate-induced conformational dynamics · Biogenic monoamines · Dopamine transporter · Norepinephrine transporter · Serotonin transporter · Cooperative sodium binding · Potassium countertransport · Chloride dependent · Coupled gating mechanisms

1 Introduction

In 1966, Oleg Jardetzky proposed a concise mechanism for how transporter proteins move solutes across the plasma membrane [1]. Although Jardetzky did not coin the term "alternating access" in his original publication, his suggested mechanism has nevertheless become known as the alternating-access model of substrate transport. To this day, our understanding of the function of the monoamine transporters (MATs) is based on the alternating-access model.

The MATs comprise the three transporters for the neurotransmitters of the biogenic monoamine class: the dopamine (DA) transporter (DAT), the norepinephrine (NE) transporter (NET), and the serotonin/5-hydroxytyramine (5-HT) transporter (SERT) [2]. They are responsible for the reuptake of these neurotransmitters back into the presynaptic neuron after release to the synapse, and the MATs thereby play a fundamental role in upholding proper neurotransmission [2].

All three transporters belong to the transporter superfamily denoted by the neurotransmitter:sodium symporters (NSSs). The transporters of the NSS superfamily are characterized by utilizing the energy of Na^+ cotransport along its electrochemical gradient, to drive the transport of substrate against its concentration gradient [3]. Another member of the NSS superfamily is the prokaryotic homolog of the MATs, the bacterial leucine transporter (LeuT), the structure of which was determined by Eric Gouaux and coworkers in 2005 [4]. Based on the high degree of sequence homology between LeuT and the conserved regions of the MATs [4, 5], their structures were proposed to share the same fold as LeuT, which became known as the LeuT-fold. The LeuT-fold is characterized by the arrangement of the first ten of the 12 transmembrane helices (TMs), with TM1–5 positioned in a pseudo-symmetrical fashion in relation to TM6–10. TM11 and TM12 sit on the outside of the transporter core structure and are not participating in the LeuT-fold (Fig. 1). Centrally located within the fold, approximately halfway through the transporter, is the central substrate-binding (S1) site to where the substrate binds during the transport event, and around the S1 site are binding sites for Na^+ and Cl^- (see Sect. 3). Given the proposed structural resemblance, LeuT became a trusted model of the MATs used to test the predictions made by the alternating-access model and to generalize its principles across DAT, NET, and SERT [6]. Finally, the structural resemblance of LeuT and the MATs was confirmed when the structures of the *drosophila melanogaster* DAT (dDAT) and human SERT (hSERT) were published in 2013 and 2016, respectively [7, 8], further warranting the use of LeuT as MAT model. Since

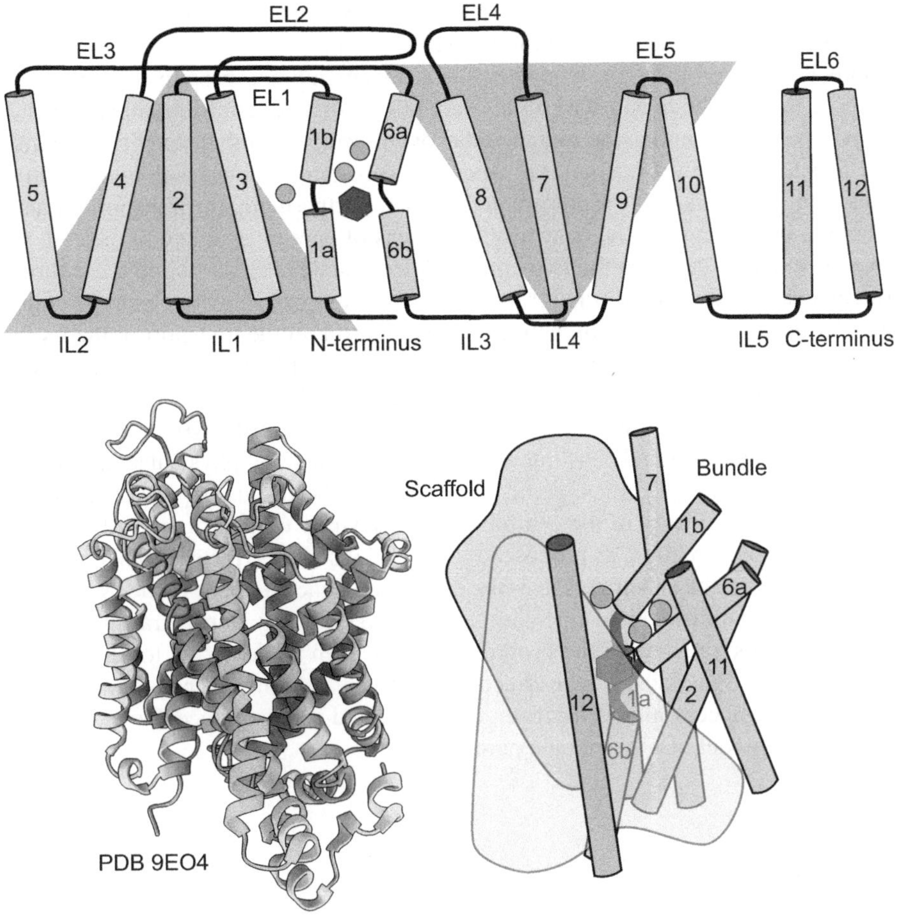

Fig. 1 The monoamine transporters display the LeuT-fold. The transmembrane helices 1–10 of the monoamine transporters acquire the LeuT-fold and are shown colored brown and light blue. The upper panel shows the transporter topology with LeuT-fold characterized by the two 5+5 helical, pseudo-symmetrical inverted repeats marked by the two inverted pink triangles. One repeat is made up by transmembrane helices 1–5 and the other by 6–10. Helices 11 and 12 in gray do not participate in the fold. Helices 1 and 6 are parted in two, termed 1a/b and 6a/b, respectively, and the substrate shown as a dark blue hexagon binds to the central substrate binding site, situated at the unwound segments of these two helices. Two Na^+, shown as orange spheres, bind near the central substrate binding site to the Na1 site, located between helix 1b and 6a, and the Na2 site between helix 1 and 3. A Cl^-, shown as a green sphere, binds to the Cl site closest to the Na1 site. The transmembrane helices 1, 2, 6, and 7, colored light blue, constitute the bundle domain where most conformational changes happen during substrate transport, whereas helices 3, 4, 5, 8, 9, and 10, colored brown, constitute the more static scaffold domain. The lower left panel shows how the helices oriented into the two domains in 3D space with helices 11 and 12 sitting outside the transporter core structure. The lower right panel shows a schematic representation of the two domains with the helices colored accordingly, and the helices of the scaffold domain are shown as brown shapes. The substrate and ions are shown at the core of the transporter

then, structures of the human DAT (hDAT), human NET (hNET), and porcine SERT have also confirmed the LeuT-fold in the MATs [9–11].

Jardetzky stated three prerequisites for a transmembrane protein to work as a pump. A transporter must (i) be able to adopt two distinct conformations that (ii) opens a pore from either the extracellular or intracellular milieu to (iii) a central binding site within the transporter to where the substrate can bind with high affinity in one conformation and low affinity in the other. Furthermore, the transporter must be able to harvest any of the conceivable sources of energy available to the cell, to drive the changes between the two conformational states [1]. Although these three principal assumptions hold true for the MATs, the case has shown to be more complex. As stated, the MATs utilize the Na^+-gradient to fuel the transport of their solutes across the plasma membrane. But they also utilize the K^+-gradient by countertransport, to increase their transport capacity. They can in fact adopt more than two distinct conformations as revealed in structures of the MATs in complex with various inhibitors and substrates, and shown also with biophysical assays probing conformational dynamics, some of which will be discussed in this chapter. Furthermore, the affinity of the central site for substrate might not depend entirely on the MAT conformation, as proposed by Jardetzky's model. Instead, other mechanisms ensure that the substrate leaves the central site, once it opens to the cytoplasm following the conformational changes that the MATs undergo during transport. This chapter presents a review of the progress that has been made on the topic of MAT conformational dynamics and the effects of substrate and ion binding, showing how the alternating-access model has been expanded and refined, since Jardetzky first proposed his model of a membrane pump.

2 The Transport Cycle

The alternating-access model is currently our best theory of how the MATs move their substrates across the plasma membrane. They do so by cycling through a series of metastable conformations, collectively referred to as the transport cycle. Although Jardetzky's membrane pump model was somewhat rudimentary, it has nevertheless laid the foundation on which the steps of the transport cycle are built. The transport cycle of the MATs has been proposed on basis of multiple functional and structural studies of DAT, NET, and SERT, as well as LeuT. Although protein structures have been crucial for our understanding of the transport cycle, these are still momentary, static snapshots of a highly dynamic process. So, to fill in the blanks, functional studies employing biophysical and pharmacological assays to report on the dynamics and kinetics of the transporters have also been widely applied. All in all, this has demonstrated how the conformational changes are orchestrated by the binding and release of substrates and ions in a highly coordinated fashion.

The transport cycle (Fig. 2) can be described in the following steps. In the absence of substrate and Na^+ (apo-state), the transporters are believed to obtain an ensemble of all possible conformations, but in an uncoordinated way [12, 13].

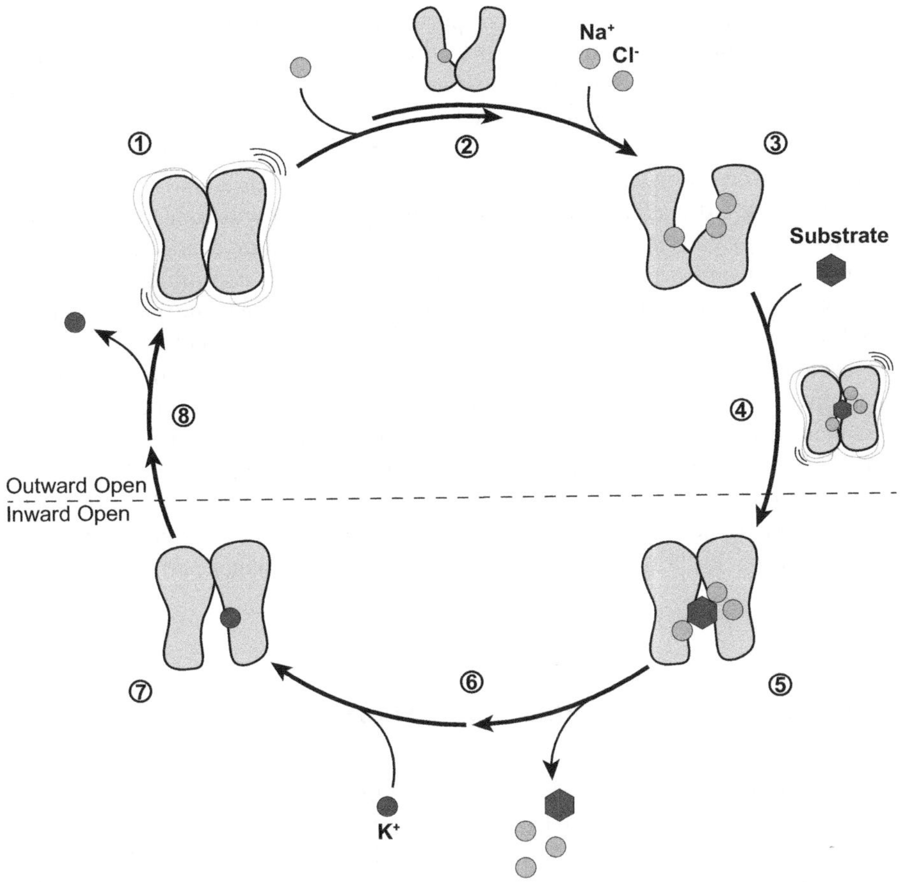

Fig. 2 The transport cycle. (1) The transporter, shown as gray shapes, in the apo-state fluctuates between the outward-facing and inward-facing conformations. (2) Binding of one Na^+, shown as orange spheres, to the Na2 site stabilizes the outward-open conformation that exposes the central substrate-binding site through the extracellular vestibule to the extracellular milieu and primes the transporter for undergoing a round of transport. (3) A second Na^+ engages the Na1 site, inducing local conformational changes around the central substrate-binding site that primes the transporter for substrate binding. A Cl^- potentially binds with the first or second Na^+, or together with the positively charged substrate. (4) Once in the outward-open, primed conformation, the substrate, shown as a dark blue hexagon, can diffuse through the extracellular permeation pathway and bind at the central substrate-binding site. Substrate binding releases the conformational tension induced by Na^+ binding and allows the transporter to fluctuate between the outward- and inward-open conformations via an occluded state. (5) When visiting the inward-open conformation, ions and the substrate can dissociate and diffuse through the intracellular permeation pathway due to the low cytoplasmic concentration of Na^+. However, the exact stoichiometry of the transported ions as well as how the substrate is released once facing the high substrate concentration of the cytoplasm is still uncertain. One possibility is that release of one or both Na^+ and/or the Cl^- triggers the release of the substrate. (6) Another possibility is that K^+, shown as a purple sphere, binding competes out one or both the Na^+ and potentially the Cl^- which then triggers substrate release. (7) The K^+-bound transporter reaches the outward-open conformation by an unclear mechanism, and (8) the K^+ is released on the extracellular side due to the low K^+ concentration

Binding of Na^+ and Cl^- primes the transporter for substrate binding by stabilizing the outward-open conformation where the centrally located S1 site is accessible from the extracellular environment. Substrate binding triggers local conformational changes that prevent access to the S1 site from the extracellular space and marks the transition to the outward-occluded conformation (see Sect. 3.3). In the outward-occluded conformation, newly formed interactions between amino acid residues constituting the extracellular gate and interactions formed with substrate and ions lower the energy barrier, which allows for global conformation change of the transporter to the inward-occluded or holo-occluded state. In this conformation, spontaneous breakage of the intracellular gate exposes the bound substrate and ions to the cytoplasm in the inward-open conformation. Presumably, Na^+ and Cl^- dissociate first due to the low intracellular Na^+ concentration, and this decreases the affinity for the substrate, causing it to dissociate. The inward-open, apo-state of the transporter must then reset itself to the outward facing conformation to finish the transport cycle. This is thought to be the rate-limiting step of the transport cycle, and the mechanism remains somewhat elusive. K^+ countertransport plays a role [14], albeit not being strictly necessary for substrate transport, but has been found to increase the uptake capacity [15–17], suggesting it to be a fundamental part of the transport cycle.

3 Substrate Binding Initiates Transport

The third principal property of Jardetzky's membrane pump is the presence of a binding site within the transporter to where the substrate can bind with an affinity that depends on the transporter conformation [1]. As evident by the structures of hSERT, dDAT, and also LeuT [4, 7, 8], the S1 site is located in the center of the transporters, and it is spatially and functionally associated with the binding sites of two Na^+ and one Cl^-. Amino acid residues situated on TM1, TM3, TM6, and TM8 compose the S1 site where their sidechains form interactions with the substrate. TM1 and TM6 are both parted in two, designated TM1a/b and TM6a/b, by an unwound segment of each helix, which leaves free backbone carbonyl and amide groups to also engage in interactions with the substrate (Fig. 1).

3.1 The Monoamine Transporters Have Low Substrate Specificity

Even though the S1 site is the primary binding site for the MAT substrates, the site is not exclusive to their cognate substrates. In cells transiently transfected with either dDAT, human DAT (hDAT) or human NET (hNET), all three MATs showed specific uptake of radiolabeled DA ([^{3}H]DA) and norepinephrine ([^{3}H]NE).

However, each of the transporters did display the highest transport affinity for their respective substrate, with the only exception being hNET. DA is transported with a higher transport affinity by hNET than is NE, but hDAT transports DA with a higher maximal velocity, i.e., higher capacity, than hNET [18]. Even radiolabeled tyramine can be transported by all three MATs, with a transport affinity comparable to that for DA and NE. Also, dDAT transports radiolabeled epinephrine [18]. The lack of substrate specificity regarding hDAT and hNET is not surprising, as both DA and NE are catecholamines with the only chemical difference being the β-hydroxyl group of NE (Fig. 3). However, also hSERT has been found to transport [^{3}H]DA [19], even though its cognate substrate, 5-HT, belongs to the indolamines (Fig. 3). Strikingly, it does so with a higher maximal velocity than for [^{3}H]5-HT uptake, albeit with a 100-fold lower transport affinity. By introducing a subtle mutation, Y95F, the transport affinity for DA increased almost seven times when compared to wild-type hSERT [19]. Interestingly, Tyr95 is located in the S1 site, and both hDAT and hNET have a phenylalanine residue at the homologous position [2], suggesting that this position confers some substrate preference.

The most important amino acid residue of the S1 site is a highly conserved aspartic acid residue (Asp79 in hDAT, Asp75 in hNET, and Asp98 in hSERT) [2] that serves as the primary recognition site for substrate binding. It is located one helical turn above Tyr95 (Phe76 in hDAT and Phe72 in hNET), and structures of the MATs in complex with DA, NE, and 5-HT [9, 20–22] have shown that all known MAT substrates are wedged in between this conserved aspartic acid residue and the aromatic moiety below. A study of the transport associated currents in hDAT and a study of transport under different ion gradients in hSERT have revealed that DA and 5-HT, respectively, are transported in their protonated state [23, 24]. The suggested protonation of substrate is backed by the fact that the acid constant of the amine renders it protonated at physiological pH [25]. This suggests an electrostatic interaction between the substrate and the conserved aspartic acid residue in the S1 site, which ensures a strong, long-range attraction between the two [26]. Removal of the negative charge in hDAT with the mutation D79A decreased the maximal transport velocity of [^{3}H]DA to less than 10%. The charge conserving mutation, D79E, retains a maximal transport velocity only around 20% [27], suggesting that the length of the

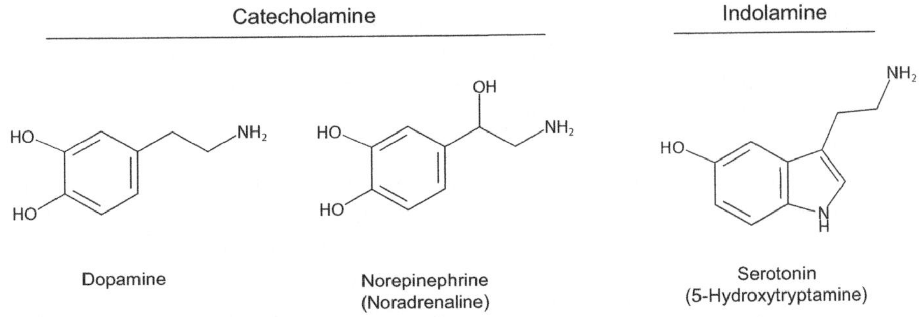

Fig. 3 Chemical structures of the biogenic monoamines

residue, and thereby the position of the negative charge in the S1 site, is not trivial for uptake. The D98A and D75A mutations have shown to completely abolish uptake in the rat SERT (rSERT) and hNET, respectively. However, rSERT-D98E retained almost 50% maximal transport velocity of the wild-type rSERT, whereas hNET-D75E showed no uptake at all. Interestingly, the rSERT-D98E had a tenfold higher affinity for gramine when compared to the wild-type rSERT. Gramine is also an indolamine substrate of SERT, but its alkylamine is one carbon shorter than that of 5-HT [28], suggesting that a certain distance between the positive and negative charges is required for optimal substrate transport. Lastly, the role of Asp79/98/75 is consolidated by the fact that LeuT, which transports amino acids, has a glycine at the homologous position. Amino acids contain a carboxyl group next to their amine group on the α-carbon which carry a negative charge [2, 5]. The mutants hDAT-D79G, rSERT-D98G, and hNET-D75G show no uptake at all [27, 28]. All in all, the position of a negative charge in the S1 site in relation to the positively charged amine of the substrate, and possibly also the aromatic moiety of Phe76/Tyr95/Phe72, is pivotal for transport of MAT substrates.

3.2 *The S1 Site Has Low Affinity for Substrate*

Binding of substrate is, logically, a prerequisite for its transport. But substrates generally bind with low affinity to the S1 site of the MATs, compared to inhibitors. The direct binding of DA to dDAT has been measured on purified protein, by displacement of [^{3}H]DA. The binding affinity of purified dDAT for DA was found to be almost 200 times lower than for the dDAT inhibitor nisoxetine [29]. The direct binding of substrate to hSERT has also been measured on purified protein by assessing the fluorescence change of a noncanonical amino acid incorporated into hSERT using genetic code expansion. The apparent affinity for 5-HT was found to be 400-fold lower than the affinity for the SERT inhibitor S-citalopram [30].

It would be reasonable to think that effective transport requires effective binding of the substrate to the transporter, reflected in a high-affinity interaction. However, the binding affinity is a ratio of the association and dissociation rates, and the association rate is ultimately limited by the diffusion rate. So, to get a high affinity interaction between the substrate and the S1 site, the dissociation rate would have to be low potentially preventing the release of substrate and thereby slowing transport. This has been studied in LeuT by comparing its binding affinity for the three amino acids, leucine, alanine, and methionine, measured by saturation binding, to its turnover number for the same amino acids measured by transport into liposomes. The authors found a negative correlation between the binding affinity of LeuT for the three amino acids and its turnover number, meaning that the tighter the substrate binds, the slower it is transported [31]. The hSERT ion currents associated with binding and transport of substrate provide some mechanistic detail to this [32]. Here the apparent affinity of hSERT for 5-HT and three analogs was investigated by looking at the transport-associated currents. The analogs varied in length, one with a shorter and two with longer alkylamines relative to 5-HT, thereby changing the position of the amine to Asp98. The apparent affinity of hSERT for the analog with the shorter alkylamine was almost 18-fold lower than for 5-HT. This decrease in affinity is similar to the increase in apparent affinity observed with gramine binding to rSERT-D98E [28]. Conversely, hSERT had more than sevenfold higher apparent affinity for the analogs with longer alkylamines. The differences in apparent affinity could be explained by the binding kinetics, measured by the hSERT ion currents. The analogs with the long alkylamines had a high association rate combined with a low dissociation rate, whereas the opposite was true for the analog with the short alkylamine. 5-HT was found to have intermediate rates of both association and dissociation compared to the analogs, but the transition rate between the inward- and outward-facing conformations of hSERT was found to be highest in presence of 5-HT. Based on molecular dynamics simulations of the system, it was suggested that the alkylamine of 5-HT has just the right length to position its amine in proximity to Asp98, while the rest of the 5-HT molecule interacted with other S1 residues when hSERT was in the occluded conformation. By satisfying the most interactions, 5-HT efficiently stabilized the occluded conformation, thereby lowering the energy barrier for the transition between outward- and inward-facing conformations making it a good substrate of hSERT compared to the analogs. This suggests that the affinity of the substrate is fine-tuned to be high enough to engage proper interactions within the S1 site, but not so high that it traps the transporter in a particular conformation of the transport cycle.

According to Jardetzky's model, the affinity for substrate should change depending on the conformation of the transporter. The affinity of hSERT for 5-HT has been measured on the purified transporter by displacing [^{3}H]ibogaine, which binds to both the inward- and outward-facing conformations of hSERT [33]. The binding affinity for 5-HT was measured in the presence of Na^+ stabilizing the outward-facing conformation, or K^+ stabilizing the inward-facing conformation (see Sect. 4) and found not to be significantly different [20]. However, the binding affinity of hSERT for 5-HT has been found to change depending on the presence of Na^+ and Cl^-. Interestingly, in the absence of Cl^-, the presence of Na^+ decreased the affinity almost twofold, whereas in the presence of Cl^-, Na^+ increased the affinity almost twofold [34]. As the affinity does not seem to depend on the identity of the cation (Na^+ or K^+) [20, 33], 5-HT can bind equally well to the S1 site in both the outward- and inward-facing conformations, deviating from Jardetzky's model. However, it has been found that when hDAT is biased toward an inward-facing conformation by mutation of the intracellular gating residue Tyr335 (see Sect. 5), the transport affinity for DA increases 20-fold, which is inconsistent with the observation in hSERT, but also Jardetzky. But, as the affinity increases when Na^+ and Cl^- are present, the S1 site could be less accessible to substrate in the absence of any ions. Perhaps hSERT spends more time in the occluded state than when Na^+ and Cl^- are present, which would agree with experiments showing that LeuT and dDAT fluctuate between conformations in the apo-state [12, 13]. Interestingly, Na^+ has the opposite effect on the 5-HT affinity in absence of Cl^-, suggesting that the negative charge of Cl^- could be balancing the positive charges of Na^+ and/or the substrate within the S1 site (see Sect. 4.3).

3.3 Substrate Binding Induces Conformational Flexibility

The effect of substrate binding on the transporter conformation has been assessed in hSERT, dDAT, and LeuT using hydrogen/deuterium exchange mass spectrometry (HDX-MS) [12, 35, 36]. In LeuT, double electron–electron resonance (DEER) spectroscopy [13], electron paramagnetic resonance (EPR) spectroscopy [37], and several fluorescence resonance energy transfer (FRET) techniques [16, 17, 38, 39] have been used to elucidate the distance changes within the transporter induced by substrate binding. Furthermore, the binding event in hSERT and hDAT has been assessed using selective cysteine accessibility method (SCAM) [40–42] and has been modeled in hDAT using homology modeling and molecular dynamic simulations [43–45].

It is evident from structural studies of the MATs that substrates can bind to the S1 site in both the outward- and inward-facing conformations of the transporters, in accordance with Jardetzky. This is reflected by the substrate-binding affinity not being different whether Na^+ or K^+ is present, at least in a purified system when the transporter does not experience any ion gradients [20, 33]. However, for hDAT, it has been shown that the DA affinity increases with the extracellular Na^+-concentration when measuring on live cells. But in membrane preparations, substrate binding was found Na^+-independent suggesting that substrate binding is dependent on the Na^+-gradient rather than the Na^+-concentration per se [46]. But how does the substrate-binding event affect the transporter conformation? Assessing the change in fluorescence upon binding of the substrates 5-HT and 3,4-methylenedioxymethamphetamine (MDMA) to hSERT, specifically labeled with a fluorescent and environmentally sensitive non-canonical amino acid at an extracellular and intracellular position, provides some insight into this. A comparison of the fluorescence "footprints" of 5-HT and MDMA to those of different inhibitors known to stabilize outward-facing (cocaine, s-citalopram, and imipramine) and inward-facing (noribogaine [33]) conformations of SERT hints at the conformations induced by the substrates. At the extracellular position, 5-HT and MDMA left footprints that were alike that of noribogaine, while being different from the inhibitors stabilizing the outward-facing conformation. This points towards that substrate binding stabilize an inward-facing conformation. However, at the intracellular position, 5-HT and MDMA left unique footprints that were also different from that of noribogaine, suggesting that they do not stabilize an inward-open conformation similar to that stabilized by noribogaine [30].

HDX-MS has been used to assess the stability of specific TM regions, by correlating the time-dependent exchange of labile backbone hydrogen atoms for the heavier isotope deuterium. For solvent-accessible and unstructured areas of the transporter, exchange occurs rapidly. When the backbone amides participate in hydrogen bonds as in an α-helix, the exchange rate slows down and is correlated to the local α-helical stability. The precise number of exchanged hydrogen atoms, as well as the region of the transporter where exchange occurred, can then be determined with mass spectrometry. HDX-MS experiments performed on LeuT revealed that most movement induced by substrate binding happens in helices composing the bundle domain [36]. These findings were mirrored by both dDAT and hSERT upon binding of DA or 5-HT, respectively, to the Na^+-bound transporter but were most pronounced in TM1 and TM6. TM6 was more stable in the presence of Na^+ compared to apo dDAT and became even more stabilized when DA was also present. However, the effect of DA on the Na^+-bound transporter was not as pronounced as the effect of Na^+ on the apo transporter and was mostly observed in TM6b. TM1a was also more stable in the presence of Na^+, whereas TM1b was less stable when compared to apo dDAT. But in both dDAT and hSERT, when comparing the Na^+ and substrate-bound state to the Na^+-bound state, the stability of TM1a was then reduced while TM1b was stabilized. This opposite stability pattern of TM1a and TM1b in the presence of both substrate an Na^+ hints at how binding of substrate to the Na^+-bound, outward-facing transporter primes it for conformational rearrangement by restricting movement on the extracellular site and inducing movement on the intracellular site [12, 35]. This is illustrated in Fig. 4. Interestingly, TM1a and TM1b line

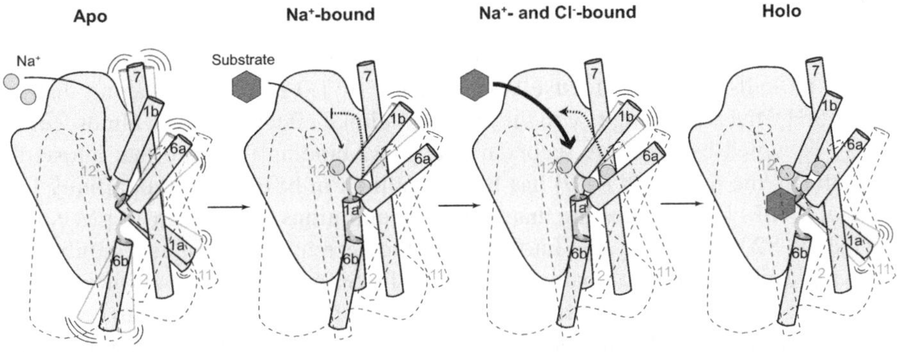

Fig. 4 Substrate binding is dependent on transporter conformation, helical flexibility, and charges near the central substrate-binding site. Schematic representation of a monoamine transporter states with the static scaffold domain shown as a brown shape and the transmembrane helices of the bundle domain shown in light blue. The outline of helices 2, 11, and 12, and half of the scaffold domain are demarcated with a broken line for better visibility. Na^+, Cl^-, and substrate are shown as orange and green spheres, and as a dark blue hexagon, respectively. In the apo-state, the transporter shows a large degree of instability of the helices comprising the bundle domain, except for helix 1b, indicative of the conformational flexibility associated with the apo-state. The Na^+-bound state shows a decreased flexibility of the helices 1a and 6b, indicative of compactification of the cytoplasmic end of the transporter and thereby stabilization of the intracellular gate. Helix 1a also moves inwardly, indicative of intracellular gate closure. In contrast, helices 6a and 7 move outwardly and are also stabilized suggesting a concomitant stabilization of the extracellular vestibule and opening of the permeation pathway. Helix 1b harboring an aspartic acid residue that is the primary recognition site for substrate is destabilized, potentially allowing for an induced fit mechanism of substrate binding. The density of positive charge in the central substrate-binding site is high with the two Na^+ of the Na^+-bound state which is not optimal for binding of the positively charged substrate. The Cl^- potentially serves to screen the positive charges of the two Na^+ and thereby promotes substrate binding to the Na^+- and Cl^--bound state. The transporter binds substrate and enters the holo-state in which stabilization and destabilization are reintroduced to helix 1b and 1a, respectively, suggestive of extracellular gate closure and intracellular gate opening

the cytoplasmic and vestibular substrate permeation pathways, and harbor residues that are involved in the intra- and extracellular gating mechanisms, respectively (see Sect. 5).

Substrate-induced movement on the intracellular site is not just evident from destabilization of TM1a but has also been reported for the intracellular part of TM5 and TM6 (TM6b) using SCAM in rSERT and hSERT, as well as hDAT. SCAM assesses the labeling degree of a specific cysteine residue with a sulfhydryl reactive compound, which correlates with the solvent accessibility of the specific cysteine residue. In rSERT, a cysteine residue inserted at position 277 in the distal, intracellular end of TM5 (rSERT-S277C) was shown to become more labile with increasing 5-HT concentrations, suggesting that the cytoplasmic substrate permeation pathway became more solvent accessible and that substrate induces an inward-open conformation [40]. This was also shown at the homologous position in hSERT, however, only at 5-HT concentrations above 10 μM [41]. Curiously, at concentrations below 10 μM, 5-HT dose-dependently protected hSERT against labeling, indicating a closed cytoplasmic permeation pathway and an inward-closed or occluded conformation of hSERT. Protection of the cytoplasmic permeation pathway by substrate has also been shown in hDAT, by labeling of the distal end of TM6b [42]. Interestingly, 5-HT-induced conformational change of hSERT has been determined to have an half-maximal effective concentration of 1.0 μM [30], meaning that at 10 μM 5-HT the site giving rise to this conformational effect should be almost completely occupied by 5-HT. A secondary substrate-binding (S2) site that allosterically affects the S1 site [47–49] has been identified in hSERT with an affinity for 5-HT of 48 μM [20], suggesting that 5-HT concentrations higher than 10 μM could engage the S2 site. Taking that into account, the decreased solvent accessibility of the cytoplasmic permeation pathway assessed by SCAM in hSERT at lower 5-HT concentrations could indicate that 5-HT binding at the S1 site induces an occluded conformation and that binding of a second molecule of 5-HT leads to opening of the cytoplasmic permeation pathway [41]. This translocation feature would be unique for hSERT, as there are no studies characterizing the function of an homologous S2 site in neither DAT nor NET, even though second molecules of DA and NE have been found in structures of hNET [9, 50].

LeuT has been suitable for assessing conformation by introduction of site-directed, molecular probes. As the transporter in contrast to its eukaryotic homologs does not contain any native cysteine residues, cysteines have been mutated into LeuT at several positions and conjugated to fluorophores and spin-labels allowing for spectroscopic studies. One such study looked at the dynamics and solvent accessibility of spin-labels by EPR at the extracellular vestibule under apo, Na^+-bound, and Na^+ and substrate-bound conditions. In accordance with the HDX-MS data, it was found that leucine and alanine decreased the solvent accessibility of the extracellular permeation pathway of the Na^+-bound LeuT. Furthermore, decreased hydration of the S1 site was observed, suggesting an overall occluded conformation. This contrasted the Na^+-bound state, where the extracellular regions fluctuated significantly more than even in the apo state [37].

Single-molecule FRET (smFRET) has been used to investigate the intracellular part of LeuT during the conformational transitions from outward-open to inward-open states. It was found that leucine did not induce movement of TM1a in the Na^+-bound LeuT. However, in the presence of a low Na^+ concentration, LeuT exhibited two FRET states indicative of TM1a movement and fluctuations between the inward-open and inward-closed conformations. Leucine then reduced the conformational heterogeneity to the FRET state corresponding to the inward-closed conformation. Taken together with the EPR data, this suggests that leucine induces the occluded conformation [38]. This occluded conformation of LeuT in the Na^+- and leucine-bound state has been underlined by precise distance measurements between sets of spin-labels at the extracellular and intracellular sides of the transporter, using DEER. Furthermore, the authors showed based on the average distances that the degree of closure of the extracellular vestibule is underestimated in the structures of the Na^+- and leucine-bound [13], emphasizing the need for complementary studies of protein dynamics to structures.

SmFRET has also been used to compare the conformational dynamics of the alanine- and leucine-bound states of the LeuT. As alanine is transported approximately 10 times faster than leucine, it better portrays the interaction of the MATs with their substrates that are transported with similar kinetic parameters. Accordingly, alanine facilitated transition to the inward-open conformation of the Na^+-bound LeuT to a larger extent than leucine. However, it did so not by stabilizing the inward-open conformation. Rather, it reduced the free energy barrier separating the open and closed states, causing increased fluctuations of LeuT that were evident from the dose-dependent reduction in time that individual LeuT protomers spent in the open and closed states. Leucine, on the other hand, increased the time spent. The total distribution of LeuT protomers between the open and closed states was almost equal regardless of the alanine concentration [39].

4 Cotransport of Ions Fuels Substrate Transport

The MATs, as the other NNS superfamily members, rely on the electrochemical potential of Na^+ across the membrane to drive the transport of their substrates. This was shown for hSERT in 1976 by Gary Rudnick [51] who performed uptake experiments on membrane vesicles from blood platelets and showed that the accumulation of 5-HT is dependent on Na^+ and Cl^-. He also demonstrated that an intact Na^+ gradient is necessary not only for 5-HT transport but also for retaining the gradient of 5-HT resulting from the transport process. In 1990, Bruce Krueger showed the same for DA uptake into synaptosomes prepared from rat caudate nucleus [52].

It is evident from structures that, like in the homolog LeuT [4], there are two sites where Na^+ can bind to the MATs [8, 21, 50], termed Na1 and Na2, respectively. Both Na^+-binding sites are spatially and functionally tightly associated with the S1 site. The Na1 site also lies in proximity to the Cl^--binding site [8–10, 21, 53]. However, it is still somewhat elusive what the role of each of the Na^+ sites is, and how they interact with the Cl^- site during transport. Furthermore, movement of Na^+ as well as the positively charged substrate across the cell membrane gives rise to transport-associated currents, which are to a certain degree balanced by cotransport of Cl^- and/or K^+ or H^+ countertransport [24, 54]. Substrate transport by hSERT has been shown to be electroneutral, and besides the positively charged substrate and one Na^+ that drives the transport, one Cl^- is cotransported and a K^+ countertransported to balance the charges [54]. However, K^+ countertransport has also been observed in dDAT, and K^+ has also been shown to affect DA transport and ligand binding in hDAT [15]. Transport by both hDAT and hNET is electrogenic [55] so if hDAT also countertransports K^+ then either Cl^- is not cotransported or two Na^+ are cotransported together with Cl^- to give a surplus of one charge. The stoichiometry and mechanisms of ion co- and countertransport remain somewhat elusive, although it has been shown by HDX-MS that K^+ induces stabilization pattern of dDAT unlike that of Na^+ [15] and promotes an outward-closed conformation of LeuT, whereas Na^+ stabilizes the outward-open conformation [17]. Besides the stoichiometric currents associated with the movement of a molecule of substrate, a transport-dependent substrate uncoupled current and a transport-independent current have been identified in hDAT and hSERT [23, 56, 57] adding further complexity to the process.

4.1 Sodium Stabilizes the Outward-Facing Conformation

Numerous studies utilizing HDX-MS, FRET, and SCAM have shown that Na^+ stabilizes the outward-facing conformation of the MATs [16, 38–42]. This conformation provides access for the substrate to the S1 site and is accompanied by tight closure of the intracellular gate, which prevents import of Na^+ uncoupled from substrate transport. When comparing the stability of the TMs in the apo-state with the Na^+-bound state of dDAT, it revealed that Na^+ causes stabilization of the intracellular gating mechanism. The increased stability was seen in TM1a, the entire TM6, the intracellular ends of TM8 and TM9, as well as IL4 which all house residues that constitute the intracellular gate (see Sect. 5) [12]. An increased stabilization will in these instances suggest a compacting of the protein, thus a closure of the gate. Concomitantly, the regions of the extracellular vestibule were destabilized, indicative of an open extracellular permeation pathway. This was apparent from the increased flexibility in TM1b, the extracellular end of TM7, and the regions of EL4 closest to it.

The accessibility of the extracellular and cytosolic permeation pathways in response to Na^+ has also been assessed using SCAM in hSERT and LeuT [58]. Here, the authors found that in the Na^+-bound state compared to the apo-state, an inserted cysteine residue in the extracellular vestibule of LeuT became more labile, whereas a cysteine residue in the cytosolic permeation pathway became less labile. Together, this suggests that Na^+ stabilizes an outward-open and inward-closed conformation. Interestingly, Na^+ and the substrate alanine showed the opposite pattern, indicating that substrate undoes the conformational changes induced by Na^+. The same findings were evident in hSERT, where a cysteine residue in the cytosolic permeation pathway became less labile in Na^+ compared with the apo-state and then more labile in Na^+ and 5-HT.

To probe for movement in the permeation pathways of LeuT, smFRET has been employed. It was shown that Na^+ dose-dependently lead to the closure of the intracellular permeation pathway, primarily by increasing the time that individual LeuT protomers spent in the closed state. Interestingly, three FRET states (high, medium, and low) were identified, indicating an intermediate conformation of the extracellular vestibule. Given that the medium FRET state was reported from the fluorophores situated on TM6a and TM12, it was interpreted as a partially closed yet outward-facing conformation of LeuT. Intriguingly, this conformation was populated upon Na^+ binding and at the expense of a profound decrease in the time spent in the low FRET state. The results suggest that Na^+ promotes a partially open, outward-facing conformation by an inward tilting of TM6a [59]. Simulations of LeuT with Na^+ bound to both the Na1 and Na2 sites, without any substrate or ligand at the S1 site, showed that when both Na^+ sites are occupied, the extracellular vestibule was slightly less open than when only the Na2 site was occupied, due to a repositioning of TM7 and EL4 [60]. These findings are supported by the observed HDX pattern of dDAT, showing a stabilization of the entire TM6 in dDAT in the Na^+-bound state. This contrasted to TM1, in which only the intracellular TM1a was stabilized. Furthermore, TM7 was destabilized and the parts of EL4 proximal and distal to TM7 were destabilized and stabilized, respectively, suggesting that these regions also react conformationally to Na^+ [12]. The authors suggested that partial closure of the extracellular permeation pathway could serve as substrate differentiation mechanism and showed that pre-incubation with Na^+ increased the association time of leucine to LeuT whereas that of alanine, which is a better substrate, was unaffected [59].

That Na^+ stabilizes the outward-facing conformation is also evident from pharmacological studies of all three MATs, where numerous inhibitors that bind to the outward-facing conformation have been shown to bind in a Na^+-dependent fashion [15, 29]. Inhibitors that bind to conformations that are not outward-open are less Na^+-dependent or completely Na^+-independent. This has been shown for ibogaine binding to hSERT [28], and the binding of atypical DAT inhibitors JHW007, modafinil, and bupropion to hDAT [52–54].

4.2 Sodium Engages an Allosteric Interaction Network

The way in which Na^+ modulates the conformation of DAT, NET, and SERT is thought to occur through an allosteric interaction network that is highly conserved among the MATs [2, 5]. The network involves TM1, TM6, and TM7, as well as EL4, and has been proposed to converge at the intracellular gate which is important for stabilizing the outward-facing conformation [61] (see Sect. 5). Indication of allosterism between the Na1 and Na2 site is apparent from direct measurements of

Na^+ binding. This has been performed on purified LeuT using $^{22}Na^+$ homologous competition binding [61, 62] and transition metal-ion FRET (tmFRET) [16], as well as on purified hSERT by looking at Na^+-dependent change in fluorescence of a site-specifically incorporated, noncanonical amino acid [30]. Also by assessing the Na^+-dependency of radiolabeled ligand binding and substrate uptake, the half-maximal effective concentration of Na^+ has been determined for purified dDAT [29], LeuT [16], and hSERT [30], as well as for cell surface-expressed hDAT [15, 63] and hSERT [30]. The Na^+ affinity has been found to be 10–30 mM for the MATs, even under the different experimental conditions that Na^+ binding has been measured. Given the concentration of Na^+ in the cerebrospinal fluid [64] and the measured Na^+ affinities, the MATs are almost completely occupied by Na^+ in their native environment of the central nervous system which is expectable as Na^+ provides the driving force for the transport. The experiments directly assessing Na^+ binding and Na^+-dependent ligand binding have revealed a cooperative interaction of Na^+ with the MATs as well as LeuT, apparent from the Hill slope being greater than 1 [65].

Mutational studies have aimed to dissect the mechanism of action and functional implications of each of the two sodium sites, and how they interact allosterically. SCAM has showed that disrupting the Na2 site by individually mutating two Na^+-coordinating residues in LeuT (LeuT-T354A and LeuT-S355A) abolished Na^+-induced closing of the cytoplasmic permeation pathway and transition to the outward-facing conformation [58]. This was confirmed by smFRET which also showed that LeuT-S355A still existed as two populations (an inward-open and a closed), but that Na^+, in contrast to in wild-type LeuT, failed to increase the population with a closed permeation pathway. These altered conformational dynamics were accompanied by a drastic decrease in the Na^+ affinity and abolished alanine uptake into liposomes [66]. Abolished substrate uptake concomitant with decreased apparent Na^+ affinity has also been observed in a disease-associated hDAT variant. Here, the Na2 site residue homologous to Thr345 in LeuT is altered from an aspartic acid residue to an asparagine residue [63]. A similar observation was seen in a homologous mutant of hSERT where the aspartic acid residue was changed to a glutamic acid residue [67]. Even these relatively conservative substitutions to residues that still can interact with Na^+ are detrimental to uptake function, suggesting that the functionality of the Na2 site is highly fine-tunned. Worth noting is that although the Na^+ affinity had decreased in the LeuT Na2 site mutants, Na^+ binding remained cooperative [66], further cementing the importance of the Na2 site. With the use of SCAM, it was found that introducing more drastic Na2 site mutations in LeuT (T345A and S355A) abolished any conformational response to the addition of alanine [58]. This suggests that Na^+ binding to the Na2 site is a prerequisite for substrate binding, or at least for the subsequent conformational change. Taken together, the studies suggest that Na^+ binding to the Na2 site closes the cytoplasmic permeation pathway which is required to establish the outward-facing conformation. Conformational changes in the extracellular vestibule/permeation pathway have been investigated in a study utilizing tmFRET to assess the conformation of LeuT-T345V. This Na2 site mutant showed to be highly biased toward the outward-closed conformation and, contrary to the wild-type LeuT, to be unresponsive to Na^+ alone [17].

Mutation of Na^+-coordinating residues of the Na1 site still enables LeuT to adopt the outward-facing conformation, suggesting that the Na1 site is not linked to the global conformational changes associated with the Na2 site. However, the Na1 site might still be responsible for conferring conformational changes which are important for the transport function. At least, the Na1 site mutants LeuT-N27S and LeuT-N286S show less than 1% uptake of alanine into liposomes. Interestingly, both mutants were conformationally responsive to Na^+ when assessed by SCAM at the cytoplasmic permeation pathway, and even so, to a larger extent than wild-type LeuT, suggesting that the intact Na2 site ensures closure of the pathway independently of the Na1 site [66]. The enhanced conformational responsiveness was concurrent with an increased Na^+ affinity, a possible explanation for the augmented response to Na^+ at the cytoplasmic permeation pathway. Surprisingly, the cooperativity of Na^+ binding that was preserved in the Na2 site mutants was abolished in the Na1 site mutants [66]. Although the conformational responsiveness at the cytoplasmic permeation pathway of the Na1 site mutants was enhanced, tmFRET probing the extracellular vestibule of the Na1 site mutant LeuT-N27Q found that it was highly biased toward the outward-open conformation and that Na^+ was unable to restore the conformation to a wild-type state [16]. This infers that the Na1 site is acting on the extracellular permeation pathway. Molecular dynamic simulations have supported this suggestion. They found the vestibule of LeuT to be more closed when Na^+ was bound at both the Na1 and Na2 site, compared to only at the Na2 site [60]. This is consistent with a study in which Na^+ was found to increase the population of LeuT in an intermediate state between fully outward-open and fully outward-closed [59], suggesting that Na^+ binding at the Na1 site is altering the extracellular vestibule where substrate is diffusing through to reach the S1 site.

Collectively, the findings described above suggest that the disruption of Na^+ binding to the Na1 and Na2 sites has different consequences and that the cooperativity of Na^+ binding is only abolished when the Na1 site is disrupted, whereas disruption of the Na2 site leads to a decreased Na^+ affinity but preserves cooperativity. But how is the cooperativity between the Na1 and the Na2 site conveyed? Simulations have suggested that both sites couple to dynamic structural motifs and microdomains that are implicated in a larger allosteric network, facilitating the conformational effects on the transporter observed upon Na^+ binding [60]. One study suggested a role of TM6 in this network and investigated the Tyr268 in LeuT, a residue located at the distal end of TM6b that has been proposed to be a key residue for facilitating the transition from the occluded to the outward-open conformation [61]. Interestingly, Tyr268 is highly conserved across the MATs and plays a pivotal role in the intracellular gating mechanism (see Sect. 5), disruption of which has been shown to bias hDAT to an inward-facing conformation [68–70]. Compellingly, LeuT-Y268A showed decreased affinity for Na^+ measured by $^{22}Na^+$ homologous competition binding which also revealed abolished cooperativity, thereby mimicking both the effects of disrupting the Na1 site (loss of cooperativity) and the Na2 site (decreased affinity). Furthermore, LeuT-Y268A was completely incapable of alanine transport into liposomes. Mutation of two other important intracellular gating residues (Arg5 and Asp369), which are also highly conserved in the MATs, also affected Na^+ affinity and uptake properties, but not cooperativity. Similar effects were seen when making the homologous mutations in hDAT [69], suggesting that this is a general mechanism of the MATs [61]. This suggests that the alanine mutation disrupts a key piece of the allosteric interaction network that the Na1 and Na2 sites couple to and that this network includes the intracellular gate. It offers a mechanistic explanation to the observed cooperative Na^+ binding in which the Na2 site could be allosterically affecting binding to the Na1 site possibly through TM1a, harboring Arg5, and Tyr268 on TM6b (Fig. 5). In this way, Na^+ binding to the Na2 site would promote the outward-open conformation, and only when the intracellular gate is shut, Na^+ binding to the Na1 site would advance the transport process.

4.3 Chloride Balances the Positive Charges of Sodium and Substrate

Close to the Na1 and S1 sites is the chloride site where Cl^- is coordinated during the transport cycle. It has been shown for both hSERT and rat DAT that 5-HT and DA transport, respectively, are Cl^--dependent with a half-maximal effective Cl^- concentration of 7.5–15 mM and a Hill slope of 1, consistent with a single Cl^- site [51, 52, 71]. The role of Cl^- is, however, more elusive than that of Na^+. Although transport seems to be dependent on Cl^-, this is not necessarily the case for inhibitor binding. Binding of the radiolabeled inhibitor [^{3}H]nisoxetine to purified dDAT has been measured directly under different Cl^- concentrations [29]. At low Na^+ concentrations, [^{3}H]nisoxetine binding was Cl^--independent, but at saturating concentrations of Na^+ [^{3}H]nisoxetine binding became Cl^--dependent. [^{3}H]nisoxetine binding was, however, Na^+-dependent regardless of the Cl^- concentration. The same has been found for [^{3}H]imipramine binding to hSERT, where Cl^- increased the affinity of imipramine, only when Na^+ was present [34], suggesting that Cl^--dependency is conferred indirectly via Na^+. Although indirect, Cl^- might still play an important role during transport, since 5-HT binds with lower affinity to hSERT in saturating concentrations of Na^+ when Cl^- is absent compared to in the absence of both ions. In the presence of both ions, the affinity for 5-HT was higher than under any other condition [34].

Asn101 has been shown to confer Cl^--dependency in hSERT [71]. Interestingly, this asparagine residue is conserved among all three MATs [2], as well as in LeuT [5], and is also involved in coordination of Na^+ at the Na1 site [4]. When this was mutated to an alanine or cysteine residue, 5-HT transport by hSERT became Cl^--independent. Additionally, the mutants reacted conformationally to 5-HT regardless of Cl^-, indicating that Cl^--dependency is related to Na^+ at the Na1 site. However, it has also been suggested that Cl^- regulates the Na2 site in DAT [72].

Mutation of a conserved Cl^--coordinating serine residue within the Cl^- sites of hDAT (Ser357) and hSERT (Ser372) to a negatively charged glutamic or aspartic acid residue makes substrate transport Cl^--independent [73, 74]. This suggests that the role of the Cl^- is to neutralize positive charges, potentially from the Na^+ at the Na1 site or from the substrate at the S1 site (Fig. 4). Convincingly, LeuT that

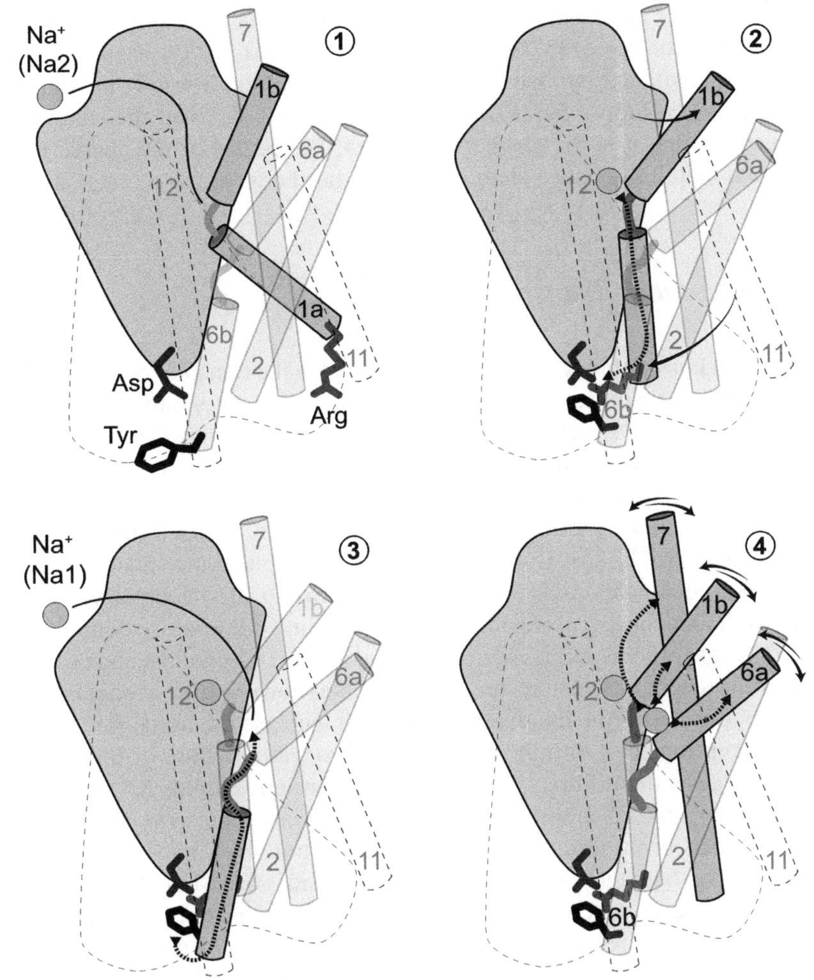

Fig. 5 Allosteric coupling of the sodium sites is mediated through the intracellular gating residues. The allosteric interaction network of the monoamine transporters conferring cooperativity of the Na^+-binding sites schematically represented with the transmembrane helices of the bundle domain colored light blue, and the scaffold domain shown as a brown shape. Half of the scaffold domain and the helices 11 and 12 are shown as outlines demarcated with broken lines for better visibility. The three intracellular gating residues, Arg5/60/56/79, Asp369/436/433/452, and Tyr268/335/332/350 in LeuT/hDAT/hNET/hSERT, are shown as sticks colored dark blue, red, and black, respectively. (1) In the transporter in the inward-open conformation shown by the broken intracellular gate and open cytoplasmic permeation pathway (2) Na^+ binding at the Na2 site induces the outward-facing conformation by engaging entire helix 1 through its Na^+-coordinating residues at the site, thereby bringing the arginine residue at the cytoplasmic end of helix 1a in proximity of the other intracellular gating residues and opening the extracellular vestibule. (3) Closure of the intracellular gate by interaction of the now close by, positively charged arginine residue with the negatively charged aspartic acid residue. The arginine residue also participates in an interaction with the aromatic moiety of the tyrosine residue, which engages helix 6b that at its proximal end coordinates the Na^+ at the Na1 site through both sidechain and backbone interactions. (4) A Na^+ can now bind at the Na1 site and induce local conformational changes to helices that line the extracellular permeation pathway and harbor residues that interact at the central substrate-binding site, thereby priming the transporter for substrate binding

transports substrates independently of Cl^- has a glutamic acid residue (E290) at the homologous position of Ser357/Ser372. The mutant LeuT-E290S has been shown to be Cl^--dependent, and by assessing direct $^{22}Na^+$ binding, it was found that Cl^--increased the Na^+ affinity of LeuT-E290S. Furthermore, Br^- had the same effect [62], suggesting that it is the charge rather than the species of the charge that is important and that this negative charge balances the charge of Na^+.

4.4 *K^+ Potentiates Transport*

It has long been known that K^+ is countertransported in the transport cycle of hSERT rendering 5-HT transport an electroneutral process. However, only Na^+ per se is required to facilitate measurable levels of 5-HT transport, but the concentrative power of hSERT has been shown to increase in the presence of an outward-directed K^+ gradient, and in the absence of K^+, the process has been shown to be electrogenic [14, 54]. Also, an outward-directed H^+-gradient has been shown to have a similar effect as K^+ on hSERT [24]. By assessing [^{3}H]5-HT uptake into platelets, it was shown that K^+ potentiates 5-HT transport by increasing the maximal velocity of hSERT but simultaneously decreasing its transport affinity. It was also demonstrated that the initial rate of transport increased with an outward K^+ gradient and decreased with an inward K^+ gradient. Interestingly, it was also shown that K^+ at equal concentrations on both sides of the membrane could also increase the initial rate of transport [14]. Taken together with the observed electroneutral nature of the transport process, it suggests that a direct interaction between intracellular K^+ and hSERT, and not necessarily the K^+-gradient, facilitates the increased transport rate. A similar phenomenon has been observed during alanine transport into liposomes by LeuT. Here, K^+ on both the intra- and extravesicular sides or just on the intravesicular side increased the concentrative potential and maximal transport velocity of LeuT, although without changing the transport affinity [16]. Mutations introduced to LeuT revealed that K^+ likely interacted with the Na1 site. However, other findings have suggested that K^+ interacts with the Na2 site in hSERT [75] which could explain the discrepancy regarding the effect of K^+ on transport affinity.

The effect of K^+ on DAT has also been assessed in both hDAT and dDAT in cells and in liposomes. Interestingly, DAT, and also NET, has been proposed to be electrogenic during substrate transport regardless of K^+. This is evident from the membrane potential dependency of hDAT and hNET during substrate transport [55] and from hDAT- and dDAT-mediated DA uptake into liposomes. When an inward-directed Na^+-gradient was imposed together with outward-directed K^+-gradient, the concentrative capacity was increased upon addition of the K^+-specific ionophore valinomycin which facilitates outward movement of K^+ to counteract the positive inward current arising from transport [10, 15]. However, K^+ countertransport has also been shown for dDAT during substrate transport into liposomes by monitoring the fluorescence of a K^+-sensitive dye on the intravesicular side. K^+ was found to efflux in response to addition of DA, and this correlated with an influx of Na^+ which was assessed, as an increase in fluorescence, in liposomes containing a Na^+-sensitive dye. The K^+ efflux was evident in the absence of an outward-directed K^+-gradient, suggesting dDAT-mediated countertransport [15]. Furthermore, application of an outward-directed K^+-gradient increased the [^{3}H]DA concentrative capacity of dDAT and, in contrast to hSERT, concomitantly increased the maximal transport velocity as well as the transport affinity. Contrasting both the findings in hSERT and LeuT, dDAT was dependent on the K^+-gradient as equimolar, intra- and extravesicular K^+ concentrations did not increase transport [15]. Taken together, this indicates that K^+ might affect the three MATs and LeuT in different ways.

5 Gating Mechanisms Facilitate Alternating Access to the S1 Site

In the previous section, it was discussed how cooperative Na^+ binding to LeuT is brought about by coupling of the Na1 and Na2 sites to an allosteric interaction network that converges at the intracellular gate (Fig. 5). This ensures that Na^+ binding stabilizes the outward-open conformation simultaneously with closure of the gate, possibly to prevent Na^+ leak uncoupled from substrate transport. The intracellular gate has also been defined in the MATs and consists of Arg60/56/79 on TM1, Tyr335/332/350 on TM6, and Asp436/433/452 on TM8 in hDAT, hNET, and hSERT, respectively, which are all highly conserved in all the MATs and in LeuT [5], and constitute a "thin" gate as opposed to the "thick" gate consisting of the transporter bulk only found on the intracellular side of the MATs [76]. The homologous residues in LeuT were seen to form an ionic interaction between the aspartic acid residue and arginine residue which was consolidated by the tyrosine residue also engaging by a pi–cation interaction with the arginine residue [4, 69].

The functionality of the intracellular gate has been extensively studied in hDAT, using molecular dynamics simulations, SCAM, and radioligand binding assays [68–70]. Several other amino acid residues, besides the already discussed arginine, aspartic acid, and tyrosine residues, are also implicated in the formation of the gate. The tyrosine residue is, however, pivotal as mutation of Tyr335 in hDAT to an alanine residue (Y335A) has been shown to drastically decrease the maximum transport velocity measured by [^{3}H]DA uptake into cells [68], demonstrating the importance of the intracellular gate in facilitating transport. The affinity of hDAT-Y335A for cocaine and the cocaine analog ((-)-2-carbomethoxy-3-(4-fluorophenyl) tropane (CFT), which both have been found to bind to hDAT in the outward-open conformation [10, 77], was decreased [68], suggesting that disruption of the intracellular gate renders hDAT in a less outward-open conformation. This is supported by a substantial decrease in the affinity for Na^+ which stabilizes the outward-open conformation [61]. Interestingly, hDAT as the only MAT harbors an endogenous Zn^{2+} binding site, to which binding of a Zn^{2+} promotes the outward-facing conformation [68, 70, 77]. Addition of Zn^{2+} to hDAT-Y335A was shown to rescue DA transport to a certain degree, and increase the affinity for cocaine and CFT [68], suggesting that disruption of the intracellular gate changes the conformational equilibrium away from the outward-facing state. Homology modeling of hDAT using the LeuT as a template showed that an interaction between Tyr335 and Glu428, as well as Arg60 and Ser334, also play a role in the intracellular gating mechanism. Simulations of wild-type LeuT showed that Arg5 interacted with both Ser267 (Ser334 in hDAT) and Asp369 (Asp436 in hDAT) simultaneously and stabilized the gate, whereas in LeuT-Y268A (Y335A in hDAT), Arg5 either interacted with Ser267 or Asp369 and the network of the gating residues was more dynamic [69], suggesting that Tyr268 positions Arg5 to interact with both residues simultaneously, while itself interacting with Gln361(Glu428 in hDAT) and thereby stabilizing the gate. Mutation of either Arg60 or Asp436 to alanine residues in hDAT decreased the maximum transport velocity, suggesting that the ionic interaction is also important for the intracellular gate. This was also shown by introducing the charge-inverted mutations (R60D+D436R), which preserves the ability to form an ionic interaction, rescuing some DA transport [69]. However, transport was not rescued to wild-type levels, underscoring the importance of Arg60 also interacting with Ser334. Using SCAM, it was shown that hDAT-R60A and -D436A obtained a more outward-closed conformation than wild-type hDAT. Zn^{2+} was able to shift the conformational equilibrium to a more outward-open state of both mutants which correlated with Zn^{2+} also enhancing DA transport, especially for hDAT-R60A. Interestingly, hDAT-R60D+D436R was found to be less outward-closed [69], consolidating the importance of the ionic interaction of the intracellular gate, as well as the effects of the gate on the conformational equilibrium.

Besides the intracellular gate, the MATs also have a thin extracellular gating mechanism. After substrate has bound to the S1 site in the outward-open conformation, the extracellular gating mechanism is the first to be engaged by the local conformational changes brought about by the binding event. The extracellular gating mechanism has also been extensively studied with radioligand binding and SCAM assays, combined with mutagenesis and molecular simulation [78]. Firstly, an extracellular thin gate was described in the structure of LeuT. The structure revealed that Arg30 in TM 1 and Asp404 in TM10 form a salt bridge which shuts the S1 site off from the extracellular environment [4]. These residues are conserved in all MATs and even over multiple species [18]. In hDAT, these residues correspond to Arg85 and Asp476, and mutation of either residue into an alanine residue (hDAT-R85A and hDAT-D476A) causes a complete loss of dopamine transport capacity as well as the capacity to bind [^{3}H]CFT [78]. Interestingly, preserving the salt bridge by inverting the charges with the R85D and D476R mutations preserves some [^{3}H]CFT binding capacity. This was also observed for [^{3}H]citalopram binding to hSERT, where the corresponding residues are Arg104 and Glu493, respectively [78]. A cysteine introduced at position 159 of TM3 in hDAT, and homologous positions in hNET and hSERT [79], is accessible to sulfhydryl-reacting compounds when the transporters are in the outward-open conformation [70]. Disruption of the extracellular gate, Arg85 and Asp476, caused higher degree of cysteine reactivity in hDAT, suggesting that the transporter is biased toward an outward-open conformation [78]. Taken together with the loss of dopamine uptake capacity, this suggests, like with the intracellular gating mechanism, a tight connection between the extracellular gating mechanism, and the conformational equilibrium of the transport cycle. In addition to the ionic interaction formed by the arginine and aspartic/glutamic acid residues of the extracellular gating mechanism, there is also evidence of interactions involving aromatic residues in the extracellular vestibule of the MATs. These interactions are not evident in the LeuT structure, but molecular dynamics suggests that they exist [80], and they are positioned more proximal to the S1 site. Based on homology modeling and structures of hDAT, it is evident that both dopamine and CFT interact tightly with the residues Asp79 of TM1, Tyr156 of TM3, and Phe320 of TM6, thereby positioning these residues to also interact with each other [21, 44, 77]. When dopamine interacts with the pivotal Asp79 residue, the aspartic acid residue can form a hydrogen bond with Tyr156 which can then pi-stack with Phe320 as well as with the aromatic moiety of dopamine. Furthermore, Trp84 next to Arg85 interacts with residues of extracellular loop 4, compacting the extracellular vestibule. This cluster of interactions composes the thin gate together with the ionic interaction between Arg85 and Asp476 [44, 45]. The aromatic residues are also conserved among all human MATs [4, 18], and their implication in gating has been validated in mutational studies of hDAT [43]. The mutant hDAT-Y156A displays no apparent dopamine uptake nor CFT binding when expressed in COS7 cells. This is consistent with simulations, as both the ability of Tyr156 to do pi-stacking and shut the gate by hydrogen bonding Asp79 is lost [43]. By mutating Tyr156 to a phenylalanine residue (hDAT-Y156F), the pi-stacking ability is preserved, but the hydrogen bond to Asp79 is abolished causing dopamine uptake velocity to decrease markedly, and the transport affinity to decrease. Interestingly, [^{3}H]CFT binding is unchanged, suggesting the importance of the gating mechanism for substrate transport but not for ligands binding to the outward-open conformation [10, 43]. Also, mutation Trp84 to a leucine residue has been shown to increase the affinity for cocaine and CFT while decreasing the affinity for DA [81], for which the outward-closed/inward-open conformation has been shown to have higher affinity [68], suggesting that the extracellular gate pushes the conformational equilibrium toward a more outward-closed state. Interestingly, Trp84 has been proposed to be coupled to the Na1 site through TM1b [10], providing a potential mechanism of interplay between Na^+ binding and the extracellular gate. To sum up, disruption of the intra- and extracellular gating mechanisms affects the conformational equilibrium of the MATs. Subsequently, this has effect on the ability of the MATs to progress through the transport cycle, evident from the abolished DA uptake in the hDAT mutants of the two gates. Based on radioligand binding experiments using CFT, and also employing the endogenous Zn^{2+} binding site of hDAT, it is suggested that disruption of the two gates has oppositely directed effects on the conformational equilibrium. Mutation of the intracellular gating residues was shown to render hDAT more outward-closed, whereas mutations of the extracellular gating residues lead to a more outward-open conformation. This proposes a potential "tug-of-war" mechanism in which interactions between the residues of the two respective gates drive the conformational equilibrium in two separate directions. It was discussed how the intracellular gate couples to the binding of Na^+ at the Na2 site via an allosteric interaction network. Binding of Na^+ at the Na1 site was proposed to rearrange the extracellular vestibule and could potentially be coupled to the extracellular gate via TM1b and Trp84 of hDAT. This would provide a potential mechanism of how the gates integrate the energy of Na^+ binding, by translating it into conformational change that ensures alternating access to the S1 site, to drive the cycling of the MATs through the transport cycle, thereby facilitating the substrate uptake.

References

1. Jardetzky O (1966) Simple allosteric model for membrane pumps. Nature 211:969–970. https://doi.org/10.1038/211969a0
2. Aggarwal S, Mortensen OV (2017) Overview of monoamine transporters. Curr Protoc Pharmacol 79:12.16.11–12.16.17. https://doi.org/10.1002/cpph.32
3. Nelson N (1998) The family of Na+/Cl− neurotransmitter transporters. J Neurochem 71:1785–1803. https://doi.org/10.1046/j.1471-4159.1998.71051785.x
4. Yamashita A, Singh SK, Kawate T, Jin Y, Gouaux E (2005) Crystal structure of a bacterial homologue of Na+/Cl--dependent neurotransmitter transporters. Nature 437:215–223. https://doi.org/10.1038/nature03978
5. Beuming T, Shi L, Javitch JA, Weinstein H (2006) A comprehensive structure-based alignment of prokaryotic and eukaryotic neurotransmitter/Na+ symporters (NSS) aids in the use of the LeuT structure to probe NSS structure and function. Mol Pharmacol 70:1630–1642. https://doi.org/10.1124/mol.106.026120
6. Penmatsa A, Gouaux E (2014) How LeuT shapes our understanding of the mechanisms of sodium-coupled neurotransmitter transporters. J Physiol 592:863–869. https://doi.org/10.1113/jphysiol.2013.259051
7. Penmatsa A, Wang KH, Gouaux E (2013) X-ray structure of dopamine transporter elucidates antidepressant mechanism. Nature 503:85–90. https://doi.org/10.1038/nature12533
8. Coleman JA, Green EM, Gouaux E (2016) X-ray structures and mechanism of the human serotonin transporter. Nature 532:334–339. https://doi.org/10.1038/nature17629
9. Hu T et al (2024) Transport and inhibition mechanisms of the human noradrenaline transporter. Nature 632:930–937. https://doi.org/10.1038/s41586-024-07638-z
10. Nielsen JC et al (2024) Structure of the human dopamine transporter in complex with cocaine. Nature 632:678–685. https://doi.org/10.1038/s41586-024-07804-3
11. Yang D, Zhao Z, Tajkhorshid E, Gouaux E (2023) Structures and membrane interactions of native serotonin transporter in complexes with psychostimulants. Proc Natl Acad Sci 120:e2304602120. https://doi.org/10.1073/pnas.2304602120
12. Nielsen AK et al (2019) Substrate-induced conformational dynamics of the dopamine transporter. Nat Commun 10:2714. https://doi.org/10.1038/s41467-019-10449-w
13. Kazmier K et al (2014) Conformational dynamics of ligand-dependent alternating access in LeuT. Nat Struct Mol Biol 21:472–479. https://doi.org/10.1038/nsmb.2816
14. Nelson PJ, Rudnick G (1979) Coupling between platelet 5-hydroxytryptamine and potassium transport. J Biol Chem 254:10084–10089
15. Schmidt SG et al (2022) The dopamine transporter antiports potassium to increase the uptake of dopamine. Nat Commun 13:2446. https://doi.org/10.1038/s41467-022-30154-5
16. Schmidt SG, Nygaard A, Mindell JA, Loland CJ (2024) Exploring the K+ binding site and its coupling to transport in the neurotransmitter:sodium symporter LeuT. elife 12:RP87985. https://doi.org/10.7554/eLife.87985

17. Billesbolle CB et al (2016) Transition metal ion FRET uncovers K(+) regulation of a neurotransmitter/sodium symporter. Nat Commun 7:12755. https://doi.org/10.1038/ncomms12755
18. Porzgen P, Park SK, Hirsh J, Sonders MS, Amara SG (2001) The antidepressant-sensitive dopamine transporter in Drosophila melanogaster: a primordial carrier for catecholamines. Mol Pharmacol 59:83–95. https://doi.org/10.1124/mol.59.1.83
19. Larsen MB et al (2011) Dopamine transport by the serotonin transporter: a mechanistically distinct mode of substrate translocation. J Neurosci 31:6605–6615. https://doi.org/10.1523/jneurosci.0576-11.2011
20. Yang D, Gouaux E (2021) Illumination of serotonin transporter mechanism and role of the allosteric site. Sci Adv 7:eabl3857. https://doi.org/10.1126/sciadv.abl3857
21. Li Y et al (2024) Dopamine reuptake and inhibitory mechanisms in human dopamine transporter. Nature 632:686–694. https://doi.org/10.1038/s41586-024-07796-0
22. Wang KH, Penmatsa A, Gouaux E (2015) Neurotransmitter and psychostimulant recognition by the dopamine transporter. Nature 521:322–327. https://doi.org/10.1038/nature14431
23. Sonders MS, Zhu SJ, Zahniser NR, Kavanaugh MP, Amara SG (1997) Multiple ionic conductances of the human dopamine transporter: the actions of dopamine and psychostimulants. J Neurosci 17:960–974. https://doi.org/10.1523/jneurosci.17-03-00960.1997
24. Keyes SR, Rudnick G (1982) Coupling of transmembrane proton gradients to platelet serotonin transport. J Biol Chem 257:1172–1176. https://doi.org/10.1016/S0021-9258(19)68170-6
25. Rudnick G, Sandtner W (2019) Serotonin transport in the 21st century. J Gen Physiol 151:1248–1264. https://doi.org/10.1085/jgp.201812066
26. Zhou HX, Pang X (2018) Electrostatic interactions in protein structure, folding, binding, and condensation. Chem Rev 118:1691–1741. https://doi.org/10.1021/acs.chemrev.7b00305
27. Kitayama S et al (1992) Dopamine transporter site-directed mutations differentially alter substrate transport and cocaine binding. Proc Natl Acad Sci USA 89:7782–7785. https://doi.org/10.1073/pnas.89.16.7782
28. Barker EL, Moore KR, Rakhshan F, Blakely RD (1999) Transmembrane domain I contributes to the permeation pathway for serotonin and ions in the serotonin transporter. J Neurosci 19:4705–4717. https://doi.org/10.1523/jneurosci.19-12-04705.1999
29. Pugh CF, DeVree BT, Schmidt SG, Loland CJ (2022) Pharmacological characterization of purified full-length Dopamine transporter from Drosophila melanogaster. Cells 11. https://doi.org/10.3390/cells11233811
30. Nygaard A, Zachariassen LG, Larsen KS, Kristensen AS, Loland CJ (2024) Fluorescent noncanonical amino acid provides insight into the human serotonin transporter. Nat Commun 15:9267. https://doi.org/10.1038/s41467-024-53584-9
31. Singh SK, Piscitelli CL, Yamashita A, Gouaux E (2008) A competitive inhibitor traps LeuT in an open-to-out conformation. Science 322:1655–1661. https://doi.org/10.1126/science.1166777
32. Gradisch R et al (2024) Ligand coupling mechanism of the human serotonin transporter differentiates substrates from inhibitors. Nat Commun 15:417. https://doi.org/10.1038/s41467-023-44637-6
33. Coleman JA et al (2019) Serotonin transporter-ibogaine complexes illuminate mechanisms of inhibition and transport. Nature 569:141–145. https://doi.org/10.1038/s41586-019-1135-1
34. Humphreys CJ, Wall SC, Rudnick G (1994) Ligand binding to the Serotonin transporter: equilibria, kinetics, and ion dependence. Biochemistry 33:9118–9125. https://doi.org/10.1021/bi00197a014
35. Moller IR et al (2019) Conformational dynamics of the human serotonin transporter during substrate and drug binding. Nat Commun 10:1687. https://doi.org/10.1038/s41467-019-09675-z
36. Merkle PS et al (2018) Substrate-modulated unwinding of transmembrane helices in the NSS transporter LeuT. Sci Adv 4:eaar6179. https://doi.org/10.1126/sciadv.aar6179
37. Claxton DP et al (2010) Ion/substrate-dependent conformational dynamics of a bacterial homolog of neurotransmitter:sodium symporters. Nat Struct Mol Biol 17:822–829. https://doi.org/10.1038/nsmb.1854

38. Zhao Y et al (2010) Single-molecule dynamics of gating in a neurotransmitter transporter homologue. Nature 465:188–193. https://doi.org/10.1038/nature09057
39. Zhao Y et al (2011) Substrate-modulated gating dynamics in a Na+-coupled neurotransmitter transporter homologue. Nature 474:109–113. https://doi.org/10.1038/nature09971
40. Zhang Y-W, Rudnick G (2006) The cytoplasmic substrate permeation pathway of Serotonin transporter*. J Biol Chem 281:36213–36220. https://doi.org/10.1074/jbc.M605468200
41. Bjerregaard H, Severinsen K, Said S, Wiborg O, Sinning S (2015) A dualistic conformational response to substrate binding in the human Serotonin transporter reveals a high affinity state for Serotonin*. J Biol Chem 290:7747–7755. https://doi.org/10.1074/jbc.M114.573477
42. Dehnes Y et al (2014) Conformational changes in dopamine transporter intracellular regions upon cocaine binding and dopamine translocation. Neurochem Int 73:4–15. https://doi.org/10.1016/j.neuint.2014.02.003
43. Beuming T et al (2008) The binding sites for cocaine and dopamine in the dopamine transporter overlap. Nat Neurosci 11:780–789. https://doi.org/10.1038/nn.2146
44. Cheng MH, Bahar I (2015) Molecular mechanism of Dopamine transport by human Dopamine transporter. Structure 23:2171–2181. https://doi.org/10.1016/j.str.2015.09.001
45. Huang X, Zhan CG (2007) How dopamine transporter interacts with dopamine: insights from molecular modeling and simulation. Biophys J 93:3627–3639. https://doi.org/10.1529/biophysj.107.110924
46. Chen N, Rickey J, Reith ME (2003) Na+ stimulates binding of dopamine to the dopamine transporter in cells but not in cell-free preparations. J Neurochem 86:678–686. https://doi.org/10.1046/j.1471-4159.2003.01889.x
47. Wennogle LP, Meyerson LR (1982) Serotonin modulates the dissociation of [3H]imipramine from human platelet recognition sites. Eur J Pharmacol 86:303–307. https://doi.org/10.1016/0014-2999(82)90333-8
48. Plenge P et al (2020) The mechanism of a high-affinity allosteric inhibitor of the serotonin transporter. Nat Commun 11:1491. https://doi.org/10.1038/s41467-020-15292-y
49. Plenge P et al (2021) The antidepressant drug vilazodone is an allosteric inhibitor of the serotonin transporter. Nat Commun 12:5063. https://doi.org/10.1038/s41467-021-25363-3
50. Tan J et al (2024) Molecular basis of human noradrenaline transporter reuptake and inhibition. Nature 632:921–929. https://doi.org/10.1038/s41586-024-07719-z
51. Rudnick G (1977) Active transport of 5-hydroxytryptamine by plasma membrane vesicles isolated from human blood platelets. J Biol Chem 252:2170–2174
52. Krueger BK (1990) Kinetics and block of Dopamine uptake in Synaptosomes from Rat Caudate nucleus. J Neurochem 55:260–267. https://doi.org/10.1111/j.1471-4159.1990.tb08847.x
53. Penmatsa A, Wang KH, Gouaux E (2015) X-ray structures of Drosophila dopamine transporter in complex with nisoxetine and reboxetine. Nat Struct Mol Biol 22:506–508. https://doi.org/10.1038/nsmb.3029
54. Rudnick G, Nelson PJ (1978) Platelet 5-hydroxytryptamine transport, an electroneutral mechanism coupled to potassium. Biochemistry 17:4739–4742. https://doi.org/10.1021/bi00615a021
55. Bhat S et al (2021) Handling of intracellular K(+) determines voltage dependence of plasmalemmal monoamine transporter function. elife 10. https://doi.org/10.7554/eLife.67996
56. Adams SV, DeFelice LJ (2003) Ionic currents in the human Serotonin transporter reveal inconsistencies in the alternating access hypothesis. Biophys J 85:1548–1559. https://doi.org/10.1016/S0006-3495(03)74587-1
57. Rodriguez-Menchaca AA, Solis E Jr, Cameron K, De Felice LJ (2012) S(+)amphetamine induces a persistent leak in the human dopamine transporter: molecular stent hypothesis. Br J Pharmacol 165:2749–2757. https://doi.org/10.1111/j.1476-5381.2011.01728.x
58. Zhang YW et al (2018) Structural elements required for coupling ion and substrate transport in the neurotransmitter transporter homolog LeuT. Proc Natl Acad Sci USA 115:E8854–E8862. https://doi.org/10.1073/pnas.1716870115

59. Terry DS et al (2018) A partially-open inward-facing intermediate conformation of LeuT is associated with Na(+) release and substrate transport. Nat Commun 9:230. https://doi.org/10.1038/s41467-017-02202-y
60. Zhao C et al (2012) Ion-controlled conformational dynamics in the outward-open transition from an occluded state of LeuT. Biophys J 103:878–888. https://doi.org/10.1016/j.bpj.2012.07.044
61. Stolzenberg S et al (2015) Mechanism of the association between Na+ binding and conformations at the intracellular gate in neurotransmitter:sodium symporters. J Biol Chem 290:13992–14003. https://doi.org/10.1074/jbc.M114.625343
62. Kantcheva AK et al (2013) Chloride binding site of neurotransmitter sodium symporters. Proc Natl Acad Sci USA 110:8489–8494. https://doi.org/10.1073/pnas.1221279110
63. Herborg F, Andreassen TF, Berlin F, Loland CJ, Gether U (2018) Neuropsychiatric disease-associated genetic variants of the dopamine transporter display heterogeneous molecular phenotypes. J Biol Chem 293:7250–7262. https://doi.org/10.1074/jbc.RA118.001753
64. Harrington MG et al (2010) Cerebrospinal fluid sodium rhythms. Cerebrospinal Fluid Res 7:3. https://doi.org/10.1186/1743-8454-7-3
65. Prinz H (2010) Hill coefficients, dose-response curves and allosteric mechanisms. J Chem Biol 3:37–44. https://doi.org/10.1007/s12154-009-0029-3
66. Tavoulari S et al (2016) Two Na+ sites control conformational change in a neurotransmitter transporter homolog. J Biol Chem 291:1456–1471. https://doi.org/10.1074/jbc.M115.692012
67. Felts B et al (2014) The two Na+ sites in the human serotonin transporter play distinct roles in the ion coupling and electrogenicity of transport. J Biol Chem 289:1825–1840. https://doi.org/10.1074/jbc.M113.504654
68. Loland CJ, Norregaard L, Litman T, Gether U (2002) Generation of an activating Zn(2+) switch in the dopamine transporter: mutation of an intracellular tyrosine constitutively alters the conformational equilibrium of the transport cycle. Proc Natl Acad Sci USA 99:1683–1688. https://doi.org/10.1073/pnas.032386299
69. Kniazeff J et al (2008) An intracellular interaction network regulates conformational transitions in the dopamine transporter. J Biol Chem 283:17691–17701. https://doi.org/10.1074/jbc.M800475200
70. Loland CJ, Granas C, Javitch JA, Gether U (2004) Identification of intracellular residues in the dopamine transporter critical for regulation of transporter conformation and cocaine binding. J Biol Chem 279:3228–3238. https://doi.org/10.1074/jbc.M304755200
71. Henry LK et al (2011) A conserved asparagine residue in transmembrane segment 1 (TM1) of serotonin transporter dictates chloride-coupled neurotransmitter transport. J Biol Chem 286:30823–30836. https://doi.org/10.1074/jbc.M111.250308
72. Borre L, Andreassen TF, Shi L, Weinstein H, Gether U (2014) The second sodium site in the dopamine transporter controls cation permeation and is regulated by chloride. J Biol Chem 289:25764–25773. https://doi.org/10.1074/jbc.M114.574269
73. Forrest LR, Tavoulari S, Zhang Y-W, Rudnick G, Honig B (2007) Identification of a chloride ion binding site in $Na^{+}Cl$-dependent transporters. Proc Natl Acad Sci 104:12761–12766. https://doi.org/10.1073/pnas.0705600104
74. Zomot E et al (2007) Mechanism of chloride interaction with neurotransmitter:sodium symporters. Nature 449:726–730. https://doi.org/10.1038/nature06133
75. Hellsberg E et al (2024) Identification of the potassium-binding site in serotonin transporter. Proc Natl Acad Sci 121:e2319384121. https://doi.org/10.1073/pnas.2319384121
76. Zhao C, Noskov SY (2013) The molecular mechanism of ion-dependent gating in secondary transporters. PLoS Comput Biol 9:e1003296. https://doi.org/10.1371/journal.pcbi.1003296
77. Srivastava DK et al (2024) Structure of the human dopamine transporter and mechanisms of inhibition. Nature 632:672–677. https://doi.org/10.1038/s41586-024-07739-9
78. Pedersen AV, Andreassen TF, Loland CJ (2014) A conserved salt bridge between transmembrane segments 1 and 10 constitutes an extracellular gate in the dopamine transporter. J Biol Chem 289:35003–35014. https://doi.org/10.1074/jbc.M114.586982

79. Chen JG, Rudnick G (2000) Permeation and gating residues in serotonin transporter. Proc Natl Acad Sci USA 97:1044–1049. https://doi.org/10.1073/pnas.97.3.1044
80. Cheng MH, Bahar I (2013) Coupled global and local changes direct substrate translocation by neurotransmitter-sodium symporter ortholog LeuT. Biophys J 105:630–639. https://doi.org/10.1016/j.bpj.2013.06.032
81. Chen N, Zhen J, Reith MEA (2004) Mutation of Trp84 and Asp313 of the dopamine transporter reveals similar mode of binding interaction for GBR12909 and benztropine as opposed to cocaine. J Neurochem 89:853–864. https://doi.org/10.1111/j.1471-4159.2004.02386.x

Unlocking the Potential of High-Quality Dopamine Transporter Pharmacological Data: Advancing Robust Machine Learning-Based QSAR Modeling

Kuo Hao Lee, Sung Joon Won, Precious Oyinloye, and Lei Shi

Contents

Abstract The dopamine transporter (DAT) plays a critical role in the central nervous system and has been implicated in numerous psychiatric disorders. The ligand-based approaches are instrumental to decipher the structure–activity relationship (SAR) of DAT ligands, especially the quantitative SAR (QSAR) modeling. By gathering and analyzing data from literature and databases, we systematically assemble a diverse range of ligands binding to DAT, aiming to discern the general features of DAT ligands and uncover the chemical space for potential novel DAT ligand scaffolds. The aggregation of DAT pharmacological activity data, particularly from databases like ChEMBL, provides a foundation for constructing robust QSAR models. The compilation and meticulous filtering of these data, establishing high-quality training data sets with specific divisions of pharmacological assays and

Kuo Hao Lee and Sung Joon Won contributed equally with all other contributors.

K. H. Lee · S. J. Won · P. Oyinloye · L. Shi (✉)
Computational Chemistry and Molecular Biophysics Section, Molecular Targets and Medications Discovery Branch, National Institute on Drug Abuse – Intramural Research Program, National Institutes of Health, Baltimore, MD, USA
e-mail: lei.shi2@nih.gov

O. V. Mortensen (ed.), *Monoamine Transporters*, Advances in Neurobiology 46,
https://doi.org/10.1007/978-3-031-96364-3_3

data types, along with the application of QSAR modeling, prove to be a promising strategy for navigating the pertinent chemical space. Through a systematic comparison of DAT QSAR models using training data sets from various ChEMBL releases, we underscore the positive impact of enhanced data set quality and increased data set size on the predictive power of DAT QSAR models.

Keywords Dopamine transporter · Psychostimulants · Structure · activity relationship (SAR) · Quantitative SAR (QSAR) modeling · Ligand-based drug design

1 The Dopamine Transporter and Its Inhibitors

Neurotransmitter:sodium symporters (NSSs) transport synaptic neurotransmitters through a Na^+ and Cl^--dependent mechanism back to the presynaptic neuron [1, 2]. Among NSSs, the monoamine transporters include the dopamine, serotonin, and the norepinephrine transporters (DAT, SERT, and NET, respectively) [3]. The cognate substrates of these transporters play key roles in the central nervous system (CNS), responsible for mood, emotion, learning, cognition, memory, sleep, and appetite. Inhibition of these transporters leads to reduced clearance of these neurotransmitters, resulting in prolonged synaptic signaling with higher intensity [3].

Specifically, dopamine is essential in the brain's reward system [4]. DAT is responsible for regulating the extracellular dopamine levels in the brain, using the energy stored in the Na^+ and Cl^- gradients to symport dopamine back to the neuron [5, 6]. Disrupting the DAT function can lead to several psychiatric disorders, including attention-deficit hyperactivity disorder (ADHD) [7, 8], bipolar disorder [9, 10], and depression [11]. Many psychostimulants, including cocaine and amphetamine, primarily target DAT. Upon entering the human brain, psychostimulants elevate extracellular dopamine levels and can have a highly addictive effect on the individuals consuming the substances, with the possibility of triggering substance use disorders (SUDs). SUDs encompass complex health conditions that include significant impairments of physiological, mental, and social functions due to consumption of substances at high doses and/or frequencies [12]. Specifically, cocaine use disorder (CUD) is characterized by the compulsive consumption of cocaine despite its adverse medical, psychological, and behavioral consequences and was found to affect more than 5 million individuals in 2019 [13]. Notably, cocaine overdose death rates increased nearly 54% from 2019 to 2021 [14]. While users experience a brief sense of intense euphoria when consuming cocaine before the effect wears off, prolonged usage is linked to the development of mental disorders (e.g., depression) and cognitive impairments [13]. Although cocaine acts nonselectively on the three monoamine transporters, substantial evidence indicates that DAT plays a key role in the development of CUD [15, 16].

Many efforts have been spent in the past decades toward developing effective medications for CUD; however, there are no approved FDA drugs for CUD. Early efforts in developing therapeutic agents for CUD have centered around cocaine analogs, which have been optimized to possess both higher affinity and improved selectivity for the DAT but decreased stimulatory effects and less abuse liability [3, 17, 18]. It was then discovered that the compounds with decreased stimulatory effects stabilized the DAT in a distinct conformation as compared to cocaine [17, 19]. These compounds were classified as "atypical" DAT inhibitors, distinguishing them from "typical" DAT inhibitors like cocaine [18–20]. Thus, cocaine stabilizes DAT in an outward-facing conformation, whereas atypical inhibitors, such as modafinil and JHW007, favor inward-facing conformations [21–23]. Notable examples of atypical DAT inhibitors include benztropine [17, 24], rimcazole [17], GBR12909 [25], and modafinil analogs [26–28].

However, translating these compounds to a human CUD treatment has been challenging [29], while the potential for discovering additional candidate atypical DAT inhibitors based on the well-characterized scaffolds is diminishing. In this work, we provide a comprehensive overview from the cheminformatics perspective in identifying novel DAT inhibitor scaffolds.

2 Ligand-Based Versus Structure-Based Drug Discovery for the NSSs

Computer-aided drug design (CADD) can be broadly categorized into protein structure-based drug design (SBDD) and ligand-based drug design (LBDD) [30]. SBDD leverages in-depth knowledge of the three-dimensional (3D) structures of the protein targets, allowing the design of small molecules that can interact optimally with the targets [31]. When the high-resolution structures of mammalian NSSs were unavailable, SBDD heavily depended on the qualities of homolog-modeling-based computational models of the targets, emphasizing the importance of accurate and reliable molecular modeling approaches [32–36]. LBDD, on the other hand, operates without the 3D structural information of the targets but relies on the analysis of the known pharmacological information between small molecules (ligands) and their targets. LBDD is particularly useful when the 3D structure of the target is either unknown or challenging to obtain [37]. Thus, while SBDD provides the target structure-guided precision, LBDD offers ligand-based adaptability, in the quest for new therapeutic agents.

In LBDD, several popular techniques have been employed, with notable examples being pharmacophore modeling and quantitative structure–activity relationship (QSAR) modeling. In pharmacophore modeling, pharmacophore refers to the spatial arrangements of structural and chemical features, usually on a chemical scaffold, which can be extracted from the active compounds of a target. The resulting model can facilitate molecular recognition to effectively identify and characterize new compounds [38–40]. QSAR modeling was first established by Corwin Hansch for its application in the field of virtual drug screening [41]. QSAR modeling focuses on establishing a quantitative relationship between the chemical structures of small compounds and their specific pharmacological activities, by building models that correlate the structural features of compounds with their observed bioactivities. The QSAR method operates on the premise that molecules sharing similar structural or physicochemical properties would demonstrate comparable biological activities [42]. Consequently, the robustness of the QSAR models depends on both the quantity and quality of available pharmacological activity and compound data, much of which have been systematically accumulated and curated in pertinent databases [43]. These models not only provide valuable insights into the essential molecular features contributing to the desired pharmacological effects but also aid in the design and optimization of novel drug candidates.

The molecular descriptors that QSAR utilizes come in many different forms, including quantitative (molecular shape) and qualitative descriptors (fingerprints) [44], and can range from 2D to 6D [45]. 2D molecular descriptors, which are the most popular, provide information on the connectivity of atoms, properties of chemical bonds, and chemical fingerprints [45]. Many commercial software, such as Mold^2 and DRAGON system, can be used to generate 2D molecular descriptors [46, 47]. 3D descriptors provide physical information, such as surface properties, molecular volume, and molecular interaction fields [48–50]. A caveat to 3D descriptors is that these additional complexities may not be necessary to improve the predictive power of QSAR models at the expense of the significantly more computational cost, as its performance is sensitive to the accuracy of predicting ligand conformation.

The early pioneering work in QSAR modeling of DAT ligands includes those of the Newman group, in constructing both 2D and 3D QSAR models, which used 2D and 3D descriptors, respectively. Specifically, the 2D QSAR study was carried out with 70 diverse inhibitors of the DAT and produced robust QSAR models, resulting in a $q^2 = 0.85$. These DAT models were then used to search through the National Cancer Institute database, yielding five candidate compounds suitable for testing new DAT scaffolds [51]. A follow-up study was conducted using 3D QSAR methods for the design of new mazindol analogs and to identify molecular interactions for optimal DAT binding. Using comparative molecular field analysis (CoMFA), each compound's steric and electrostatic potential fields were calculated and used as features to generate robust 3D QSAR models [52].

In recent years, there has been extensive applications of machine learning (ML) and deep learning (DL) techniques in the construction of QSAR models [53–55]. These advanced computational techniques offer powerful tools for extracting intricate patterns and relationships from complex data sets, beyond the traditional linear regression based QSAR modeling approaches (see below). By leveraging ML and DL algorithms, significantly enhanced accuracy and predictive capabilities of QSAR models have been achieved [56–60].

3 Databases Containing Pharmacological Activity Information of the DAT

Constructing QSAR models involves consideration of many factors, and one of the initial and pivotal aspects is the collection of available and reliable data from relevant databases. We found that the publicly accessible databases containing pharmacological activity information of the DAT include ChEMBL [61], DrugBank [62], binding database (BindingDB) [63, 64], therapeutic target database (TTD) [65], psychoactive drug screening program (PDSP) [66], and PubChem [67]. These databases compile data on biomolecules, protein targets, compound characteristics, ADMET properties, binding interactions, functional assays, and more. Typically, these databases are maintained by nonprofit organizations and undergo regular updates. While most databases primarily focus on drug-related information and associated targets, BindingDB also offers 3D insights into protein targets and ligands.

To evaluate the available pharmacological data that can be used to construct QSAR models for DAT, we compared pharmacological data of DAT from these databases (data retrieved in December, 2023). Note that in this process, it is essential to consider the number of unique compounds when comparing across databases by removing redundant information. If such information was not found in the database, we employed Morgan fingerprints and Tanimoto similarity measurements to calculate the pairwise similarities to identify the distinct compounds within each database. In addition to human DAT (UniProt ID Q01959), the rattus norvegicus (UniProt ID P23977), bos taurus (UniProt ID P23977), and mus musculus DAT (UniProt ID Q61327) are also included.

- ChEMBL release 33 (May 2023) provides 2,399,743 compounds, 20,334,684 activities, 1,610,596 assays, 15,398 targets, and 88,630 publications [61]. We found 14,102 pharmacological activity data related to DAT, and among them, we identified 7496 unique compounds, with clear indications of being curated by experts.
- DrugBank Online (version 5.1.10, released 2023-01-04) contains 15,325 drug entries of FDA-approved drugs or experimental drugs going through the FDA approval process [62]. Upon querying for the DAT information in DrugBank, our search revealed 51 drugs specifically targeting DAT. Nevertheless, our comparison of data from other databases and literature suggests that additional drugs listed by DrugBank may also bind to DAT (see Table S1).
- BindingDB collects experimental data of protein-small molecule interaction [63, 64]. It provides both 2D- and 3D-information of proteins and ligands. Since its initial launch in the year 2000, BindingDB has accumulated 2.8 million binding data for more than 9000 targets and over 1.2 million compounds (released 2023-11-30). Regarding DAT, we found 10,650 pharmacological data points, encompassing 5724 unique compounds.

- TTD provides information of known therapeutic protein targets and the corresponding drugs [65]. The latest update (2024) includes 3730 targets and 39,862 drugs. We found 3578 pharmacological data points directed at DAT. Within that data set, some compound structures are unavailable, making it a challenge to identify unique compounds directly.
- PDSP provides data assessing pharmacological and functional effects at CNS receptors, channels, and transporters [66]. We found 1234 instances of pharmacological data associated with DAT, which are related to 327 unique compounds.
- PubChem is one of the largest chemical databases of publicly accessible chemicals [67]. There are hundreds of data sources connected to PubChem, containing over 116 M compounds, 309 M substances, 229 M bioactivities, 36 M publications, and 38 M patterns. When focusing on DAT specifically, we found a total of 9108 pharmacological activities, related to 6021 distinct compounds. However, the level of expert curation on these DAT activities is not clear.

Among these six databases, ChEMBL is clearly outstanding, as it amasses the largest number of unique compounds and offers an extensive repository of expert-curated bioactivity information. Consequently, it was our primary choice for sieving training data for the development of a robust DAT QSAR model [68]. However, we note that BindingDB is another important resource for DAT pharmacological activity data.

To illustrate the trend of the evolution and the accumulation of DAT pharmacological data over the past four decades, we queried ChEMBL 33 and compiled the retrieved DAT pharmacological activity data along the years (Fig. 1, see its legend for query criteria). The first publication curated by ChEMBL on DAT studied amphetamine analogs and tested their inhibition of dopamine uptake in rats [69]. Along the years, the number of relevant publications and the pharmacological activity data have been accumulating substantially (Fig. 1a, b). In particular, the number of publications had a large spike in 2008 (55 publications), possibly attributed to the availability of the crystal structure of a bacterial NSS homolog, the leucine transporter (LeuT) in 2005 [70]), which allowed homolog modeling of the DAT and other NSSs with a high-resolution experimentally determined structure as the template. Integrated with molecular dynamics simulations and uptake experiments, the LeuT-based homology DAT models have been used to understand the substrate binding mode and the molecular mechanism of inhibitor interactions with DAT [71–74]. Importantly, the LeuT-based homology DAT models have provided valuable guidance to facilitate the drug discovery [36, 75]. However, we did not detect that the publication of the drosophila DAT structure in 2013 [76] resulted in another obvious large accumulation in the relevant publications, although the pharmacological activity data and the number of new unique compounds somewhat increased in 2014 as compared to those in 2013 (Fig. 1b, c).

Notably, whereas the number of publications appears to be plateaued in recent years (Fig. 1a), between 2008–2010 and 2019–2020, there has been an obvious decline of novel DAT inhibitors being discovered and the associated pharmacological characterizations (Fig. 1b, c). Even though there is a noticeable increase of

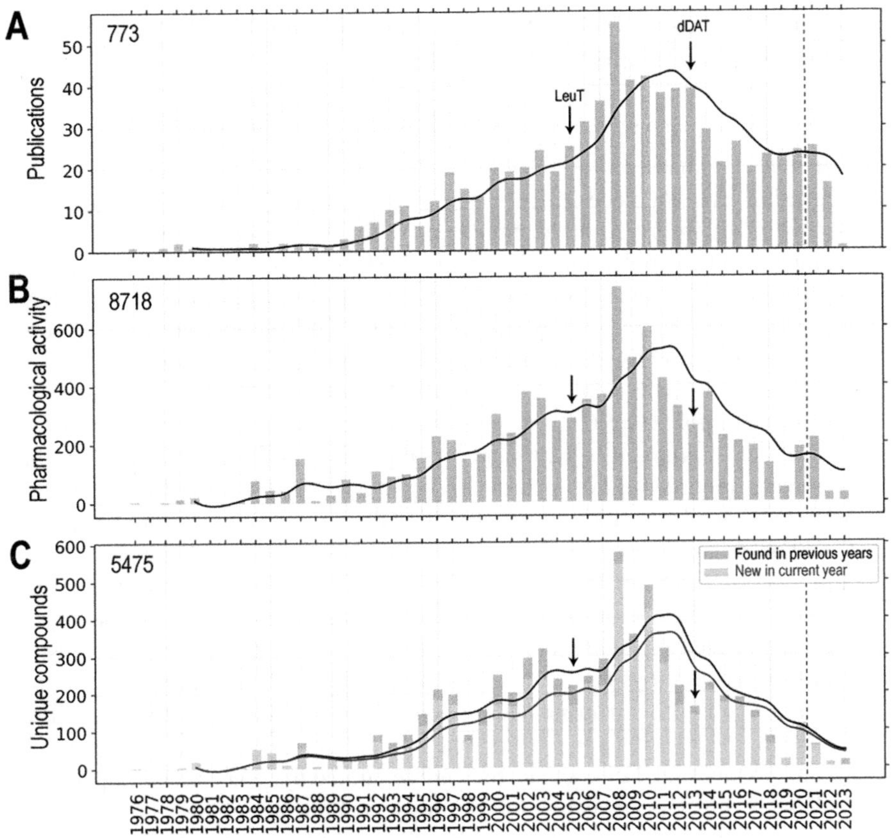

Fig. 1 Statistics of DAT activity data from ChEMBL 33. The DAT data set was queried and retrieved from entries in a locally installed instance of ChEMBL 33 (May 2023 release). We employed the same query criteria as in our previous study to filter the DAT pharmacological data set [68]. The number of publications on DAT ligands (**a**), the number of related pharmacological activity data (**b**), and the number of unique DAT ligands (**c**) were plotted along the years. If a compound has <0.999 pairwise similarity to all the other compound in the data set, we consider it as a unique compound. For panel C, the new unique compound in the current year is colored wheat, and the unique compounds found in previous years are in light gray. The total number of publications, pharmacological activity data, and unique compounds are 773, 8718, and 5475 (with pKi or pIC50 $\geq$ 4), respectively. The data sets in each panel are separated by dash lines between 2020 and 2021, to indicated what we observed that the curation in ChEMBL can be delayed for more than 2 years

related publications in 2021, the increase of novel DAT ligands was minimal (note that to indicate the delay of ChEMBL data curation, which we found could be more than 2 years, a dashed line was added to separate the data of year 2021–2023 from early years in Fig. 1).

4 Some of Representative DAT Ligands Specifically and Not Specifically Developed for DAT

When targeting a specific protein, such as DAT, for therapeutic purpose, it is critical to understand the pharmacology beyond the target. Conversely, some high-affinity DAT ligands were not necessarily specifically developed for DAT. To evaluate whether the SAR information accumulated by the field focusing on DAT ligand development can be supplemented by other efforts, we gathered and analyzed a set of representative DAT ligands from both well-respected review articles on the DAT inhibitors [21, 22, 77] and DrugBank [62], which collects and curates the compounds that have progressed significantly along the drug development processes. We identified those with relatively high affinities using dopamine as a reference (with either pKi, pIC50, pEC50, or pKd > 5.06) to narrow down to a total of 61 ligands (Table S1). Among the 61 ligands, 48 ligands can be found in ChEMBL, but only 46 have recorded DAT pharmacological activity therein. In addition to ChEMBL, 52 ligands can be found in BindingDB, and 38 ligands can be found in PDSP, emphasizing the necessity to source the data beyond an individual database. To collect adequate data for Table S1, we also had to search the literature extensively.

To characterize the chemical features of these 61 representative DAT ligands, we clustered them based on their chemical structures (see Fig. 2 legend for the clustering algorithm we employed). Cocaine and eight other DAT ligands were clustered together and share both a tropane and a phenyl ring. However, a noticeable difference between cocaine and other compounds is the linkage distance between the tropane and phenyl ring. For cocaine, they are separated by a carboxyl group, but for others, they are directly connected. Interestingly, cocaine has the least binding affinity compared to the other compounds in this cluster. Indeed, compared to the affinities at SERT and NET, cocaine is not a DAT selective ligand, while RTI-55, RTI-82, MFZ 2-24, ioflupane I-123, and WIN 35,428 have been used as radiolabeled probes in the binding assays for DAT [78–82]. Among them, we could not find activity data at SERT or NET for RTI-82, MFZ 2-24 and ioflupane I-123, suggesting that these cocaine analogs may have been primarily designed for DAT. Altropane is the only DAT selective ligand in this cluster and has been used as an imaging probe to monitor DAT [83]. Troparil (WIN 35,065-2) is a common inhibitor for the three monoamine transporters and has exhibited a stronger affinity at DAT compared to that of cocaine [19, 84]. Tesofensine, another common inhibitor for the three transporters, was investigated for its potential applications in obesity-related conditions, Parkinson's disease, and Alzheimer's disease [85].

In the cluster containing methamphetamine, amphetamine, and three other ligands, the common feature among them is a propylbenzene scaffold linked to a protonated amine (Fig. 2b). Note that most ligands in this cluster are DAT substrates except for sydnocarb. Amphetamine has been used medically to aid in treating ADHD [86, 87] and in enhancing cognitive functions [88]. Due to the size and

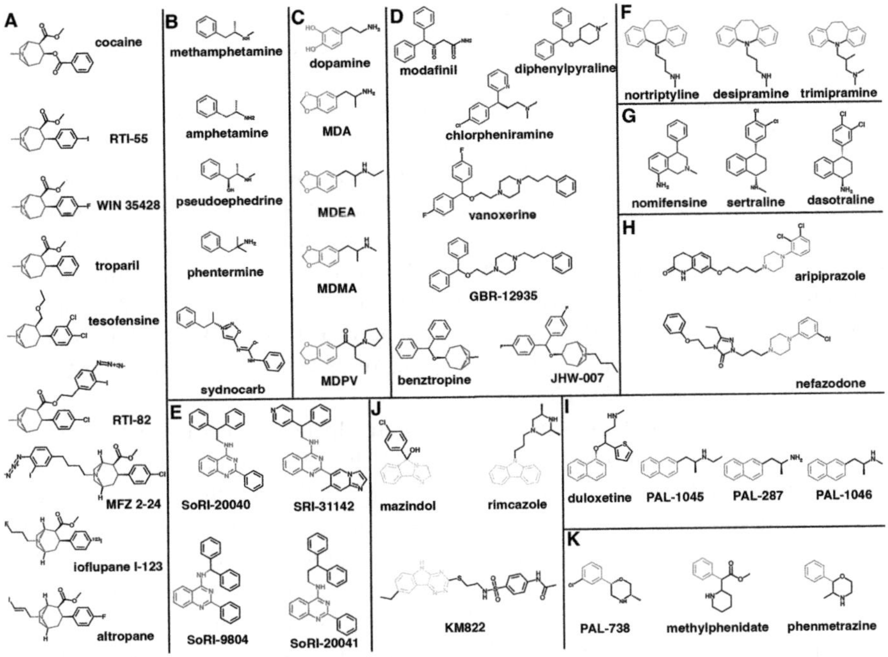

Fig. 2 Clustering of representative DAT ligands. A total of 61 well-known DAT inhibitors were collected from literature [21, 22, 77] and DrugBank [62]. To characterize these DAT ligands, we employed the hierarchical clustering approach implemented in the Schrodinger Maestro suite (version 2023-3). We used linear fingerprints, Tanimoto similarity, and average linkage method. Note that three pairs of compounds (armodafinil and modafinil, methylphenidate and dexmethylphenidate, and amphetamine and dextroamphetamine) are enantiomers, and we only use modafinil, methylphenidate, and amphetamine when carrying out the clustering. The single-member clusters are shown in Fig. S1

structure of amphetamine analogs, they can cross through the blood–brain barrier (BBB) [89]. However, it has been found that methamphetamine disrupts the BBB and affects its structural integrity and permeability [90]. Methamphetamine is a recreational psychostimulant drug known for its strong neurotoxic and addictive effects [91]. Methamphetamine use disorder (MUD) exhibits a wide spectrum of adverse outcomes from hallucinatory ideation to self-injurious behaviors [92]. Notably, MUD has been linked to severe cardiac complications [92]. Both methamphetamine and amphetamine function as potent releasers that increase the extracellular concentrations of key neurotransmitters such as dopamine and serotonin in the brain [92, 93]. Sydnocarb functions as a noncompetitive inhibitor of DAT and is used for the treatment of schizophrenia and depression but has been linked to having psychostimulant effects [94, 95].

Dopamine is an endogenous substrate of DAT. It can be clustered together with several benzodioxole compounds, including 3,4-methylenedioxymethamphetamine (MDMA), 3,4-methylenedioxypyrovalerone (MDPV), 3,4-methylenedioxy-N-ethylamphetamine (MDEA), and 3,4-methylenedioxyamphetamine (MDA) (Fig. 2c). MDMA, commonly known as "ecstasy," acts on both DAT and SERT and can have stimulatory effects and alter perception and induces hallucinatory experiences [96–98]. MDEA functions as a partial releaser of DAT [99], while MDA has been shown to act as a substrate for all three monoamine NSSs [100]. MDPV, a synthetic cathinone typically seen in "bath salts," is a psychostimulant drug characterized by its heightened selectivity for DAT and increased potency when compared to that of cocaine [101]. Studies have shown that MDPV is a DAT inhibitor but not a substrate [102], whereas all the other members of this cluster showed at least some substrate properties for DAT [100, 103].

A few well-recognized atypical inhibitors were clustered together including benztropine, JHW-007, vanoxerine, modafinil, GBR-12935, diphenylpyraline, and chlorpheniramine (Fig. 2d). In this cluster, diphenylmethane is the common moiety. Among them, modafinil is a therapeutic drug involved in treating excessive sleepiness in narcolepsy patients by increasing extracellular levels of dopamine [27, 104]. Modafinil has two enantiomers (S- and R-modafinil), and it has been shown that the R-enantiomer (also known as armodafinil) has ~3-fold higher affinity than S-enantiomer at DAT [27], while the R-enantiomer is a DAT selective ligand, when comparing the affinities at DAT, SERT, and NET [105]. Vanoxerine (GBR12909), a piperazine derivative and another atypical DAT inhibitor, was initially developed as a therapeutic treatment in combatting cocaine addiction, but it was later discontinued due to its heart-related side effects [25, 106]. Diphenylpyraline functions as an antihistamine but has also been shown to inhibit DAT. Previous research suggested a potential for it to reduce the effects of cocaine [107]. Chlorpheniramine (chlorphenamine) is an antihistamine as well and displayed a high affinity for SERT and a relatively low affinity to DAT [108]. In addition to the diphenylmethane moiety, JHW-007 and benztropine also possess a tropane ring. Benztropine, formally used as a drug for Parkinson's disease, shared noticeable structural similarities with cocaine. It has been shown to function as an atypical DAT inhibitor and utilized as a pharmacological agent for CUD research [17, 25, 109]. JHW-007 is also an atypical DAT inhibitor and has been demonstrated to possess reduced psychostimulant effects compared to cocaine [110]. Interestingly, except for diphenylpyraline and chlorpheniramine, the other five ligands in cluster D are selective for DAT over SERT and DAT based on the data that we could collect (Table S1).

Some known allosteric inhibitors of DAT were clustered in one group (Fig. 2e). Specifically, SRI-31142 is a 4-quinazolinamine derivative that could inhibit the actions of cocaine; however, further research is needed to fully explore its therapeutic potential [111]. Three other 4-quinazolinamine analogs, SoRI-9804, SoRI-20040, and SoRI-20041, are clustered with SRI-31142 and have been found to be partial inhibitors and allosteric modulators for DAT [112, 113].

While clusters A to E mentioned above have garnered more attentions for their interactions with DAT, the remaining clusters of ligands were mainly developed for use as antidepressants, appetite suppressants, and antipsychotics, which, however, have the off-target effects at DAT. In particular, two clusters of DAT inhibitors were obviously developed to treat depression. They have either tricyclic or bicyclic ring moiety. In the cluster F (Fig. 2f), desipramine, nortriptyline, and trimipramine are tricyclic antidepressants (TCAs). Both desipramine and nortriptyline are FDA-approved drugs to treat depression, due to their actions at NET and/or SERT [114], and have been shown to have an inhibiting effect on DAT as well [108]. Trimipramine has been shown to be a DAT inhibitor but is more potent at SERT [108]. Nomifensine, sertraline, and dasotraline share a bicyclic ring moiety attached to a toluene group (Fig. 2g). Sertraline, a selective serotonin reuptake inhibitor (SSRI) that also inhibits DAT, serves as an antidepressant as well [115]. Dasotraline is a dual DAT and NET inhibitor, which is being studied for its therapeutic potential in treating ADHD [116].

Two other clusters also include antidepressants. One cluster shares phenylpiperazine as the common feature (Fig. 2h). In this cluster, aripiprazole is an atypical antipsychotic drug that primarily targets dopamine D2 receptor with a high affinity [117], but can also bind to both DAT and SERT, with only slightly higher affinity at SERT [118, 119]. Nefazodone is an atypical antidepressant and has similar affinities in the three monoamine NSSs [120]. Duloxetine, PAL-287, PAL-1045, and PAL-1046 share a naphthalene moiety and are clustered together (Fig. 2i). Duloxetine, a dual NET and SERT inhibitor, has exhibited the potential to serve as an antidepressant based on previous research findings [121], but demonstrated a weaker affinity to DAT [122]. In addition to duloxetine, the other three compounds in cluster I are all substrates. PAL-287 functions as a releaser for DAT, SERT, and NET and has been shown to suppress the effects of cocaine [99, 123] (Table S1). PAL-1045 tends to act as a partial releaser for both DAT and SERT [99, 124], whereas PAL-1046 functions as a full substrate for DAT [99].

Two clusters include appetite suppressants. Phenmetrazine is an anorectic drug used in the mid-1900s to promote the release of dopamine in the brain [125] and was clustered together with methylphenidate and PAL-738. All three ligands in this cluster share a toluene moiety either linked to a 3-methylmorpholine moiety or carboxyl group (Fig. 2k). However, the phenylacetic acid moiety of methylphenidate, which acts as a typical DAT inhibitor, is common to cocaine, but not to the other two members of this cluster. Although methylphenidate serves as a therapeutic drug for ADHD, it also exhibits psychostimulant effects [126]. PAL-738, on the other hand, serves as a partial releaser for DAT [99]. The next cluster includes mazindol, which is another appetite suppressant, and a triple reuptake inhibitor of DAT, NET, and SERT [127, 128]. It can be clustered together with rimcazole and KM822, and they share a similar tricyclic ring moiety (Fig. 2j). Rimcazole has been observed to inhibit the cocaine binding at DAT [129]. KM822 is a noncompetitive inhibitor of DAT and has been shown to reduce the affinity of cocaine to DAT [35].

Ten compounds have distinct scaffolds and are the sole representatives in their own respective clusters in this set of 61 ligands (Fig. S1). In particular, bupropion functions as a dual DAT and NET inhibitor and is used for the treatment of ADHD, depression, and smoking cessation [130]. Tamoxifen functions as both a DAT inhibitor with some atypical property without psychostimulant effect and a selective estrogen receptor modulator, which is typically prescribed for the treatment of estrogen-related breast cancer [131]. While having a high affinity at SERT, ibogaine operates as a competitive inhibitor of DAT as well and is a psychostimulant under investigation for its potential to treat SUDs [132].

In summary, while the potential for leveraging the "traditional" DAT inhibitor scaffolds is diminishing, our analysis indicated that ample chemical space can be explored for developing the DAT inhibitors with desired pharmacological profiles. However, the accumulation and availability of large quantity of DAT pharmacological data present both challenges and opportunities.

5 Curation and Filtering the DAT Pharmacological Data Sets

In order to assemble high-quality training data sets for QSAR modeling, adequate understanding and careful curation of the raw data retrieved from the databases are essential. One important issue is to differentiate the half maximal inhibitory concentration (IC50) and inhibition constant (Ki) in measuring the binding potencies of the ligand at a protein target. Whereas Ki can be derived from IC50 based on the Cheng-Prusoff equation, Ki is a more accurate representation of binding affinity than IC50, because IC50 can be influenced by the methods used for measurement [133]. Based on the Cheng-Prusoff equation, Ki values are always expected to be smaller than the IC50 values. However, if the same experimental approaches are employed, the trends of Ki and IC50 should be similar. Additionally, for transporter proteins, we can separate the data according to the assay type, either the radiolabeled inhibitor binding assay (referred to as "binding" below) or uptake inhibition (referred to as "uptake" below), as they represent inhibitory potencies of related but different biological processes [68].

Thus, the DAT pharmacological data retrieved from ChEMBL can be divided into four distinct sets: uptake IC50, uptake Ki, binding IC50, and binding Ki data sets (Table 1). Notably, among the four distinct data sets, there are some compounds that appeared in more than one set (termed as "overlapping" compounds). By analyzing the trends among these compounds, we validate the points described above. First, in both the uptake pKi versus uptake IC50 and the binding pKi versus binding IC50 plots, we can observe that pKi values have a trend of being larger than IC50s, confirming the deduction from the Cheng-Prusoff equation described above (Fig. 3c, d). As expected, there is a strong correlation between uptake and binding

Table 1 Queried pharmacological activity data and training data sets of the DAT ligands from different ChEMBL releases

			Uptake		Binding	
Version	Release date	Queried data	Ki	IC50	Ki	IC50
ChEMBL33	May 2023	14102 (7496)	366 (366)	646 (646)	1166 (1166)	482 (482)
ChEMBL31	Aug 2022	14099 (7470)	366 (366)	646 (646)	1166 (1166)	482 (482)
ChEMBL29	Jul 2021	13915 (7326)	366 (366)	646 (646)	1131 (1131)	482 (482)
ChEMBL27	May 2020	13456 (7049)	366 (366)	606 (606)	1113 (1113)	482 (482)
ChEMBL25	Mar 2019	13273 (6935)	350 (350)	593 (593)	1189 (1189)	556 (556)

The numbers of the unique compounds for each data set are shown in the brackets. Note that although some ChEMBL releases have the same numbers for the final training data sets, the value of pharmacological activity data may be different due to the measurement of the same compounds from new studies collected in new ChEMBL releases

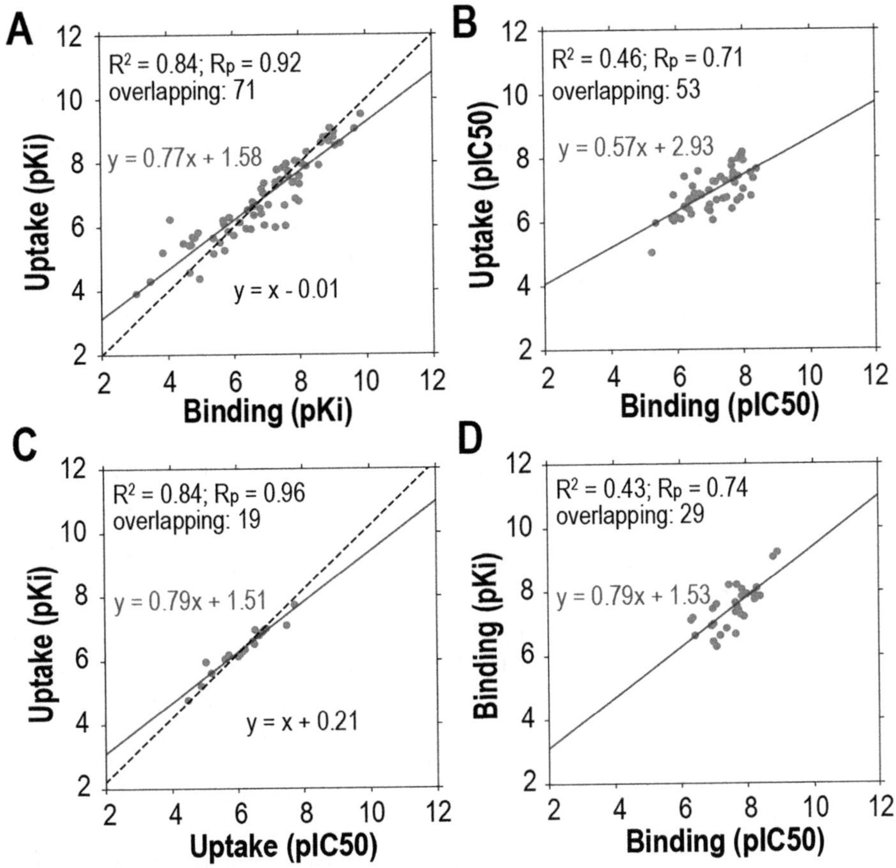

Fig. 3 Comparative analysis of DAT pharmacological data sets with correlation metrics and linear regressions. The DAT pharmacological data can be divided into four sets, uptake pKi, uptake pIC50, binding pKi, and binding pIC50. The overlapping compounds between different data sets were extracted for comparisons. The correlation of determination (R^2) and the Pearson coefficient correlation (Rp), as well as the number of the overlapping compounds (referred as "overlapping"), are indicated at the top left corner of each panel for the indicated comparisons. The red lines are the linear regressions of the indicated data sets; the black dotted lines are the linear regressions with the slope restrained to 1. Note that slope-restrained regression results were not shown for panels B and D, due to poor goodness-of-fit

pKi values, as well as between uptake and binding IC50 values (Fig. 3a, b). However, due to limited available data, it is not conclusive whether uptake pKi or pIC50 values tend to be larger than those of binding (Fig. 3a, b).

Hence, it is not advisable to combine either Ki and IC50, or uptake and binding data sets in building the QSAR models of DAT and likely other transporter proteins. As there are more Ki data than IC50 data in the binding data sets, we mainly focus on using the binding Ki data for DAT QSAR modeling.

6 ML-Based QSAR Models Using DAT Binding Data Set

Machine learning (ML) and deep learning (DL) have been extensively applied in biological and pharmaceutical research, including protein structure prediction [134, 135], absorption, distribution, metabolism, excretion, and toxicity (ADMET) profiling [136, 137], QSAR modeling [58, 68], de novo drug design [138–140], and the blood–brain barrier (BBB) permeability prediction [141]. With the accumulation of large amount of pharmacological data in the databases such as ChEMBL and the drastic improvement of computation hardware in the recent decade, especially the applications of graphics processing units (GPUs) in scientific computing, ML and DL have risen to become the primary approaches in QSAR modeling, by handling the big data sets as well as the high volume of chemical features that can be generated for each compound.

Among many ML algorithms, XGBoost and RF have gained increasing popularity for their strong prediction performances, relative ease in usage, robustness with adjustable hyperparameters, and interpretability [59, 60, 142]. A comparison of ML- and DL-based QSAR models in the human *ether-à-go-go-* related gene (hERG) indicates that XGBoost provides the best prediction results [143]. XGBoost was found to be an excellent choice for both large and small data sets for QSAR modeling [144]. When utilizing the DAT binding data set, XGBoost-trained models exhibited superior performance [68].

To evaluate impact of the additional data on the quality of the QSAR models, we performed a benchmark analysis of the DAT binding Ki data sets retrieved from different ChEMBL releases. Across the ChEMBL releases from the past 5 years, we observed a noticeable enhancement in the coefficient of determination (R^2) value from 0.72 to 0.77 (Table 2). Similarly, mean square error (MSE), Pearson correlation coefficient, and Spearman's rank correlation coefficient followed this positive trend, with the latest ChEMBL release yielding the most favorable benchmarks.

Note that while overall volume of the queried data continued to grow from ChEMBL 25 to ChEMBL 33, there was no significant increase in the number of unique compounds within the final training data sets (Table 1), which may partially be due to the downward trend of medicinal chemistry efforts on DAT (Fig. 1), and potentially on NSSs in general. Notably, there were changes of curation criteria between ChEMBL 25 and ChEMBL 27, resulting in a reduction in the number of inhibitor's Ki and IC50 data sets. These changes in ChEMBL led to an enhancement

Table 2 Benchmarks of the DAT QSAR models trained with the XGBoost algorithm

	Binding Ki							
Release	R^2		MSE		Rp		Rs	
ChEMBL33	0.77	0.03	0.35	0.04	0.88	0.02	0.86	0.02
ChEMBL31	0.77	0.04	0.37	0.06	0.88	0.02	0.86	0.03
ChEMBL29	0.75	0.03	0.42	0.07	0.87	0.02	0.85	0.02
ChEMBL27	0.74	0.03	0.43	0.06	0.86	0.02	0.85	0.02
ChEMBL25	0.72	0.04	0.43	0.08	0.85	0.03	0.85	0.03

We constructed 10 distinct QSAR models for each indicated ChEMBL release to mitigate any stochastic effects. For each reported metric, the first value is the average across the 10 QSAR models, which is followed by the standard deviation

in the data set's quality, resulting in a 0.02 increase in the R^2 value (Table 2). Even though there was only a modest increase of 53 data points in the binding Ki training data set from ChEMBL 27 to ChEMBL 33, this change contributed to an overall benchmark improvement of 0.03 in R^2. Thus, even relatively moderate addition, update, and refinement of data may play a role in enhancing the quality of QSAR models.

While extensive ML/DL efforts have been made on various aspects of the NSSs [43, 145, 146], to our knowledge, no DL-based DAT QSAR research has been published. Because proper applications of DL may require larger data sets, we speculate that there are still not yet enough data points available in the public database to build robust DL-based DAT QSAR models. Indeed, ML-based models can use a small training data to still produce robust predictive QSAR models [147] and have been seen to outperform DL-based models in various cases [148].

7 Potential Applications and Challenge

Virtual screening provides a rapid and low-cost compounds screening in the early stage of drug discovery, by searching potential hits from databases [149, 150]. Such techniques have been applied in identifying new compounds in SERT [151–153]. The application of the DAT QSAR model in virtual screening may similarly explore a large compound library and to identify hit compounds and novel scaffolds. In addition, in a medicinal chemistry synthesis campaign, the predicted Ki of candidate compounds by robust DAT QSAR models can help to prioritize the candidate compounds to be synthesized. The iterative prediction, synthesis, and pharmacological measurements are expected to both improve the quality of the QSAR models and improve the efficiency of synthesis campaign.

The screening process can also involve conducting counter- or synergistic-screening with other targets. In particular, we have established the protocol to screen the ligands for the DAT but against the human ether-à-go-go-related gene (hERG)

[68]. The increasing concerns regarding the risks associated with hERG binding have led to a notable surge in the availability of hERG binding affinity data in both public databases and scientific literature. The prediction by the QSAR models can be instrumental in enabling the identification of potential medication candidates characterized by specific attributes, e.g., to aid in pinpointing high-affinity DAT inhibitors with minimal hERG affinity.

At physiological pH of 7.4, dopamine exists predominantly in its cationic form [154]. pH-induced changes between dopamine and DAT could alter their interactions and conformations. Specifically, the protonation of the amine group not only influences the conformation of dopamine but also introduces a positive charge, enhancing the interaction between the bound dopamine and negatively charged residues in DAT, such as Asp79 in the central binding pocket [155]. In comparison, modafinil is not in a charged form when it is bound in DAT [27]. Therefore, considering the proper protonation state of the ligands accordingly should improve the quality of the training data set, by providing a critical chemical context for the models to consider when making bioactivity predictions. However, while different algorithms and software packages are available to predict pKa, such as Epik [156] and MolGpka [157], which may generate decent pKa predictions in water, it is generally a challenging problem to predict the pKa of the ligands in the binding pocket of the target protein.

8 Conclusion

DAT is a key therapeutic target of SUDs. By collecting and analyzing the representative DAT ligands, we reveal that DAT binds to a variety of compounds, while there is still a large chemical space to be explored for novel DAT ligand scaffolds. The compilation of extensive DAT pharmacological activity data sets, coupled with the power of QSAR modeling, presents a promising avenue for exploring such a space.

Acknowledgments Support for this research was provided by the National Institute on Drug Abuse–Intramural Research Program, Z1A DA000606 (L.S.).

Declarations of Competing Interests No potential conflict of interest was reported by the authors.

Appendix

Table S1 The pharmacological activities of some representative DAT ligands

Name[a]	DrugBank ID	DAT (activity)	SERT (activity)	NET (activity)	DAT-SERT (Δactivity)	DAT-NET (Δactivity)	Selectivity	Cluster ID	Reference (DAT)
Vanoxerine[a]	DB03701	10.22	7.62	7.10	2.60	3.12	DAT selective	D	[158]
RTI-55[a]		9.32	9.42	8.85	−0.10	0.47		A	[159]
SRI-31142[a]		8.72					DAT only	E	[160]
Dasotraline	DB12305	8.70	7.85	8.40	0.85	0.30		G	[161]
MDPV[a]		8.68					DAT only	C	[162]
GBR-12935[a]		8.43	6.21	6.20	2.22	2.23	DAT selective	D	[163]
Altropane	DB04947	8.18	6.74		1.44			A	[83]
Dextroamphetamine	DB01576	8.18	8.32	8.21	−0.14	−0.03		B	[164]
Nomifensine[a]	DB04821	8.15	6.10	8.42	2.05	−0.27		G	[165]
Benztropine[a]	DB00245	8.10	5.29	5.86	2.81	2.24	DAT selective	D	[51]
Tesofensine	DB06156	8.10	7.96	8.49	0.14	−0.39		A	[161]
Sydnocarb[a]		8.08	5.00	5.82	3.08	2.26	DAT selective	B	[166]
PAL-1046[a]		8.00					DAT only	I	[99]
JHW-007[a]		7.98	5.76	5.88	2.22	2.10	DAT selective	D	[167]
PAL-287[a]		7.90	8.47	7.96	−0.57	−0.06		I	[168]
RTI-82[a]		7.85					DAT only	A	[169]
Sertraline	DB01104	7.60	10.12	6.80	−2.52	0.80	SERT selective	G	[120]
Ioflupane I-123	DB08824	7.54					DAT only	A	
MFZ 2-24[a]		7.48					DAT only	A	[170]
Mazindol[a]	DB00579	7.35	6.61	9.10	0.74	−1.75	NET selective	J	[171]
PAL-1045[a]		7.34					DAT only	I	[99]
PAL-738[a]		7.24					DAT only	K	[99]
MRS7292[a]		6.90					DAT only	L	[172]
Dexmethylphenidate	DB06701	6.79	8.00	6.69	−1.21	0.10	SERT selective	K	[173]
WIN 35428[a]		6.77	7.52	5.96	−0.75	0.81		A	[174]
Methylphenidate[a]	DB00422	6.72	5.01	6.66	1.71	0.06		K	[175]
Rimcazole[a]		6.65	6.08	5.67	0.57	0.98		J	[176]
Duloxetine	DB00476	6.62	9.10	8.13	−2.48	−1.51		I	[177]
Troparil[a]		6.59	6.64	6.59	−0.05	0.00		A	[174]

Nefazodone	DB01149	6.44	6.86	6.44	−0.42	0.00		H	[120]
Venlafaxine	DB00285	6.44	8.42	6.86	−1.98	−0.42	SERT selective	L	[120]
Bupropion[a]	DB01156	6.43	4.33	5.16	2.10	1.27	DAT selective	L	[51]
Trodusquemine	DB06333	6.40					DAT only	L	[178]
Diphenylpyraline	DB01146	6.38					DAT only	D	[179]
Methamphetamine[a]	DB01577	6.34	4.50	6.96	1.84	−0.62		B	[180]
Sibutramine	DB01105	6.30	5.96	5.25	0.34	1.05		L	
SoRI-9804[a]		6.28					DAT only	E	[181]
MDEA[a]	DB01566[b]	6.21					DAT only	C	[20]
Amphetamine[a]	DB00182	6.19	4.41	7.15	1.78	−0.96		B	[182]
Cocaine[a]	DB00907	6.19	6.85	5.80	−0.66	0.39		A	[174]
Chlorpheniramine	DB01114	6.13	7.89	5.31	−1.76	0.82	SERT selective	D	
Phenmetrazine[a]	DB00830	6.12					DAT only	K	[183]
MDMA[a]	DB01454	6.05	6.02	6.46	0.03	−0.41		C	[164]
Nisoxetine[a]	DB09186[b]	5.94	8.85	8.82	−2.91	−2.88		L	[184]
Nortriptyline[a]	DB00540[b]	5.88	8.16	8.50	−2.28	−2.62		F	
Tamoxifen[a]	DB00675[b]	5.83	5.91	5.84	−0.08	−0.01		L	
Phentermine	DB00191	5.80	4.86	6.61	0.94	−0.81		B	[185]
SoRI-20041[a]		5.75					DAT only	E	[113]
Pseudoephedrine	DB00852	5.68					DAT only	B	[186]
MDA[a]	DB01509[b]	5.62	5.64	6.15	−0.02	−0.53		C	[164]
Desipramine[a]	DB01151[b]	5.52	7.70	9.22	−2.18	−3.70	NET selective	F	[187]
Aripiprazole	DB01238	5.49	7.50		−2.01			H	[188]
KM822[a]		5.44					DAT only	J	[35]
Trimipramine	DB00726	5.42	6.83	5.61	−1.41	−0.19	SERT selective	F	[108]
Ibogaine[a]		5.39	6.23		−0.84			L	[189]
Amoxapine	DB00543	5.37	7.24	7.80	−1.87	−2.43		L	[108]
SoRI-20040[a]		5.27					DAT only	E	[181]
Modafinil[a]	DB00745	5.19	<3.3(N.D.)	4.45		0.74		D	[190]
Armodafinil	DB06413	5.18	3.63	3.77	1.55	1.41	DAT selective	D	[105]
Escitalopram	DB01175	5.09	9.05	4.98	−3.96	0.11	SERT selective	L	[191]
Dopamine[a]	DB00988	5.06	3.00	5.28	2.06	−0.22		C	[174]

For each DAT ligand, we first attempted to retrieve its pharmacological activity data for DAT, SERT, and NET from ChEMBL 33, and then supplemented by literature search. When a compound has pKi, pKd, and pIC50 data in ChEMBL, we chose to report either pKi or pKd; then if more than one pKi, pKd, or pIC50 data points are available, we chose to report the highest one. For the ligand that we found having all DAT, SERT, and NET activity data from the same publication in our literature search, we replaced the data retrieved from ChEMBL with the ones from the literature search. Additional information for each compound including other references we collected can be found on https://github.com/NIDA-IRP-CCMB/QSAR_DAT-hERG. The Δactivity was then calculated to detect the DAT, SERT, and NET selective ligands, with Δactivity >= 1 for one transporter over both the other two transporters. The yellow in reference (DAT) indicates that the entry in ChEMBL has activity data, but no publication record. The gray means the best pKi has been annotation wrongly in ChEMBL, and the second best pKi was used in the table

[a]DAT ligands are collected from the well-respected reviews [21, 22, 77]

[b]DAT ligands can be found in DrugBank, but not listed as targeting DAT in DrugBank.

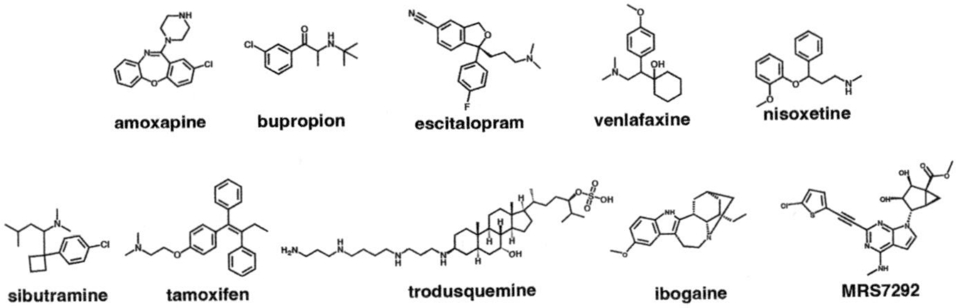

Fig. S1 Single-member clusters

References

1. Chen NH, Reith ME, Quick MW (2004) Synaptic uptake and beyond: the sodium- and chloride-dependent neurotransmitter transporter family SLC6. Pflugers Arch 447(5):519–531. https://doi.org/10.1007/s00424-003-1064-5. From NLM Medline
2. Beuming T, Shi L, Javitch JA, Weinstein H (2006) A comprehensive structure-based alignment of prokaryotic and eukaryotic neurotransmitter/Na+ symporters (NSS) aids in the use of the LeuT structure to probe NSS structure and function. Mol Pharmacol 70(5):1630–1642. https://doi.org/10.1124/mol.106.026120
3. Aggarwal S, Mortensen OV (2017) Overview of monoamine transporters. Curr Protoc Pharmacol 79:12 16 11–12 16 17. https://doi.org/10.1002/cpph.32. From NLM Medline
4. Juarez Olguin H, Calderon Guzman D, Hernandez Garcia E, Barragan Mejia G (2016) The role of Dopamine and its dysfunction as a consequence of oxidative stress. Oxidative Med Cell Longev 2016:9730467. https://doi.org/10.1155/2016/9730467. From NLM Medline
5. Stolzenberg S, Quick M, Zhao C, Gotfryd K, Khelashvili G, Gether U, Loland CJ, Javitch JA, Noskov S, Weinstein H et al (2015) Mechanism of the association between Na+ binding and conformations at the intracellular gate in neurotransmitter:sodium symporters. J Biol Chem 290(22):13992–14003. https://doi.org/10.1074/jbc.M114.625343
6. Borre L, Andreassen TF, Shi L, Weinstein H, Gether U (2014) The second sodium site in the dopamine transporter controls cation permeation and is regulated by chloride. J Biol Chem 289(37):25764–25773. https://doi.org/10.1074/jbc.M114.574269
7. Cook EH Jr, Stein MA, Krasowski MD, Cox NJ, Olkon DM, Kieffer JE, Leventhal BL (1995) Association of attention-deficit disorder and the dopamine transporter gene. Am J Hum Genet 56(4):993–998. From NLM Medline
8. Waldman ID, Rowe DC, Abramowitz A, Kozel ST, Mohr JH, Sherman SL, Cleveland HH, Sanders ML, Gard JM, Stever C (1998) Association and linkage of the dopamine transporter gene and attention-deficit hyperactivity disorder in children: heterogeneity owing to diagnostic subtype and severity. Am J Hum Genet 63(6):1767–1776. https://doi.org/10.1086/302132. From NLM Medline
9. Greenwood TA, Schork NJ, Eskin E, Kelsoe JR (2006) Identification of additional variants within the human dopamine transporter gene provides further evidence for an association with bipolar disorder in two independent samples. Mol Psychiatry 11(2):125–133, 115. https://doi.org/10.1038/sj.mp.4001764. From NLM Medline
10. Ashok AH, Marques TR, Jauhar S, Nour MM, Goodwin GM, Young AH, Howes OD (2017) The dopamine hypothesis of bipolar affective disorder: the state of the art and implications for treatment. Mol Psychiatry 22(5):666–679. https://doi.org/10.1038/mp.2017.16. From NLM Medline

11. Tamura T, Sugihara G, Okita K, Mukai Y, Matsuda H, Shiwaku H, Takagi S, Daisaki H, Tateishi U, Takahashi H (2022) Dopamine dysfunction in depression: application of texture analysis to dopamine transporter single-photon emission computed tomography imaging. Transl Psychiatry 12(1):309. https://doi.org/10.1038/s41398-022-02080-z. From NLM Medline
12. McLellan AT (2017) Substance misuse and substance use disorders: why do they matter in healthcare? Trans Am Clin Climatol Assoc 128:112–130. From NLM Medline
13. Schwartz EKC, Wolkowicz NR, De Aquino JP, MacLean RR, Sofuoglu M (2022) Cocaine Use Disorder (CUD): current clinical perspectives. Subst Abus Rehabil 13:25–46. https://doi.org/10.2147/SAR.S337338. From NLM PubMed-not-MEDLINE
14. Abuse, N. I. o. D. Drug overdose death rates (2023). https://nida.nih.gov/research-topics/trends-statistics/overdose-death-rates. Accessed 16 Dec 2023
15. Giros B, Jaber M, Jones SR, Wightman RM, Caron MG (1996) Hyperlocomotion and indifference to cocaine and amphetamine in mice lacking the dopamine transporter. Nature 379(6566):606–612. https://doi.org/10.1038/379606a0. From NLM Medline
16. Chen R, Tilley MR, Wei H, Zhou F, Zhou FM, Ching S, Quan N, Stephens RL, Hill ER, Nottoli T et al (2006) Abolished cocaine reward in mice with a cocaine-insensitive dopamine transporter. Proc Natl Acad Sci USA 103(24):9333–9338. https://doi.org/10.1073/pnas.0600905103. From NLM Medline
17. Newman AH, Kulkarni S (2002) Probes for the dopamine transporter: new leads toward a cocaine-abuse therapeutic--a focus on analogues of benztropine and rimcazole. Med Res Rev 22(5):429–464. https://doi.org/10.1002/med.10014. From NLM Medline
18. Newman AH, Katz JL (2009) Atypical Dopamine uptake inhibitors that provide clues about Cocaine's mechanism at the Dopamine transporter. Top Med Chem Ser 4:95–129. https://doi.org/10.1007/7355_2008_027
19. Loland CJ, Desai RI, Zou MF, Cao J, Grundt P, Gerstbrein K, Sitte HH, Newman AH, Katz JL, Gether U (2008) Relationship between conformational changes in the dopamine transporter and cocaine-like subjective effects of uptake inhibitors. Mol Pharmacol 73(3):813–823. https://doi.org/10.1124/mol.107.039800. From NLM Medline
20. Reith ME, Blough BE, Hong WC, Jones KT, Schmitt KC, Baumann MH, Partilla JS, Rothman RB, Katz JL (2015) Behavioral, biological, and chemical perspectives on atypical agents targeting the dopamine transporter. Drug Alcohol Depend 147:1–19. https://doi.org/10.1016/j.drugalcdep.2014.12.005. From NLM Medline
21. Schmitt KC, Reith ME (2011) The atypical stimulant and nootropic modafinil interacts with the dopamine transporter in a different manner than classical cocaine-like inhibitors. PLoS One 6(10):e25790. https://doi.org/10.1371/journal.pone.0025790. From NLM Medline
22. Nepal B, Das S, Reith ME, Kortagere S (2023) Overview of the structure and function of the dopamine transporter and its protein interactions. Front Physiol 14:1150355. https://doi.org/10.3389/fphys.2023.1150355. From NLM PubMed-not-MEDLINE
23. Abramyan AM, Stolzenberg S, Li Z, Loland CJ, Noe F, Shi L (2017) The isomeric preference of an atypical Dopamine transporter inhibitor contributes to its selection of the transporter conformation. ACS Chem Neurosci 8(8):1735–1746. https://doi.org/10.1021/acschemneuro.7b00094. From NLM Medline
24. Zou MF, Cao J, Abramyan AM, Kopajtic T, Zanettini C, Guthrie DA, Rais R, Slusher BS, Shi L, Loland CJ et al (2017) Structure-activity relationship studies on a series of 3alpha-[Bis(4-fluorophenyl)methoxy]tropanes and 3alpha-[Bis(4-fluorophenyl)methylamino]tropanes as novel atypical Dopamine transporter (DAT) inhibitors for the treatment of cocaine use disorders. J Med Chem 60(24):10172–10187. https://doi.org/10.1021/acs.jmedchem.7b01454. From NLM Medline
25. Rothman RB, Baumann MH, Prisinzano TE, Newman AH (2008) Dopamine transport inhibitors based on GBR12909 and benztropine as potential medications to treat cocaine addiction. Biochem Pharmacol 75(1):2–16. https://doi.org/10.1016/j.bcp.2007.08.007. From NLM Medline

26. Zhou J, He R, Johnson KM, Ye Y, Kozikowski AP (2004) Piperidine-based nocaine/modafinil hybrid ligands as highly potent monoamine transporter inhibitors: efficient drug discovery by rational lead hybridization. J Med Chem 47(24):5821–5824. https://doi.org/10.1021/jm040117o. From NLM Medline
27. Loland CJ, Mereu M, Okunola OM, Cao J, Prisinzano TE, Mazier S, Kopajtic T, Shi L, Katz JL, Tanda G et al (2012) R-modafinil (armodafinil): a unique dopamine uptake inhibitor and potential medication for psychostimulant abuse. Biol Psychiatry 72(5):405–413. https://doi.org/10.1016/j.biopsych.2012.03.022. From NLM Medline
28. Okunola-Bakare OM, Cao J, Kopajtic T, Katz JL, Loland CJ, Shi L, Newman AH (2014) Elucidation of structural elements for selectivity across monoamine transporters: novel 2-[(diphenylmethyl)sulfinyl]acetamide (modafinil) analogues. J Med Chem 57(3):1000–1013. https://doi.org/10.1021/jm401754x
29. Newman AH, Ku T, Jordan CJ, Bonifazi A, Xi ZX (2021) New drugs, old targets: tweaking the Dopamine system to treat psychostimulant use disorders. Annu Rev Pharmacol Toxicol 61(1):609–628. https://doi.org/10.1146/annurev-pharmtox-030220-124205. (acccessed 2023/12/16). From NLM Medline
30. Pratuangdejkul J, Schneider B, Launay JM, Kellermann O, Manivet P (2008) Computational approaches for the study of serotonin and its membrane transporter SERT: implications for drug design in neurological sciences. Curr Med Chem 15(30):3214–3227. https://doi.org/10.2174/092986708786848523. From NLM Medline
31. Nolan TL, Geffert LM, Kolber BJ, Madura JD, Surratt CK (2014) Discovery of novel-scaffold monoamine transporter ligands via in silico screening with the S1 pocket of the serotonin transporter. ACS Chem Neurosci 5(9):784–792. https://doi.org/10.1021/cn500133b. From NLM Medline
32. Manepalli S, Geffert LM, Surratt CK, Madura JD (2011) Discovery of novel selective serotonin reuptake inhibitors through development of a protein-based pharmacophore. J Chem Inf Model 51(9):2417–2426. https://doi.org/10.1021/ci200280m. From NLM Medline
33. Simmons KJ, Gotfryd K, Billesbolle CB, Loland CJ, Gether U, Fishwick CW, Johnson AP (2013) A virtual high-throughput screening approach to the discovery of novel inhibitors of the bacterial leucine transporter, LeuT. Mol Membr Biol 30(2):184–194. https://doi.org/10.3109/09687688.2012.710341. From NLM Medline
34. Schlessinger A, Geier E, Fan H, Irwin JJ, Shoichet BK, Giacomini KM, Sali A (2011) Structure-based discovery of prescription drugs that interact with the norepinephrine transporter, NET. Proc Natl Acad Sci USA 108(38):15810–15815. https://doi.org/10.1073/pnas.1106030108. From NLM Medline
35. Aggarwal S, Liu X, Rice C, Menell P, Clark PJ, Paparoidamis N, Xiao YC, Salvino JM, Fontana ACK, Espana RA et al (2019) Identification of a novel allosteric modulator of the human Dopamine transporter. ACS Chem Neurosci 10(8):3718–3730. https://doi.org/10.1021/acschemneuro.9b00262. From NLM Medline
36. Indarte M, Madura JD, Surratt CK (2008) Dopamine transporter comparative molecular modeling and binding site prediction using the LeuT(Aa) leucine transporter as a template. Proteins 70(3):1033–1046. https://doi.org/10.1002/prot.21598. From NLM Medline
37. Vemula D, Jayasurya P, Sushmitha V, Kumar YN, Bhandari V (2023) CADD, AI and ML in drug discovery: a comprehensive review. Eur J Pharm Sci 181:106324. https://doi.org/10.1016/j.ejps.2022.106324. From NLM Medline
38. Voet A, Banwell EF, Sahu KK, Heddle JG, Zhang KY (2013) Protein interface pharmacophore mapping tools for small molecule protein: protein interaction inhibitor discovery. Curr Top Med Chem 13(9):989–1001. https://doi.org/10.2174/1568026611313090003. From NLM Medline
39. Khedkar SA, Malde AK, Coutinho EC, Srivastava S (2007) Pharmacophore modeling in drug discovery and development: an overview. Med Chem 3(2):187–197. https://doi.org/10.2174/157340607780059521. From NLM Medline

40. Christensen HS, Boye SV, Thinggaard J, Sinning S, Wiborg O, Schiott B, Bols M (2007) QSAR studies and pharmacophore identification for arylsubstituted cycloalkenecarboxylic acid methyl esters with affinity for the human dopamine transporter. Bioorg Med Chem 15(15):5262–5274. https://doi.org/10.1016/j.bmc.2007.05.015. From NLM Medline
41. Hansch C, Maloney PP, Fujita T (1962) Correlation of biological activity of phenoxyacetic acids with Hammett substituent constants and partition coefficients. Nature 194(4824):178. https://doi.org/10.1038/194178b0
42. Verma RP, Hansch C (2009) Camptothecins: a SAR/QSAR study. Chem Rev 109(1):213–235. https://doi.org/10.1021/cr0780210. From NLM Medline
43. Gao K, Chen D, Robison AJ, Wei GW (2021) Proteome-informed machine learning studies of cocaine addiction. J Phys Chem Lett 12(45):11122–11134. https://doi.org/10.1021/acs.jpclett.1c03133. From NLM Medline
44. Danishuddin, Khan AU (2016) Descriptors and their selection methods in QSAR analysis: paradigm for drug design. Drug Discov Today 21(8):1291–1302. https://doi.org/10.1016/j.drudis.2016.06.013. From NLM Medline
45. Mao J, Akhtar J, Zhang X, Sun L, Guan S, Li X, Chen G, Liu J, Jeon HN, Kim MS et al (2021) Comprehensive strategies of machine-learning-based quantitative structure-activity relationship models. iScience 24(9):103052. https://doi.org/10.1016/j.isci.2021.103052. From NLM PubMed-not-MEDLINE
46. Hong H, Xie Q, Ge W, Qian F, Fang H, Shi L, Su Z, Perkins R, Tong W (2008) Mold(2), molecular descriptors from 2D structures for chemoinformatics and toxicoinformatics. J Chem Inf Model 48(7):1337–1344. https://doi.org/10.1021/ci800038f. From NLM Medline
47. Mauri A, Consonni V, Pavan M, Todeschini R (2006) Dragon software: an easy approach to molecular descriptor calculations. Match-Commun Math Co 56(2):237–248, Article. Scopus
48. Cho YS, No KT, Cho KH (2012) yaInChI: modified InChI string scheme for line notation of chemical structures. SAR QSAR Environ Res 23(3–4):237–255. https://doi.org/10.1080/1062936X.2012.657677. From NLM Medline
49. Choi SY, Shin JH, Ryu CK, Nam KY, No KT, Park Choo HY (2006) The development of 3D-QSAR study and recursive partitioning of heterocyclic quinone derivatives with antifungal activity. Bioorg Med Chem 14(5):1608–1617. https://doi.org/10.1016/j.bmc.2005.10.010. From NLM Medline
50. Hayakawa D, Sawada N, Watanabe Y, Gouda H (2020) A molecular interaction field describing nonconventional intermolecular interactions and its application to protein-ligand interaction prediction. J Mol Graph Model 96:107515. https://doi.org/10.1016/j.jmgm.2019.107515. From NLM Medline
51. Hoffman BT, Kopajtic T, Katz JL, Newman AH (2000) 2D QSAR modeling and preliminary database searching for dopamine transporter inhibitors using genetic algorithm variable selection of Molconn Z descriptors. J Med Chem 43(22):4151–4159. https://doi.org/10.1021/jm990472s. From NLM Medline
52. Kulkarni SS, Newman AH, Houlihan WJ (2002) Three-dimensional quantitative structure-activity relationships of mazindol analogues at the dopamine transporter. J Med Chem 45(19):4119–4127. https://doi.org/10.1021/jm0102093. From NLM Medline
53. Zhang X, Mao J, Wei M, Qi Y, Zhang JZH (2022) HergSPred: accurate classification of hERG blockers/nonblockers with machine-learning models. J Chem Inf Model 62(8):1830–1839. https://doi.org/10.1021/acs.jcim.2c00256. From NLM Medline
54. Soares TA, Nunes-Alves A, Mazzolari A, Ruggiu F, Wei GW, Merz K (2022) The (Re)-evolution of Quantitative Structure-Activity Relationship (QSAR) studies propelled by the surge of machine learning methods. J Chem Inf Model 62(22):5317–5320. https://doi.org/10.1021/acs.jcim.2c01422. From NLM Medline
55. Alsenan SA, Al-Turaiki IM, Hafez AM (2020) Feature extraction methods in quantitative structure–activity relationship modeling: a comparative study. IEEE Access 8:78737–78752. https://doi.org/10.1109/access.2020.2990375

56. Tropsha A, Isayev O, Varnek A, Schneider G, Cherkasov A (2023) Integrating QSAR modelling and deep learning in drug discovery: the emergence of deep QSAR. Nat Rev Drug Discov 23:141. https://doi.org/10.1038/s41573-023-00832-0. From NLM Publisher
57. Yao XJ, Panaye A, Doucet JP, Zhang RS, Chen HF, Liu MC, Hu ZD, Fan BT (2004) Comparative study of QSAR/QSPR correlations using support vector machines, radial basis function neural networks, and multiple linear regression. J Chem Inf Comput Sci 44(4):1257–1266. https://doi.org/10.1021/ci049965i. From NLM Medline
58. Ma J, Sheridan RP, Liaw A, Dahl GE, Svetnik V (2015) Deep neural nets as a method for quantitative structure-activity relationships. J Chem Inf Model 55(2):263–274. https://doi.org/10.1021/ci500747n. From NLM Medline
59. Sheridan RP, Wang WM, Liaw A, Ma J, Gifford EM (2016) Extreme gradient boosting as a method for quantitative structure-activity relationships. J Chem Inf Model 56(12):2353–2360. https://doi.org/10.1021/acs.jcim.6b00591. From NLM Medline
60. Svetnik V, Liaw A, Tong C, Culberson JC, Sheridan RP, Feuston BP (2003) Random forest: a classification and regression tool for compound classification and QSAR modeling. J Chem Inf Comput Sci 43(6):1947–1958. https://doi.org/10.1021/ci034160g. From NLM PubMed-not-MEDLINE
61. Gaulton A, Bellis LJ, Bento AP, Chambers J, Davies M, Hersey A, Light Y, McGlinchey S, Michalovich D, Al-Lazikani B et al (2012) ChEMBL: a large-scale bioactivity database for drug discovery. Nucleic Acids Res 40(Database issue):D1100–1107. https://doi.org/10.1093/nar/gkr777. From NLM Medline
62. Wishart DS, Feunang YD, Guo AC, Lo EJ, Marcu A, Grant JR, Sajed T, Johnson D, Li C, Sayeeda Z et al (2018) DrugBank 5.0: a major update to the DrugBank database for 2018. Nucleic Acids Res 46(D1):D1074–D1082. https://doi.org/10.1093/nar/gkx1037. From NLM Medline
63. Gilson MK, Liu T, Baitaluk M, Nicola G, Hwang L, Chong J (2016) BindingDB in 2015: a public database for medicinal chemistry, computational chemistry and systems pharmacology. Nucleic Acids Res 44(D1):D1045–D1053. https://doi.org/10.1093/nar/gkv1072. From NLM Medline
64. Liu T, Lin Y, Wen X, Jorissen RN, Gilson MK (2007) BindingDB: a web-accessible database of experimentally determined protein-ligand binding affinities. Nucleic Acids Res 35(Database issue):D198–201. https://doi.org/10.1093/nar/gkl999. From NLM Medline
65. Zhou Y, Zhang Y, Zhao D, Yu X, Shen X, Zhou Y, Wang S, Qiu Y, Chen Y, Zhu F (2024) TTD: Therapeutic Target Database describing target druggability information. Nucleic Acids Res 52(D1):D1465–D1477. https://doi.org/10.1093/nar/gkad751. From NLM Medline
66. Besnard J, Ruda GF, Setola V, Abecassis K, Rodriguiz RM, Huang XP, Norval S, Sassano MF, Shin AI, Webster LA et al (2012) Automated design of ligands to polypharmacological profiles. Nature 492(7428):215–220. https://doi.org/10.1038/nature11691. From NLM Medline
67. Kim S, Chen J, Cheng T, Gindulyte A, He J, He S, Li Q, Shoemaker BA, Thiessen PA, Yu B et al (2023) PubChem 2023 update. Nucleic Acids Res 51(D1):D1373–D1380. https://doi.org/10.1093/nar/gkac956. From NLM Medline
68. Lee KH, Fant AD, Guo J, Guan A, Jung J, Kudaibergenova M, Miranda WE, Ku T, Cao J, Wacker S et al (2021) Toward reducing hERG affinities for DAT inhibitors with a combined machine learning and molecular modeling approach. J Chem Inf Model 61(9):4266–4279. https://doi.org/10.1021/acs.jcim.1c00856. From NLM Medline
69. Tuomisto J, Tuomisto L, Pazdernik TL (1976) Conformationally rigid amphetamine analogs as inhibitors of monoamine uptake by brain synaptosomes. J Med Chem 19(5):725–727. https://doi.org/10.1021/jm00227a030. From NLM Medline
70. Yamashita A, Singh SK, Kawate T, Jin Y, Gouaux E (2005) Crystal structure of a bacterial homologue of Na+/Cl--dependent neurotransmitter transporters. Nature 437(7056):215–223. https://doi.org/10.1038/nature03978. From NLM Medline

71. Koldso H, Christiansen AB, Sinning S, Schiott B (2013) Comparative modeling of the human monoamine transporters: similarities in substrate binding. ACS Chem Neurosci 4(2):295–309. https://doi.org/10.1021/cn300148r. From NLM Medline
72. Severinsen K, Koldso H, Thorup KA, Schjoth-Eskesen C, Moller PT, Wiborg O, Jensen HH, Sinning S, Schiott B (2014) Binding of mazindol and analogs to the human serotonin and dopamine transporters. Mol Pharmacol 85(2):208–217. https://doi.org/10.1124/mol.113.088922. From NLM Medline
73. Shan J, Javitch JA, Shi L, Weinstein H (2011) The substrate-driven transition to an inward-facing conformation in the functional mechanism of the dopamine transporter. PLoS One 6(1):e16350. https://doi.org/10.1371/journal.pone.0016350. From NLM Medline
74. Cheng MH, Bahar I (2015) Molecular mechanism of Dopamine transport by human Dopamine transporter. Structure 23(11):2171–2181. https://doi.org/10.1016/j.str.2015.09.001
75. Ravna AW, Sylte I, Dahl SG (2009) Structure and localisation of drug binding sites on neurotransmitter transporters. J Mol Model 15(10):1155–1164. https://doi.org/10.1007/s00894-009-0478-1. From NLM Medline
76. Penmatsa A, Wang KH, Gouaux E (2013) X-ray structure of dopamine transporter elucidates antidepressant mechanism. Nature 503(7474):85–90. https://doi.org/10.1038/nature12533. From NLM Medline
77. Schmitt KC, Rothman RB, Reith ME (2013) Nonclassical pharmacology of the dopamine transporter: atypical inhibitors, allosteric modulators, and partial substrates. J Pharmacol Exp Ther 346(1):2–10. https://doi.org/10.1124/jpet.111.191056. From NLM Medline
78. Gatley SJ, Volkow ND, Chen R, Fowler JS, Carroll FI, Kuhar MJ (1996) Displacement of RTI-55 from the dopamine transporter by cocaine. Eur J Pharmacol 296(2):145–151. https://doi.org/10.1016/0014-2999(95)00698-2. From NLM Medline
79. Coffey LL, Reith ME (1994) [3H]WIN 35,428 binding to the dopamine uptake carrier. I. Effect of tonicity and buffer composition. J Neurosci Methods 51(1):23–30. https://doi.org/10.1016/0165-0270(94)90022-1. From NLM Medline
80. Dahal RA, Pramod AB, Sharma B, Krout D, Foster JD, Cha JH, Cao J, Newman AH, Lever JR, Vaughan RA et al (2014) Computational and biochemical docking of the irreversible cocaine analog RTI 82 directly demonstrates ligand positioning in the dopamine transporter central substrate-binding site. J Biol Chem 289(43):29712–29727. https://doi.org/10.1074/jbc.M114.571521. From NLM Medline
81. Parnas ML, Gaffaney JD, Zou MF, Lever JR, Newman AH, Vaughan RA (2008) Labeling of dopamine transporter transmembrane domain 1 with the tropane ligand N-[4-(4-azido-3-[125I]iodophenyl)butyl]-2beta-carbomethoxy-3beta-(4-chlorophenyl)tropane implicates proximity of cocaine and substrate active sites. Mol Pharmacol 73(4):1141–1150. https://doi.org/10.1124/mol.107.043679. From NLM Medline
82. Booij J, Andringa G, Rijks LJ, Vermeulen RJ, De Bruin K, Boer GJ, Janssen AG, Van Royen EA (1997) [123I]FP-CIT binds to the dopamine transporter as assessed by biodistribution studies in rats and SPECT studies in MPTP-lesioned monkeys. Synapse 27(3):183–190. https://doi.org/10.1002/(SICI)1098-2396(199711)27:3<183::AID-SYN4>3.0.CO;2-9. From NLM Medline
83. Madras BK, Meltzer PC, Liang AY, Elmaleh DR, Babich J, Fischman AJ (1998) Altropane, a SPECT or PET imaging probe for dopamine neurons: I. Dopamine transporter binding in primate brain. Synapse 29(2):93–104. https://doi.org/10.1002/(SICI)1098-2396(199806)29:2<93::AID-SYN1>3.0.CO;2-5. From NLM Medline
84. Kimmel HL, Carroll FI, Kuhar MJ (2001) Locomotor stimulant effects of novel phenyltropanes in the mouse. Drug Alcohol Depend 65(1):25–36. https://doi.org/10.1016/s0376-8716(01)00144-2. From NLM Medline
85. Lehr T, Staab A, Tillmann C, Nielsen EO, Trommeshauser D, Schaefer HG, Kloft C (2008) Contribution of the active metabolite M1 to the pharmacological activity of tesofensine in vivo: a pharmacokinetic-pharmacodynamic modelling approach. Br J Pharmacol 153(1):164–174. https://doi.org/10.1038/sj.bjp.0707539. From NLM Medline

86. Elia J, Ambrosini PJ, Rapoport JL (1999) Treatment of attention-deficit-hyperactivity disorder. N Engl J Med 340(10):780–788. https://doi.org/10.1056/NEJM199903113401007. From NLM Medline
87. Kutcher S, Aman M, Brooks SJ, Buitelaar J, van Daalen E, Fegert J, Findling RL, Fisman S, Greenhill LL, Huss M et al (2004) International consensus statement on attention-deficit/hyperactivity disorder (ADHD) and disruptive behaviour disorders (DBDs): clinical implications and treatment practice suggestions. Eur Neuropsychopharmacol 14(1):11–28. https://doi.org/10.1016/s0924-977x(03)00045-2. From NLM Medline
88. Barch DM, Carter CS (2005) Amphetamine improves cognitive function in medicated individuals with schizophrenia and in healthy volunteers. Schizophr Res 77(1):43–58. https://doi.org/10.1016/j.schres.2004.12.019. From NLM Medline
89. Berman SM, Kuczenski R, McCracken JT, London ED (2009) Potential adverse effects of amphetamine treatment on brain and behavior: a review. Mol Psychiatry 14(2):123–142. https://doi.org/10.1038/mp.2008.90. From NLM Medline
90. Northrop NA, Yamamoto BK (2015) Methamphetamine effects on blood-brain barrier structure and function. Front Neurosci 9:69. https://doi.org/10.3389/fnins.2015.00069. From NLM PubMed-not-MEDLINE
91. McCann UD, Wong DF, Yokoi F, Villemagne V, Dannals RF, Ricaurte GA (1998) Reduced striatal dopamine transporter density in abstinent methamphetamine and methcathinone users: evidence from positron emission tomography studies with [11C]WIN-35,428. J Neurosci 18(20):8417–8422. https://doi.org/10.1523/JNEUROSCI.18-20-08417.1998. From NLM Medline
92. Paulus MP, Stewart JL (2020) Neurobiology, clinical presentation, and treatment of methamphetamine use disorder: a review. JAMA Psychiatry 77(9):959–966. https://doi.org/10.1001/jamapsychiatry.2020.0246. From NLM Medline
93. Martin D, Le JK (2023) Amphetamine [Updated 2023 Jul 31]. In: StatPearls [Internet]. https://www.ncbi.nlm.nih.gov/books/NBK556103/. Accessed
94. Erdo SL, Kiss B, Rosdy B (1981) Inhibition of dopamine uptake by a new psychostimulant mesocarb (Sydnocarb). Pol J Pharmacol Pharm 33(2):141–147. From NLM Medline
95. Aggarwal S, Cheng MH, Salvino JM, Bahar I, Mortensen OV (2021) Functional characterization of the Dopaminergic psychostimulant Sydnocarb as an allosteric modulator of the human dopamine transporter. Biomedicines 9(6). https://doi.org/10.3390/biomedicines9060634. From NLM PubMed-not-MEDLINE
96. Kalant H (2001) The pharmacology and toxicology of "ecstasy" (MDMA) and related drugs. CMAJ 165(7):917–928. From NLM Medline
97. Steinkellner T, Freissmuth M, Sitte HH, Montgomery T (2011) The ugly side of amphetamines: short- and long-term toxicity of 3,4-methylenedioxymethamphetamine (MDMA, 'Ecstasy'), methamphetamine and D-amphetamine. Biol Chem 392(1–2):103–115. https://doi.org/10.1515/BC.2011.016. From NLM Medline
98. Hagino Y, Takamatsu Y, Yamamoto H, Iwamura T, Murphy DL, Uhl GR, Sora I, Ikeda K (2011) Effects of MDMA on extracellular Dopamine and Serotonin levels in mice lacking Dopamine and/or Serotonin transporters. Curr Neuropharmacol 9(1):91–95. https://doi.org/10.2174/157015911795017254. From NLM PubMed-not-MEDLINE
99. Rothman RB, Partilla JS, Baumann MH, Lightfoot-Siordia C, Blough BE (2012) Studies of the biogenic amine transporters. 14. Identification of low-efficacy "partial" substrates for the biogenic amine transporters. J Pharmacol Exp Ther 341(1):251–262. https://doi.org/10.1124/jpet.111.188946. From NLM Medline
100. Rothman RB, Baumann MH (2006) Therapeutic potential of monoamine transporter substrates. Curr Top Med Chem 6(17):1845–1859. https://doi.org/10.2174/156802606778249766. From NLM Medline
101. Lopez-Arnau R, Duart-Castells L, Aster B, Camarasa J, Escubedo E, Pubill D (2019) Effects of MDPV on dopamine transporter regulation in male rats. Comparison with cocaine.

Psychopharmacology 236(3):925–938. https://doi.org/10.1007/s00213-018-5052-z. From NLM Medline

102. Baumann MH, Partilla JS, Lehner KR, Thorndike EB, Hoffman AF, Holy M, Rothman RB, Goldberg SR, Lupica CR, Sitte HH et al (2013) Powerful cocaine-like actions of 3,4-methylenedioxypyrovalerone (MDPV), a principal constituent of psychoactive 'bath salts' products. Neuropsychopharmacology 38(4):552–562. https://doi.org/10.1038/npp.2012.204. From NLM Medline
103. Baumann MH, Wang X, Rothman RB (2007) 3,4-Methylenedioxymethamphetamine (MDMA) neurotoxicity in rats: a reappraisal of past and present findings. Psychopharmacology 189(4):407–424. https://doi.org/10.1007/s00213-006-0322-6. From NLM Medline
104. Volkow ND, Fowler JS, Logan J, Alexoff D, Zhu W, Telang F, Wang GJ, Jayne M, Hooker JM, Wong C et al (2009) Effects of modafinil on dopamine and dopamine transporters in the male human brain: clinical implications. JAMA 301(11):1148–1154. https://doi.org/10.1001/jama.2009.351. From NLM Medline
105. Shanmugasundaram B, Aher YD, Aradska J, Ilic M, Daba Feyissa D, Kalaba P, Aher NY, Dragacevic V, Saber Marouf B, Langer T et al (2017) R-Modafinil exerts weak effects on spatial memory acquisition and dentate gyrus synaptic plasticity. PLoS One 12(6):e0179675. https://doi.org/10.1371/journal.pone.0179675. From NLM Medline
106. Rotolo RA, Dragacevic V, Kalaba P, Urban E, Zehl M, Roller A, Wackerlig J, Langer T, Pistis M, De Luca MA et al (2019) The novel atypical Dopamine uptake inhibitor (S)-CE-123 partially reverses the effort-related effects of the Dopamine depleting agent Tetrabenazine and increases progressive ratio responding. Front Pharmacol 10:682. https://doi.org/10.3389/fphar.2019.00682. From NLM PubMed-not-MEDLINE
107. Oleson EB, Ferris MJ, Espana RA, Harp J, Jones SR (2012) Effects of the histamine H(1) receptor antagonist and benztropine analog diphenylpyraline on dopamine uptake, locomotion and reward. Eur J Pharmacol 683(1–3):161–165. https://doi.org/10.1016/j.ejphar.2012.03.003. From NLM Medline
108. Tatsumi M, Groshan K, Blakely RD, Richelson E (1997) Pharmacological profile of antidepressants and related compounds at human monoamine transporters. Eur J Pharmacol 340(2–3):249–258. https://doi.org/10.1016/s0014-2999(97)01393-9. From NLM Medline
109. Newman AH, Agoston GE (1998) Novel benztropine [3a-(diphenylmethoxy)tropane] analogs as probes for the dopamine transporter. Curr Med Chem 5(4):305–319. From NLM Medline
110. Kopajtic TA, Liu Y, Surratt CK, Donovan DM, Newman AH, Katz JL (2010) Dopamine transporter-dependent and -independent striatal binding of the benztropine analog JHW 007, a cocaine antagonist with low abuse liability. J Pharmacol Exp Ther 335(3):703–714. https://doi.org/10.1124/jpet.110.171629. From NLM Medline
111. Moerke MJ, Ananthan S, Banks ML, Eltit JM, Freitas KC, Johnson AR, Saini SK, Steele TWE, Negus SS (2018) Interactions between Cocaine and the putative allosteric Dopamine transporter ligand SRI-31142. J Pharmacol Exp Ther 367(2):222–233. https://doi.org/10.1124/jpet.118.250902. From NLM Medline
112. Rothman RB, Dersch CM, Carroll FI, Ananthan S (2002) Studies of the biogenic amine transporters. VIII: identification of a novel partial inhibitor of dopamine uptake and dopamine transporter binding. Synapse 43(4):268–274. https://doi.org/10.1002/syn.10046. From NLM Medline
113. Pariser JJ, Partilla JS, Dersch CM, Ananthan S, Rothman RB (2008) Studies of the biogenic amine transporters. 12. Identification of novel partial inhibitors of amphetamine-induced dopamine release. J Pharmacol Exp Ther 326(1):286–295. https://doi.org/10.1124/jpet.108.139675. From NLM Medline
114. Owens MJ, Morgan WN, Plott SJ, Nemeroff CB (1997) Neurotransmitter receptor and transporter binding profile of antidepressants and their metabolites. J Pharmacol Exp Ther 283(3):1305–1322. From NLM Medline

115. Goodnick PJ, Goldstein BJ (1998) Selective serotonin reuptake inhibitors in affective disorders--I. Basic pharmacology. J Psychopharmacol 12(3 Suppl B):S5–20. https://doi.org/10.1177/026988119801200302 1. From NLM Medline
116. Koblan KS, Hopkins SC, Sarma K, Jin F, Goldman R, Kollins SH, Loebel A (2015) Dasotraline for the treatment of attention-deficit/hyperactivity disorder: a randomized, double-blind, placebo-controlled, proof-of-concept trial in adults. Neuropsychopharmacology 40(12):2745–2752. https://doi.org/10.1038/npp.2015.124. From NLM Medline
117. Lawler CP, Prioleau C, Lewis MM, Mak C, Jiang D, Schetz JA, Gonzalez AM, Sibley DR, Mailman RB (1999) Interactions of the novel antipsychotic aripiprazole (OPC-14597) with dopamine and serotonin receptor subtypes. Neuropsychopharmacology 20(6):612–627. https://doi.org/10.1016/S0893-133X(98)00099-2. From NLM Medline
118. Davies MA, Sheffler DJ, Roth BL (2004) Aripiprazole: a novel atypical antipsychotic drug with a uniquely robust pharmacology. CNS Drug Rev 10(4):317–336. https://doi.org/10.1111/j.1527-3458.2004.tb00030.x. From NLM Medline
119. Duggal HS (2003) Aripiprazole-induced improvement in tardive dyskinesia. Can J Psychiatr 48(11):771–772. https://doi.org/10.1177/070674370304801116. From NLM Medline
120. Carlier PR, Lo MM, Lo PC, Richelson E, Tatsumi M, Reynolds IJ, Sharma TA (1998) Synthesis of a potent wide-spectrum serotonin-, norepinephrine-, dopamine-reuptake inhibitor (SNDRI) and a species-selective dopamine-reuptake inhibitor based on the gamma-amino alcohol functional group. Bioorg Med Chem Lett 8(5):487–492. https://doi.org/10.1016/s0960-894x(98)00062-6. From NLM Medline
121. Karpa KD, Cavanaugh JE, Lakoski JM (2002) Duloxetine pharmacology: profile of a dual monoamine modulator. CNS Drug Rev 8(4):361–376. https://doi.org/10.1111/j.1527-3458.2002.tb00234.x. From NLM Medline
122. Cashman JR, Ghirmai S (2009) Inhibition of serotonin and norepinephrine reuptake and inhibition of phosphodiesterase by multi-target inhibitors as potential agents for depression. Bioorg Med Chem 17(19):6890–6897. https://doi.org/10.1016/j.bmc.2009.08.025. From NLM Medline
123. Rothman RB, Blough BE, Woolverton WL, Anderson KG, Negus SS, Mello NK, Roth BL, Baumann MH (2005) Development of a rationally designed, low abuse potential, biogenic amine releaser that suppresses cocaine self-administration. J Pharmacol Exp Ther 313(3):1361–1369. https://doi.org/10.1124/jpet.104.082503. From NLM Medline
124. Hasenhuetl PS, Bhat S, Freissmuth M, Sandtner W (2019) Functional selectivity and partial efficacy at the Monoamine transporters: a unified model of allosteric modulation and amphetamine-induced substrate release. Mol Pharmacol 95(3):303–312. https://doi.org/10.1124/mol.118.114793. From NLM Medline
125. Rothman RB, Katsnelson M, Vu N, Partilla JS, Dersch CM, Blough BE, Baumann MH (2002) Interaction of the anorectic medication, phendimetrazine, and its metabolites with monoamine transporters in rat brain. Eur J Pharmacol 447(1):51–57. https://doi.org/10.1016/s0014-2999(02)01830-7. From NLM Medline
126. Fone KC, Nutt DJ (2005) Stimulants: use and abuse in the treatment of attention deficit hyperactivity disorder. Curr Opin Pharmacol 5(1):87–93. https://doi.org/10.1016/j.coph.2004.10.001. From NLM Medline
127. Marco P, Maddalena M, Silvana B, Erika G, Maria N (2022) Attention deficit hyperactivity disorder. In: Comprehensive pharmacology, pp 256–285
128. Raffel DM, Chen W (2004) Binding of [3H]mazindol to cardiac norepinephrine transporters: kinetic and equilibrium studies. Naunyn Schmiedeberg's Arch Pharmacol 370(1):9–16. https://doi.org/10.1007/s00210-004-0949-y. From NLM Medline
129. Hiranita T, Soto PL, Kohut SJ, Kopajtic T, Cao J, Newman AH, Tanda G, Katz JL (2011) Decreases in cocaine self-administration with dual inhibition of the dopamine transporter and sigma receptors. J Pharmacol Exp Ther 339(2):662–677. https://doi.org/10.1124/jpet.111.185025. From NLM Medline

130. Clark A, Tate B, Urban B, Schroeder R, Gennuso S, Ahmadzadeh S, McGregor D, Girma B, Shekoohi S, Kaye AD (2023) Bupropion mediated effects on depression, attention deficit hyperactivity disorder, and smoking cessation. Health Psychol Res 11:81043. https://doi.org/10.52965/001c.81043. From NLM PubMed-not-MEDLINE
131. Mikelman SR, Guptaroy B, Schmitt KC, Jones KT, Zhen J, Reith MEA, Gnegy ME (2018) Tamoxifen directly interacts with the Dopamine transporter. J Pharmacol Exp Ther 367(1):119–128. https://doi.org/10.1124/jpet.118.248179. From NLM Medline
132. Bulling S, Schicker K, Zhang YW, Steinkellner T, Stockner T, Gruber CW, Boehm S, Freissmuth M, Rudnick G, Sitte HH et al (2012) The mechanistic basis for noncompetitive ibogaine inhibition of serotonin and dopamine transporters. J Biol Chem 287(22):18524–18534. https://doi.org/10.1074/jbc.M112.343681. From NLM Medline
133. Cheng Y, Prusoff WH (1973) Relationship between the inhibition constant (K1) and the concentration of inhibitor which causes 50 per cent inhibition (I50) of an enzymatic reaction. Biochem Pharmacol 22(23):3099–3108. https://doi.org/10.1016/0006-2952(73)90196-2. From NLM Medline
134. Jumper J, Evans R, Pritzel A, Green T, Figurnov M, Ronneberger O, Tunyasuvunakool K, Bates R, Zidek A, Potapenko A et al (2021) Highly accurate protein structure prediction with AlphaFold. Nature 596(7873):583–589. https://doi.org/10.1038/s41586-021-03819-2. From NLM Medline
135. AlQuraishi M (2021) Machine learning in protein structure prediction. Curr Opin Chem Biol 65:1–8. https://doi.org/10.1016/j.cbpa.2021.04.005. From NLM Medline
136. Wenzel J, Matter H, Schmidt F (2019) Predictive multitask deep neural network models for ADME-Tox properties: learning from large data sets. J Chem Inf Model 59(3):1253–1268. https://doi.org/10.1021/acs.jcim.8b00785. From NLM Medline
137. Wu K, Wei GW (2018) Quantitative toxicity prediction using topology based multitask deep neural networks. J Chem Inf Model 58(2):520–531. https://doi.org/10.1021/acs.jcim.7b00558. From NLM Medline
138. Salas-Estrada L, Provasi D, Qiu X, Kaniskan HU, Huang XP, DiBerto JF, Lamim Ribeiro JM, Jin J, Roth BL, Filizola M (2023) De Novo design of kappa-opioid receptor antagonists using a generative deep-learning framework. J Chem Inf Model 63(16):5056–5065. https://doi.org/10.1021/acs.jcim.3c00651. From NLM Medline
139. Skinnider MA, Wang F, Pasin D, Greiner R, Foster LJ, Dalsgaard PW, Wishart DS (2021) A deep generative model enables automated structure elucidation of novel psychoactive substances. Nat Mach Intell 3(11):973-+. https://doi.org/10.1038/s42256-021-00407-x
140. Atance SR, Diez JV, Engkvist O, Olsson S, Mercado R (2022) De Novo drug design using reinforcement learning with graph-based deep generative models. J Chem Inf Model 62(20):4863–4872. https://doi.org/10.1021/acs.jcim.2c00838. From NLM Medline
141. Kumar R, Sharma A, Alexiou A, Bilgrami AL, Kamal MA, Ashraf GM (2022) DeePred-BBB: a blood brain barrier permeability prediction model with improved accuracy. Front Neurosci 16:858126. https://doi.org/10.3389/fnins.2022.858126. From NLM PubMed-not-MEDLINE
142. Xu Y (2022) Deep neural networks for QSAR. Methods Mol Biol 2390:233–260. https://doi.org/10.1007/978-1-0716-1787-8_10. From NLM Medline
143. Siramshetty VB, Nguyen DT, Martinez NJ, Southall NT, Simeonov A, Zakharov AV (2020) Critical assessment of artificial intelligence methods for prediction of hERG channel inhibition in the "Big Data" era. J Chem Inf Model 60(12):6007–6019. https://doi.org/10.1021/acs.jcim.0c00884. From NLM Medline
144. Wu Z, Zhu M, Kang Y, Leung EL, Lei T, Shen C, Jiang D, Wang Z, Cao D, Hou T (2021) Do we need different machine learning algorithms for QSAR modeling? A comprehensive assessment of 16 machine learning algorithms on 14 QSAR data sets. Brief Bioinform 22(4). https://doi.org/10.1093/bib/bbaa321. From NLM Medline
145. Feng H, Gao K, Chen D, Shen L, Robison AJ, Ellsworth E, Wei GW (2022) Machine learning analysis of cocaine addiction informed by DAT, SERT, and NET-based interactome networks.

J Chem Theory Comput 18(4):2703–2719. https://doi.org/10.1021/acs.jctc.2c00002. From NLM Medline
146. Zhu Z, Dou B, Cao Y, Jiang J, Zhu Y, Chen D, Feng H, Liu J, Zhang B, Zhou T et al (2023) TIDAL: topology-inferred drug addiction learning. J Chem Inf Model 63(5):1472–1489. https://doi.org/10.1021/acs.jcim.3c00046. From NLM Medline
147. Altae-Tran H, Ramsundar B, Pappu AS, Pande V (2017) Low data drug discovery with one-shot learning. ACS Cent Sci 3(4):283–293. https://doi.org/10.1021/acscentsci.6b00367. From NLM PubMed-not-MEDLINE
148. Shwartz-Ziv R, Armon A (2022) Tabular data: deep learning is not all you need. Inf Fusion 81:84–90. https://doi.org/10.1016/j.inffus.2021.11.011
149. Garcia-Hernandez C, Fernandez A, Serratosa F (2019) Ligand-based virtual screening using graph edit distance as molecular similarity measure. J Chem Inf Model 59(4):1410–1421. https://doi.org/10.1021/acs.jcim.8b00820. From NLM Medline
150. Cereto-Massague A, Ojeda MJ, Valls C, Mulero M, Garcia-Vallve S, Pujadas G (2015) Molecular fingerprint similarity search in virtual screening. Methods 71:58–63. https://doi.org/10.1016/j.ymeth.2014.08.005. From NLM Medline
151. Gabrielsen M, Kurczab R, Siwek A, Wolak M, Ravna AW, Kristiansen K, Kufareva I, Abagyan R, Nowak G, Chilmonczyk Z et al (2014) Identification of novel serotonin transporter compounds by virtual screening. J Chem Inf Model 54(3):933–943. https://doi.org/10.1021/ci400742s. From NLM Medline
152. Wang P, Yang F, Yang H, Xu X, Liu D, Xue W, Zhu F (2015) Identification of dual active agents targeting 5-HT1A and SERT by combinatorial virtual screening methods. Biomed Mater Eng 26(Suppl 1):S2233–S2239. https://doi.org/10.3233/BME-151529. From NLM Medline
153. Erol I, Aksoydan B, Kantarcioglu I, Salmas RE, Durdagi S (2017) Identification of novel serotonin reuptake inhibitors targeting central and allosteric binding sites: a virtual screening and molecular dynamics simulations study. J Mol Graph Model 74:193–202. https://doi.org/10.1016/j.jmgm.2017.02.001. From NLM Medline
154. Berfield JL, Wang LC, Reith ME (1999) Which form of dopamine is the substrate for the human dopamine transporter: the cationic or the uncharged species? J Biol Chem 274(8):4876–4882. https://doi.org/10.1074/jbc.274.8.4876. From NLM Medline
155. Chen N, Reith ME (2000) Structure and function of the dopamine transporter. Eur J Pharmacol 405(1–3):329–339. https://doi.org/10.1016/s0014-2999(00)00563-x. From NLM Medline
156. Shelley JC, Cholleti A, Frye LL, Greenwood JR, Timlin MR, Uchimaya M (2007) Epik: a software program for pK(a) prediction and protonation state generation for drug-like molecules. J Comput Aided Mol Des 21(12):681–691. https://doi.org/10.1007/s10822-007-9133-z. From NLM Medline
157. Pan X, Wang H, Li C, Zhang JZH, Ji C (2021) MolGpka: a web server for small molecule pK(a) prediction using a graph-convolutional neural network. J Chem Inf Model 61(7):3159–3165. https://doi.org/10.1021/acs.jcim.1c00075. From NLM Medline
158. Neumeyer JL, Tamagnan G, Wang S, Gao Y, Milius RA, Kula NS, Baldessarini RJ (1996) N-substituted analogs of 2 beta-carbomethoxy-3 beta- (4′-iodophenyl)tropane (beta-CIT) with selective affinity to dopamine or serotonin transporters in rat forebrain. J Med Chem 39(2):543–548. https://doi.org/10.1021/jm9505324. From NLM Medline
159. Bois F, Baldwin RM, Kula NS, Baldessarini RJ, Innis RB, Tamagnan G (2004) Synthesis and monoamine transporter affinity of 3′-analogs of 2-beta-carbomethoxy-3-beta-(4′-iodophenyl) tropane (beta-CIT). Bioorg Med Chem Lett 14(9):2117–2120. https://doi.org/10.1016/j.bmcl.2004.02.043. From NLM Medline
160. Rothman RB, Ananthan S, Partilla JS, Saini SK, Moukha-Chafiq O, Pathak V, Baumann MH (2015) Studies of the biogenic amine transporters 15. Identification of novel allosteric dopamine transporter ligands with nanomolar potency. J Pharmacol Exp Ther 353(3):529–538. https://doi.org/10.1124/jpet.114.222299. From NLM Medline

161. Subbaiah MAM (2018) Triple reuptake inhibitors as potential therapeutics for depression and other disorders: design paradigm and developmental challenges. J Med Chem 61(6):2133–2165. https://doi.org/10.1021/acs.jmedchem.6b01827. From NLM Medline
162. Glennon RA (2017) The 2014 Philip S. Portoghese Medicinal Chemistry Lectureship: the "Phenylalkylaminome" with a focus on selected drugs of abuse. J Med Chem 60(7):2605–2628. https://doi.org/10.1021/acs.jmedchem.7b00085. From NLM Medline
163. Hsin LW, Chang LT, Rothman RB, Dersch CM, Jacobson AE, Rice KC (2008) Design and synthesis of 2- and 3-substituted-3-phenylpropyl analogs of 1-[2-[bis(4-fluorophenyl)methoxy]ethyl]-4-(3-phenylpropyl)piperazine and 1-[2-(diphenylmethoxy)ethyl]-4-(3-phenylpropyl)piperazine: role of amino, fluoro, hydroxyl, methoxyl, methyl, methylene, and oxo substituents on affinity for the dopamine and serotonin transporters. J Med Chem 51(9):2795–2806. https://doi.org/10.1021/jm701270n. From NLM Medline
164. Arunotayanun W, Dalley JW, Huang XP, Setola V, Treble R, Iversen L, Roth BL, Gibbons S (2013) An analysis of the synthetic tryptamines AMT and 5-MeO-DALT: emerging 'Novel Psychoactive Drugs'. Bioorg Med Chem Lett 23(11):3411–3415. https://doi.org/10.1016/j.bmcl.2013.03.066. From NLM Medline
165. Brown DG, Bernstein PR, Wu Y, Urbanek RA, Becker CW, Throner SR, Dembofsky BT, Steelman GB, Lazor LA, Scott CW et al (2013) Azepines and piperidines with dual norepinephrine dopamine uptake inhibition and antidepressant activity. ACS Med Chem Lett 4(1):46–51. https://doi.org/10.1021/ml300262e. From NLM PubMed-not-MEDLINE
166. Gruner JA, Mathiasen JR, Flood DG, Gasior M (2011) Characterization of pharmacological and wake-promoting properties of the dopaminergic stimulant sydnocarb in rats. J Pharmacol Exp Ther 337(2):380–390. https://doi.org/10.1124/jpet.111.178947. From NLM Medline
167. Cao J, Slack RD, Bakare OM, Burzynski C, Rais R, Slusher BS, Kopajtic T, Bonifazi A, Ellenberger MP, Yano H et al (2016) Novel and high affinity 2-[(Diphenylmethyl)sulfinyl]acetamide (Modafinil) analogues as atypical Dopamine transporter inhibitors. J Med Chem 59(23):10676–10691. https://doi.org/10.1021/acs.jmedchem.6b01373. From NLM Medline
168. Decker AM, Partilla JS, Baumann MH, Rothman RB, Blough BE (2016) The biogenic amine transporter activity of vinylogous amphetamine analogs. Medchemcomm 7(8):1657–1663. https://doi.org/10.1039/c6md00245e
169. Mavel S, Mincheva Z, Meheux N, Carcenac Y, Guilloteau D, Abarbri M, Emond P (2012) QSAR study and synthesis of new phenyltropanes as ligands of the dopamine transporter (DAT). Bioorg Med Chem 20(4):1388–1395. https://doi.org/10.1016/j.bmc.2012.01.014. From NLM Medline
170. Zou MF, Kopajtic T, Katz JL, Wirtz S, Justice JB Jr, Newman AH (2001) Novel tropane-based irreversible ligands for the dopamine transporter. J Med Chem 44(25):4453–4461. https://doi.org/10.1021/jm0101904. From NLM Medline
171. Orjales A, Mosquera R, Toledo A, Pumar MC, Garcia N, Cortizo L, Labeaga L, Innerarity A (2003) Syntheses and binding studies of new [(aryl)(aryloxy)methyl]piperidine derivatives and related compounds as potential antidepressant drugs with high affinity for serotonin (5-HT) and norepinephrine (NE) transporters. J Med Chem 46(25):5512–5532. https://doi.org/10.1021/jm0309349. From NLM Medline
172. Tosh DK, Janowsky A, Eshleman AJ, Warnick E, Gao ZG, Chen Z, Gizewski E, Auchampach JA, Salvemini D, Jacobson KA (2017) Scaffold repurposing of nucleosides (Adenosine Receptor agonists): enhanced activity at the human Dopamine and Norepinephrine sodium symporters. J Med Chem 60(7):3109–3123. https://doi.org/10.1021/acs.jmedchem.7b00141. From NLM Medline
173. Williard RL, Middaugh LD, Zhu HJ, Patrick KS (2007) Methylphenidate and its ethanol transesterification metabolite ethylphenidate: brain disposition, monoamine transporters and motor activity. Behav Pharmacol 18(1):39–51. https://doi.org/10.1097/FBP.0b013e3280143226. From NLM Medline

174. Ritz MC, Cone EJ, Kuhar MJ (1990) Cocaine inhibition of ligand binding at dopamine, norepinephrine and serotonin transporters: a structure-activity study. Life Sci 46(9):635–645. https://doi.org/10.1016/0024-3205(90)90132-b. From NLM Medline
175. Froimowitz M, Gu Y, Dakin LA, Nagafuji PM, Kelley CJ, Parrish D, Deschamps JR, Janowsky A (2007) Slow-onset, long-duration, alkyl analogues of methylphenidate with enhanced selectivity for the dopamine transporter. J Med Chem 50(2):219–232. https://doi.org/10.1021/jm0608614. From NLM Medline
176. Husbands SM, Izenwasser S, Kopajtic T, Bowen WD, Vilner BJ, Katz JL, Newman AH (1999) Structure-activity relationships at the monoamine transporters and sigma receptors for a novel series of 9-[3-(cis-3, 5-dimethyl-1-piperazinyl)propyl]carbazole (rimcazole) analogues. J Med Chem 42(21):4446–4455. https://doi.org/10.1021/jm9902943. From NLM Medline
177. Kuo F, Gillespie TA, Kulanthaivel P, Lantz RJ, Ma TW, Nelson DL, Threlkeld PG, Wheeler WJ, Yi P, Zmijewski M (2004) Synthesis and biological activity of some known and putative duloxetine metabolites. Bioorg Med Chem Lett 14(13):3481–3486. https://doi.org/10.1016/j.bmcl.2004.04.066. From NLM Medline
178. Lantz KA, Hart SG, Planey SL, Roitman MF, Ruiz-White IA, Wolfe HR, McLane MP (2010) Inhibition of PTP1B by trodusquemine (MSI-1436) causes fat-specific weight loss in diet-induced obese mice. Obesity (Silver Spring) 18(8):1516–1523. https://doi.org/10.1038/oby.2009.444. From NLM Medline
179. Lapa GB, Byrd GD, Lapa AA, Budygin EA, Childers SR, Jones SR, Harp JJ (2005) The synthesis and biological evaluation of dopamine transporter inhibiting activity of substituted diphenylmethoxypiperidines. Bioorg Med Chem Lett 15(22):4915–4918. https://doi.org/10.1016/j.bmcl.2005.08.028. From NLM Medline
180. Andersen PH (1989) The dopamine inhibitor GBR 12909: selectivity and molecular mechanism of action. Eur J Pharmacol 166(3):493–504. https://doi.org/10.1016/0014-2999(89)90363-4. From NLM Medline
181. Ananthan S, Saini SK, Khare R, Clayton SD, Dersch CM, Rothman RB (2002) Identification of a novel partial inhibitor of dopamine transporter among 4-substituted 2-phenylquinazolines. Bioorg Med Chem Lett 12(16):2225–2228. https://doi.org/10.1016/s0960-894x(02)00348-7. From NLM Medline
182. Han DD, Gu HH (2006) Comparison of the monoamine transporters from human and mouse in their sensitivities to psychostimulant drugs. BMC Pharmacol 6:6. https://doi.org/10.1186/1471-2210-6-6. From NLM Medline
183. Solis E Jr, Suyama JA, Lazenka MF, DeFelice LJ, Negus SS, Blough BE, Banks ML (2016) Dissociable effects of the prodrug phendimetrazine and its metabolite phenmetrazine at dopamine transporters. Sci Rep 6:31385. https://doi.org/10.1038/srep31385. From NLM Medline
184. Han M, Song C, Jeong N, Hahn HG (2014) Exploration of 3-Aminoazetidines as triple reuptake inhibitors by bioisosteric modification of 3-alpha-Oxyazetidine. ACS Med Chem Lett 5(9):999–1004. https://doi.org/10.1021/ml500187a. From NLM PubMed-not-MEDLINE
185. Rothman RB, Baumann MH, Dersch CM, Romero DV, Rice KC, Carroll FI, Partilla JS (2001) Amphetamine-type central nervous system stimulants release norepinephrine more potently than they release dopamine and serotonin. Synapse 39(1):32–41. https://doi.org/10.1002/1098-2396(20010101)39:1<32::AID-SYN5>3.0.CO;2-3. From NLM Medline
186. Rothman RB, Vu N, Partilla JS, Roth BL, Hufeisen SJ, Compton-Toth BA, Birkes J, Young R, Glennon RA (2003) In vitro characterization of ephedrine-related stereoisomers at biogenic amine transporters and the receptorome reveals selective actions as norepinephrine transporter substrates. J Pharmacol Exp Ther 307(1):138–145. https://doi.org/10.1124/jpet.103.053975. From NLM Medline
187. Carroll FI, Howell LL, Kuhar MJ (1999) Pharmacotherapies for treatment of cocaine abuse: preclinical aspects. J Med Chem 42(15):2721–2736. https://doi.org/10.1021/jm9706729. From NLM Medline

188. Shapiro DA, Renock S, Arrington E, Chiodo LA, Liu LX, Sibley DR, Roth BL, Mailman R (2003) Aripiprazole, a novel atypical antipsychotic drug with a unique and robust pharmacology. Neuropsychopharmacology 28(8):1400–1411. https://doi.org/10.1038/sj.npp.1300203. From NLM Medline
189. Efange SM, Mash DC, Khare AB, Ouyang Q (1998) Modified ibogaine fragments: synthesis and preliminary pharmacological characterization of 3-ethyl-5-phenyl-1,2,3,4,5,6-hexahydroazepino[4,5-b]benzothiophenes. J Med Chem 41(23):4486–4491. https://doi.org/10.1021/jm980156y. From NLM Medline
190. Madras BK, Xie Z, Lin Z, Jassen A, Panas H, Lynch L, Johnson R, Livni E, Spencer TJ, Bonab AA et al (2006) Modafinil occupies dopamine and norepinephrine transporters in vivo and modulates the transporters and trace amine activity in vitro. J Pharmacol Exp Ther 319(2):561–569. https://doi.org/10.1124/jpet.106.106583. From NLM Medline
191. Zhang P, Cyriac G, Kopajtic T, Zhao Y, Javitch JA, Katz JL, Newman AH (2010) Structure-activity relationships for a novel series of citalopram (1-(3-(dimethylamino)propyl)-1-(4-fluorophenyl)-1,3-dihydroisobenzofuran-5-carbonitrile) analogues at monoamine transporters. J Med Chem 53(16):6112–6121. https://doi.org/10.1021/jm1005034. From NLM Medline

Regulation of Dopamine Transporter Surface Expression

Christina Besada and Ole Valente Mortensen

Contents

Abstract The dopamine transporter (DAT) is a plasma membrane protein expressed in dopamine (DA) neurons of the central nervous system and is critical for regulating DA neurotransmission. The DAT is responsible for the reuptake of released DA back into the presynaptic neuron, resulting in the termination of DA transmission. This process also recycles the DA back into the dopaminergic neuron for subsequent release. DAT is the target of psychostimulants including cocaine and amphetamines and has been associated with several neuropsychiatric disorders. Only DAT proteins located on the plasma membrane can remove DA from the extracellular space, and the number of DAT proteins on the cell-surface therefore determines the efficiency of DA clearance. As a result, regulating DAT surface expression is a critical means to regulating the magnitude and duration of DA neurotransmission. This chapter will discuss the different processes and proteins that have been shown to affect DAT surface expression and discuss the relevance to normal DA physiology and diseases that involve aberrant DA signaling.

Keywords Dopamine transporter · Psychostimulants · Cocaine · Amphetamine · Endocytosis · Membrane protein trafficking

C. Besada · O. V. Mortensen (✉)
Department of Pharmacology and Physiology, Drexel University College of Medicine, Philadelphia, PA, USA
e-mail: ovm23@drexel.edu

O. V. Mortensen (ed.), *Monoamine Transporters*, Advances in Neurobiology 46,
https://doi.org/10.1007/978-3-031-96364-3_4

1 Introduction

The monoamine neurotransmitter dopamine (DA) functions as a modulatory neurotransmitter and is important for behaviors such as motivation, reward, movement, and learning [100]. Regulating DA signaling is therefore central for proper functioning of these processes. The dopamine transporter (DAT) is a plasma membrane protein expressed in dopamine (DA) neurons of the central nervous system [1, 61]. The DAT is responsible for the reuptake of released DA back into the presynaptic neuron, resulting in the termination of DA transmission. This process also recycles the DA back into the dopaminergic neuron for subsequent release. DAT is the target of psychostimulants including cocaine and amphetamines and has been associated with several neuropsychiatric disorders including substance use disorders, ADHD, autism, and bipolar disorder [92]. Only DAT proteins located on the plasma membrane can remove DA from the extracellular space. Therefore, the number of DAT proteins on the cell surface determines the efficiency of DA clearance, and DAT surface expression is directly proportional to DA reuptake. However, DAT does not remain stationary at the cell surface, as it dynamically cycles to and from the cell surface. Regulating DAT surface expression is therefore a critical way of regulating the magnitude and duration of DA neurotransmission. Transporter expression at the surface of the cell occurs by delivery of newly translated and processed DAT to the cell surface, redistribution of the transporter from the cell membrane to intracellular compartments, and recycling of internalized DAT proteins back to the cell surface in processes known as trafficking. Broadening our knowledge of how DA neurons regulate transporter surface expression and trafficking and thereby DA reuptake will aid in a better understanding of the physiological role of DAT and could lead to the development of novel therapeutics. This chapter will discuss the different processes and proteins that affect DAT surface expression and trafficking.

2 The Trafficking Cycle

Membrane proteins are synthesized on the ribosomal machinery at the endoplasmic reticulum (ER) of cells and following translation immediately inserted into membranes. From the endoplasmic reticulum, membrane proteins may travel in vesicles to the Golgi complex, or directly to the cell surface [7]. During these steps, the membrane proteins can be posttranslationally modified. The membrane proteins will reside at the cell membrane performing their respective function until internalized through a process known as endocytosis [80]. Endocytosis is a fundamental process by which cells absorb extracellular metabolites, hormones, and peptides. In addition, the cells also use endocytosis to internalize proteins residing in the plasma membrane to regulate the surface expression levels of critical membrane proteins, such as neurotransmitter transporters like DAT. As a result, the cell can tightly regulate the composition and number of proteins on the cell membrane surface through these two processes of surface delivery and endocytosis. Within the cell, there are several types of endocytosis including phagocytosis, pinocytosis, caveolae-associated, and clathrin-mediated endocytosis with the two latter types having been implicated in DAT endocytosis.

In caveolae-associated and clathrin-mediated endocytosis, the membrane proteins to be internalized reside in the plasma membrane that is pinched off through invagination into endocytic vesicles and inserted into the cell. In clathrin-mediated endocytosis, clathrin is responsible for the formation and coating of the endocytic vesicle [94]. During this process, clathrin-coated vesicles containing molecules known as "cargo" bud off from the plasma membrane and are internalized into the cell. The first step is nucleation, in which the cargo is gathered into the internalizing endocytic vesicle. Then, propagation occurs, in which the coated pit with its cargo invaginates into the cell. Then, budding/scission of the endocytic vesicle occurs in which the vesicle is fully separated from the plasma membrane and is inserted into the cell. When the vesicle buds off from the membrane, this scission is mediated by the protein dynamin [87]. Finally, uncoating and clathrin recycling is the last step of this process in which the clathrin coat is removed from the vesicle.

Once a material (luminal and membrane) is internalized, it can go through different fates—mainly either degradation or recycling back to the cell membrane (Fig. 1).

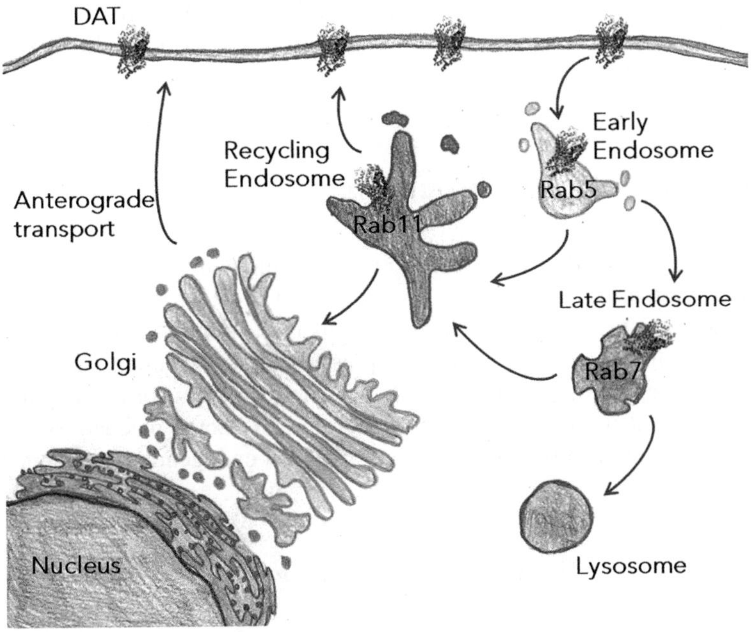

Fig. 1 Regulation of dopamine transporter (DAT) trafficking. Newly synthesized DAT is transported from the Golgi to the plasma membrane through anterograde transport. DAT is internalized through endocytosis to early endosomes and sorted to recycling endosomes for delivery back to the plasma membrane or late endosomes and lysosomes for degradation. Some Rab protein markers for endosomal organelles are included. Illustration by Isabella Karklin Mortensen

The balance between endocytosis, degradation, and recycling is a critical mechanism for controlling protein expression on the plasma membrane. The vesicle's fate is determined by the endocytic compartment that it associates with. There are three main compartments of endocytosis: early endosomes, late endosomes and lysosomes for degradation, and recycling endosomes for recycling. Early endosomes act as the first compartment for endocytosis, where vesicles internalizing from the cell surface are received. The early endosome compartment serves as a sorting station and is responsible for sorting molecules into late endosomes or recycling endosomes. When early endosomes mature into late endosomes, they fuse with lysosomes that provide an acidic environment and hydrolases and proteases responsible for degrading the intravesicular material [53]. When early endosomes mature into recycling endosomes, the intravesicular cargo is returned to the plasma membrane.

In order to ensure proper flow of vesicular traffic, transport vesicles must have a way to indicate the correct target membrane with which to fuse. The main way this is done is through Rab GTPases that serve as master regulators and provide traffic control and regulation of transfer of the endocytic vesicles to the proper organelles [51, 84]. In addition to serving as regulators of the trafficking pathway, some of these Rab GTPases are also used in the field of membrane trafficking as markers for specific types of endocytic compartments and organelles to distinguish between the different phases of the endocytic pathways. For example, Rab5 is a key factor in regulating early endocytosis and localizes to early endosomes where it recruits its effectors. Rab7 localizes to late endosomes, and Rab11 is associated with recycling endosomes (Fig. 1).

3 DAT Trafficking and Regulation

We want to preface the following review of our present understanding of the regulation of DAT surface expression, with the fact that most of the studies described below of mechanisms of DAT trafficking and its impact on DA function have been conducted in cell lines heterogeneously expressing DAT, in primary cultures of DA neurons, or in preparations from brain regions where the tissue processing results in some disruption of the DA neurons. Therefore, most known mechanisms of DAT trafficking have not been fully elucidated in its intact endogenous environment.

3.1 DAT Surface Delivery and Targeting

Following synthesis of DAT in the endoplasmic reticulum (ER), DAT is translocated through the ER membrane and packaged into vesicles. It then undergoes anterograde transport to the Golgi where posttranslational modifications such as glycosylation are introduced [42, 64], and DAT is ultimately trafficked to the cell membrane [126]. Electron microscopy studies have shown that DAT is present in the ER in the soma and proximal dendrites of DA neurons in ventral tegmental area (VTA) and substantia nigra [46, 83].

Several studies have also demonstrated that dimerization/oligomerization of DAT takes place in the ER and that it is important for the proper trafficking of DAT through the organelles to the surface. The oligomerization of DAT is thought to stabilize and thereby facilitate the delivery of DAT to the surface [44, 106, 114]. Interestingly a small molecule AIM100 has been found to induce this oligomerization and regulate DAT surface expression [108]. The DAT oligomerization process has also been shown to be regulated as it is affected by the E3 ubiquitin ligase, parkin, through its direct interaction with DAT [54, 79]. In these studies, it was found that the overexpression of parkin in cell lines resulted in increased transporter activity and cell surface expression of DAT [54]. Furthermore, it was found that parkin produced this effect by targeting misfolded DAT proteins for degradation.

Other proteins have also been identified to play a role in delivering DAT to the surface, and these include studies on the surface delivery machinery demonstrating that the Sec24D protein, a component of the coat protein complex II that promotes the formation of transport vesicles from the ER, supports surface delivery and axonal targeting of DAT [112]. The secretory carrier membrane protein 2 (SCAMP2) is a member of a family of proteins that have been proposed to be involved in exocytosis, and when overexpressed in a cell line transfected with DAT, it caused a reduction in transporter activity and a decrease in DAT surface expression levels [31]. Syntaxin1A, which is part of the SNARE complexes driving synaptic vesicular fusion, has been found to interact directly with DAT, and this interaction appears to have functional consequences as cleavage of syntaxin1A results in increased DAT activity [16]. In the same study, it was also shown that overexpression of syntaxin1A resulted in decreased DAT activity and transporter surface expression.

Following arrival at the surface, it has been established that DAT is highly mobile in the plasma membrane of dopaminergic neurons to facilitate proper targeting and bidirectional movement along neuronal extensions was demonstrated [27]. When DAT is residing in the plasma membrane, it has also been shown that membrane lipids are important for DAT activity as cholesterol directly interacts with DAT [85] and other studies suggest that cholesterol and cholesterol-containing lipid rafts might play a role in regulating DAT location and function at the cell surface [32, 48]. Interestingly, it was found that the C-terminal PDZ domain-binding sequence found in DAT is important through unknown protein interactions to be involved in the proper targeting of DAT to neuronal processes [93, 105].

3.2 *Constitutive DAT Endocytosis*

Following delivery to the cell surface, DAT can be internalized, and numerous studies have determined that this internalization can occur through two mechanisms: a constitutive and a regulated type of endocytosis. In constitutive endocytosis, DAT is continuously internalized, sorted into endosomal compartments, and recycled back to the cell surface or targeted for degradation. In vitro studies utilizing rat adrenal pheochromocytoma PC12 cells stably expressing DAT have shown that the cell surface half-life of DAT is about 13 minutes. Furthermore, the basal surface DAT endocytosis rate is 3–5% per minute [68]. Studies on characterizing where DAT localizes following endocytosis have shown that DAT colocalizes with several of the Rab proteins involved in intracellular trafficking noted above (Fig. 1). In both in vitro and ex vivo studies, it was shown that internalized DAT proteins partially colocalize with several endosomal markers including EEA1, Rab4, Rab5, Rab7, Rab11, and the Vps35 retromer complex component [26, 49, 68, 124]. It was found that DAT is transferred from Rab5 containing early endosomes to downstream recycling endosomes containing the Rab11 marker. A constitutively active Rab11 mutant upregulates DAT activity and trafficking to the surface [37] further demonstrating that recycling of DAT is Rab11 dependent. In two separate studies employing neuronal cultures, constitutive internalization of DAT was also demonstrated and DAT was found in both axonal and somatodendritic compartments [26, 109]. Furthermore, in cultured neurons, DAT was found to colocalize with Eps15, a marker of clathrin-coated pits [109]. Finally, in studies using antibody feeding assays or fluorescent cocaine analogs to visualize DAT in primary neuronal cultures, it was demonstrated that internalized DAT is targeted to both recycling and degradative pathways such as recycling and late endosomes and lysosomes [26, 27, 49, 90, 91].

Regarding mechanisms involved in regulating constitutive endocytosis, it has been demonstrated that a particular sequence motif within the DAT is involved in mediating basal trafficking. The DAT is a 12 transmembrane protein in which its amino terminus (N-terminus) and carboxy terminus (C-terminus) both reside in the intracellular space. Studies have revealed that the C-terminus residues Leucine 591, tyrosine 593, and isoleucine 595 are required for basal DAT endocytosis. Mutations within this motif resulted in an increased DAT endocytic rate and decreases recycling to the cell surface [9, 47]. Additionally, residues within the N-terminus also play a significant role in endocytosis. Residues 60–65 were shown to be critical for the retention of DAT in the plasma membrane and prevent rapid constitutive endocytosis. Mutations or removal of these residues also lead to retention of DAT in early and recycling endosomes [110].

3.3 Regulated DAT Endocytosis

In addition to the constitutive internalization described above, it has been established that DAT internalization also occurs through regulated processes to redistribute the transporter and modulate DA reuptake activity.

3.3.1 Protein Kinase C-Elicited DAT Internalization

Early in vitro systems utilizing various cell lines have shown that treatment with phorbol esters, activators of protein kinase C (PKC), increases the internalization of DAT. In canine kidney MDCK cells, it was shown that PKC activation resulted in the internalization and endocytosis of DAT in a dynamin-dependent manner through clathrin-coated pits and these studies found that following internalization, DAT was targeted for degradation in lysosomes [24, 78]. Similarly, phorbol ester–induced PKC activation in porcine endothelial and human Hela cells results in a clathrin- and dynamin-dependent decrease in cell surface expression DAT and an accumulation of endosomal DAT ultimately leading to increased degradation of DAT [74, 107]. On the other hand, it was found in PC12 cells that DAT following PKC activation was also internalized in these cells, but different from the other cell studies, DAT was not degraded in PC12 cells but instead targeted to early and recycling endosomes to be recycled back to the surface [72]. Interestingly, regulated DAT trafficking appears to be dependent on DAT expression levels as results have shown that PKC activation does not significantly change the endocytic rate of DAT when DAT is expressed at high levels; however, PKC activation does increase levels of DAT endocytosis when DAT is expressed at low and moderate levels in PC12 cells [68].

At least 11 isoforms of PKC exist [104], and among all PKC isoforms, PKCβ (PKCβI and PKCβII) is likely to be the most important contributor to the regulated endocytosis of DAT. This is supported by the use of pharmacological tools in the studies mentioned above to inhibit PCK isoforms. In ex vivo studies, it was also shown that in PKCβ knockout mice, constitutive DAT trafficking is altered resulting in reduced surface DAT expression levels and reduced dopamine uptake [20].

In more physiological preparations such as primary midbrain cultures, it was found that PKC activation resulted in inconsistent results as internalization similar to what is observed in vitro was found in some studies [22, 49], but in other studies, no internalization following PKC activation was observed [27, 90]. This could be a result of differences in culturing and assay conditions. In brain slices, a region-specific difference was found as PKC activation only decreased DAT surface expression in the ventral striatum but not the dorsal striatum [29]. Although PKC-induced loss of surface DAT can be reproduced in ex vivo striatal and mid-brain slices, the total level of DAT is not significantly reduced suggesting most DAT in endogenous systems is targeted for recycling and not degradation as it was observed in some in vitro systems [22, 29, 38].

Some cellular processes that can regulate DAT endocytosis have been identified and suggest that there are regulatory mechanisms that possibly act as an endocytic brake in order to attenuate DAT endocytosis. The protein Rit2 [81], which is a neuronal GTPase, and the cytosolic tyrosine kinase TNK2/ACK1 physically associate with DAT on the cell surface, and these interactions normally stabilize DAT surface expression in the plasma membrane. Following PKC activation, these interactions are disrupted, and this was proposed to release an endocytic brake resulting in increased DAT internalization [29, 81, 123].

Finally, DAT has been found to interact with the lipid raft marker flotillin-1 [22]. In these studies, two findings suggest that flotillin-1 is involved in the PKC-induced internalization of DAT: PKC activation increased the direct interaction between flotillin-1 and DAT, and following flotillin-1 knockdown, PKC-mediated downregulation of DAT was diminished [22].

Recently, ex vivo brain slice and primary cell culture studies have pursued the role of endogenous G-protein-coupled receptors (GPCR) in activating PKC signaling and causing DAT internalization. One study found in brain slices that following an upregulation of DAT surface expression that resulted from increased DA neuronal activity, the activation of metabotropic glutamate receptor mGluR5 and associated Gq-mediated signaling resulted in the retrieval and internalization of DAT in a PKC- and Rit2-dependent manner [56]. Another study in primary cultures of DA neurons and brain slices found that employing carbachol to stimulate M_1/M_5 muscarinic receptors that also signal through Gq signaling resulted in a downregulation of DAT surface expression and this downregulation of DAT was inhibited by a PKC inhibitor [116].

3.3.2 Amphetamine-Elicited DAT Internalization

As mentioned above, DAT is the target of psychostimulants such as cocaine and amphetamine. Although their mechanisms differ, both cocaine and amphetamine exposure result in increased DA concentration in the synaptic cleft. Cocaine produces psychoactive effects by blocking and inhibiting DAT. This prevents presynaptic reuptake of the DA molecules, leading to increased levels of DA within the synaptic cleft. Amphetamines like amphetamine and methamphetamine also exert their rewarding and reinforcing effects by elevating extracellular DA and prolonging DA signaling in the striatum. However, amphetamine exerts its effects in a different manner compared to cocaine as it is a substrate of DAT and competes with DA at DAT also inhibiting DA reuptake. However, as amphetamine also enters the cell, it can cause the release of DA from synaptic vesicles through the reversal of the transport function of the vesicular monoamine transporter VMAT2. The DA released into the intracellular space is then reverse transported to the extracellular space by DAT. Taken together, these two effects of amphetamines, DAT inhibition and DA release, ultimately result in increased extracellular levels of DA. It is important to note that the elevated DA levels in an intact system resulting from psychostimulant exposure in addition to direct effects of cocaine and amphetamine on DAT trafficking also will activate DA D2 autoreceptors that mediate feedback mechanism for regulating extracellular DA and associated signaling as described below.

Several in vitro and in vivo experiments have demonstrated that amphetamine exposure can produce effects on DAT surface expression depending on the length of exposure. Here we will describe effects of exposure times longer than 10 minutes, and the effects of briefer exposures will be described below (Sect. 3.5.1). Longer time amphetamine exposure was found to reduce levels of cell-surface DAT by accelerating the rate of endocytosis [43, 59, 97]. It was shown that this effect was dependent on active transporter function and dependent on dynamin. This effect was initially observed in heterologous expression system [98] but has also been demonstrated in primary DA neuron cultures [121] and ex vivo brain slices [28]. A recent study found that in animals that received repeated administration of amphetamine, the last administration of amphetamine resulted in an increase of DAT localization with intracellular endosomal compartments. This effect was only observed in male mice suggesting sex-dependent regulation of the processes involved in amphetamine-elicited DAT internalization [4]. In other studies, it was found that insulin can interfere with this process and that it depends on Akt and CaMKII kinase signaling [39, 119]. However, despite PKC being activated by amphetamine, this process does not appear to involve PKC signaling [8].

On the other hand, studies have established a role for other signaling pathways including protein kinase A, RhoA-mediated dynamin-dependent endocytosis, and Rac1 signaling in this process [121]. Interestingly, it has been established that the trace amine GPCR TAAR1 plays a direct role in this process. The TAAR1 receptor is an unusual receptor being located intracellularly, and it was demonstrated that TAAR1 is the receptor responding to amphetamine mediating the activation of RhoA signaling resulting in the amphetamine-elicited DAT internalization [117].

3.4 Posttranslational Modifications Involved in Regulating DAT Surface Expression

3.4.1 Phosphorylation

Phosphorylation of DAT as a mechanism for the PKC-mediated regulation of DAT surface expression has been heavily pursued as DAT possesses several potential sites for the phosphorylation by protein kinases. Proteomic analysis of rat DAT has shown that 90% of phosphorylation occurs on serine residues, and 10% occurs on threonine residues. These phosphorylation sites are present throughout both the N- and C-terminal domains, as well as intracellular loops [33]. Peptide mapping studies have further shown that the N-terminus is the major site of basal, protein kinase C (PKC)-stimulated, and amphetamine-stimulated phosphorylation [15, 33], and several of the serine residues in the N-terminus of DAT have been identified as targets for phosphorylation by PKC [33, 75]. Upon PKC activation, Ser7 phosphorylation results in decreased V_{max} of DA uptake suggesting decreased DAT surface expression [75]. These transporter kinetics results suggest a role of direct PKC-mediated phosphorylation of DAT in the regulation of DAT surface expression, but other studies have eliminated the idea that direct phosphorylation of DAT drives the PKC-mediated increase in DAT internalization. In a study, all potential phosphorylation sites in the DAT N-terminus were deleted. This resulted in the elimination of PKC-mediated phosphorylation, but it did not diminish the level of DAT internalization following PKC activation [41]. This suggests that phosphorylation of DAT may not be solely and directly responsible for the internalization of DAT, but rather, there are more complicated mechanisms and processes at play.

3.4.2 Palmitoylation

Palmitoylation is another posttranslational modification that has been explored for a role in the regulation of DAT surface expression and activity. Palmitoylation results in palmitic acid being added to a cysteine in DAT via a thioester linkage. Palmitoylation is a dynamic and reversible process that can be catalyzed by palmitoyl acyltransferases (PATs) to add, or palmitoyl-protein thioesterases (PPTs) to remove the modification [34]. Studies using synaptosomes have shown that palmitoylation of DAT promotes DA transport capacity through increased transport kinetics [34]. More specifically, palmitoylation increases the V_{max} of DA uptake suggesting an increase in surface expression and it also decreases DAT degradation. Conversely, depalmitoylation enhances DA efflux and decreases the V_{max} of DA uptake [57, 76]. Furthermore, treatment with PKC inhibitors increases DAT palmitoylation [34], suggesting that there is cross-talk between DAT phosphorylation and palmitoylation in order to regulate DAT activity. It has also been shown that increased phosphorylation of DAT via PKC activation decreases DAT palmitoylation and these processes appear to be involved in DAT membrane mobility [103] and transport kinetics [76], but overall these two complementary processes do not appear to be directly involved in regulating DAT surface expression.

3.4.3 Ubiquitylation

As mentioned above, PKC activation in HeLa and MDCK cells results in degradation of DAT in lysosomes, and this is presumably by mechanisms involving the interaction of ubiquitinylated DAT with complexes in multivesicular bodies (MVBs) and incorporation of DAT into internal vesicles of MVBs [24, 74]. Significantly, localization of DAT in MVBs has also been observed within the soma of substantia nigra DA neurons [46]. Ubiquitylation is a regulated process that is involved in targeting unstable membrane and cytosolic proteins for degradation via the proteosome. Ubiquitin is covalently linked to its target through a three-step process: initiation is carried out by a ubiquitin-activating enzyme known as E1; then, ubiquitin structurally associates with a carrier protein known as E2; finally, this structure couples with a protein-ligase known as E3 to modify the target protein [23]. There are three lysine residues on DAT that are targets for ubiquitylation that are all located in the N-terminal domain: lysine 19, lysine 27, and lysine 35. Ubiquitinylation of DAT has been demonstrated in both heterologous expression studies in cell lines and in native tissue expressing endogenous DAT [49, 78, 109]. Studies utilizing DAT mutants have led to the discovery that ubiquitylation is required for PKC-dependent and clathrin-mediated internalization, supporting the notion that ubiquitylation is important for the internalization step [24, 73, 74, 78]. The process of ubiquitylation of DAT has been shown to be mediated by the E3 ubiquitin ligase NEDD4-2 and is an essential component of PKC-dependent ubiquitination and DAT endocytosis [109]. In our previous study identifying proteins involved in PKC-dependent regulation of DAT, we found that the ERK1/2 MAP kinase phosphatase MKP3 attenuates the PKC-dependent internalization of DAT. Interestingly, MKP3 overexpression did not diminish PKC-induced ubiquitylation suggesting the mechanism of regulated DAT internalization is more complex than just a direct result of the ubiquitylation of DAT [78].

3.4.4 SUMOylation

The small ubiquitin-like modifier (SUMO) is an important regulator of protein function and stability. The process of SUMOylation is similar to ubiquitylation and involves the reversible covalent attachment of the SUMO peptide to a single or multiple lysine residues on the target protein. There are four paralogs of the SUMO protein, and SUMO1-3 are more heavily expressed in the brain. SUMO conjugation to target proteins is carried out by a conjugating enzyme known as Ubc9, in combination with E3 ligases [99]. Ubc9 is mostly expressed in the nucleus and cytoplasm, but there have been accumulating studies demonstrating a role of Ubc9 within the plasma membrane. Regarding DAT regulation, DAT was found to be associated with SUMO1, as opposed to the other paralogs [12]. Studies utilizing overexpression of SUMO1 and increased Ubc9 activity have shown that SUMO1 and Ubc9 enhance DAT stability and prevent its degradation. Conversely, knockdown of Ubc9 resulted in decreased levels of DAT in the plasma membrane and enhanced DAT degradation [12]. These results demonstrate that SUMOylation is significantly involved in regulating DAT turnover and trafficking. Furthermore, these studies demonstrated that SUMO1 overexpression decreased DAT ubiquitination, indicating that there is a complex system of reciprocal regulation between ubiquitylation and SUMOylation in influencing DAT expression and degradation.

3.5 DAT Recycling

Following internalization, DAT can be targeted for degradation and traffic through late endosomes to lysosomes, but it has also been established that it can traffic to recycling endosomes for delivery back to the cell surface. Several processes have been implicated in regulating this important mechanism of modulating DAT surface expression.

3.5.1 Psychostimulant-Induced DAT Recycling

As described above, studies have established that longer (>10 minutes) amphetamine treatment leads to increased rate of internalization of DAT. On the other hand, several studies have found that recycling and delivery of DAT to the surface can be regulated by a brief exposure of amphetamine as this was found to produce an upregulation of DAT surface expression [55]. Length of exposure is important as this finding is different from the effect of longer exposures of amphetamine described above that results in downregulation of DAT surface expression (Sect. 3.3.2). It was found that within 30 s–1 min of amphetamine exposure, DAT surface expression in rat striatal synaptosomes was transiently elevated [55]. This effect was dependent on protein kinase Cβ (PKCβ), and it has been speculated that amphetamine itself activates PKC [15, 20, 36]. It was also found that this acute amphetamine treatment induced recycling of DAT rather than reducing endocytosis [55, 109]. Interestingly, dopamine itself caused similar upregulation of DAT surface expression, but this could be mediated by DA receptors as mentioned below [36].

As mentioned above, both amphetamine and dopamine are substrates of DAT, and this appears to be important for this fast upregulation of DAT as the presence of DAT inhibitors blocked the effect. However, cocaine, a DAT inhibitor, has also been shown to upregulate DAT cell surface levels but through longer exposures in cells in culture and synaptosomes [25, 66]. However, the mechanism in which cocaine produces these effects is not fully known, but it is thought that cocaine may also increase DAT recruitment out of recycling endosomes and increase the recycling of DAT back to the surface [21, 25].

3.5.2 Receptor-Mediated DAT Recycling

The elevated DA resulting from psychostimulant exposure that was found to affect DAT surface expression described above has in several studies implicated the dopamine D2 receptor in mediating some of these effects. This receptor acts both as a postsynaptic receptor mediating downstream signaling of DA and as presynaptic auto-receptor providing a feedback mechanism for regulating DA signaling. As an auto-receptor, it is expressed on the same DA neurons that DATs are expressed on. Early in vitro studies in heterologous DAT expressing cell lines have demonstrated D2 receptor-mediated regulation of DAT and that the activation of D2 receptor resulted in an upregulation of DAT activity [6]. A role of the D2 receptor in regulating DAT was also demonstrated in mouse striatal synaptosomes where it was found that the D2 receptor is involved in upregulating DAT surface expression. It was also found in these studies that the D2-mediated upregulation of DAT surface expression required PKCβ [19] similar to what was found for amphetamine-elicited upregulation. In further support of a role of the D2 receptor in regulating DAT, it has been demonstrated that the receptor and DAT can form an interacting complex [63].

Interestingly, in ex vivo brain slice studies, it has been determined that the D2 receptor-mediated regulation of DAT is region specific as DAT regulation resulting from the D2 receptor activation by a D2 receptor agonist was found in the dorsal striatum but not in the ventral striatum [40]. In a recent study, ex vivo brain slices were also employed to show that D2 receptor activation resulted in a rapid within minutes upregulation of DAT surface expression in both ventral and dorsal striatum, and these studies also demonstrated that this effect was dependent on Vps35 containing recycling endosomes that are involved in recycling membrane proteins back to the surface [56].

Finally, the D2 receptor is not the only GPCR that has been found to regulate DAT surface expression as a study on the kappa opioid receptor demonstrated that when this receptor was activated by salvinorin A, it also caused increased DA uptake and DAT surface expression in several preparations, including cell lines and striatal synaptosome preparations [58].

3.5.3 Role of ERK1/2 in DAT Recycling and Surface Expression

The abovementioned studies also demonstrated that the D2 receptor-mediated regulation of DAT in the in vitro studies in cell lines was mediated through the activation of ERK1/2 kinases [6]. The ERK1/2 map kinase signaling pathway has been studied in several additional studies that have demonstrated that the inhibition of ERK1/2 results in downregulation of DAT activity in both heterologous expression studies in cell lines and in native tissue and cultures expressing endogenous DAT [6, 50, 65, 77, 123, 127]. Interestingly, it has been demonstrated that ERK1/2 potentially phosphorylates DAT at residue threonine 53 and this phosphorylation has been proposed to affect DAT activity and interaction with psychostimulants such as cocaine and amphetamine [17, 35]. Amphetamine and PKC activation have also been shown to increase the phosphorylation of DAT at this site [17, 35]. In a recent animal study employing a mouse knock-in (KI) model where this residue was mutated to alanine, it was found that this caused DAT to not respond to ERK1/2 inhibition and DAT was less sensitive to inhibition by cocaine [89]. Finally, the study found that the KI animals with DAT threonine 53 mutated displayed an attenuated response to cocaine-elicited hyperlocomotion [89]. However, despite these significant findings of the role of this residue in regulating DAT activity, these studies have not demonstrated a direct role of the phosphorylation of this residue in regulating DAT surface expression. Studies by several groups to explore this are currently under way. Our group has also pursued the involvement of ERK1/2 signaling in regulating DAT activity as we have identified a role of the ERK1/2 phosphatase, MKP3, in inhibiting PKC-dependent endocytosis as described above [78]. We have continued these studies on MKP3 and recently demonstrated that overexpression of MKP3 in vivo in rat VTA DA neurons results in increased surface expression of DAT and decreased phosphorylation of threonine 53 in DAT. Similar to the KI studies on the threonine 53 residue in DAT mentioned above, we found that the decrease in threonine 53 phosphorylation resulting from MKP3 overexpression correlated with a diminished sensitivity of DAT to cocaine inhibition. Finally, we also found that overexpression of MPK3 attenuates the motivation to seek cocaine in a rat self-administration model.

This could suggest a role of the increased surface expression of DAT and decreased sensitivity of DAT to cocaine in driving this behavior [5].

3.6 *Other Protein Interactions*

Additional protein interactions with DAT different from the ones described above have been identified, and some of these might play a role in DAT trafficking. However, most of the studies have not clearly established what parts of the trafficking pathways are affected by the interaction. Early studies established that PDZ domain containing proteins interact with DAT and could play a role in regulating membrane targeting and surface expression of DAT. These include the PDZ domain containing protein PICK1 [115], and overexpression of PICK1 was shown to increase DAT surface expression in cell lines and could play a role in targeting DAT to proper synaptic locations. The focal adhesion protein Hic-5 [11] when overexpressed was shown to decrease DAT surface expression. Finally, it was shown that the PDZ domain containing protein Piccolo [14] when overexpressed slowed down DAT endocytosis.

Of relevance to Parkinson's disease, the protein synuclein that is mutated in some cases of Parkinson's disease and that forms the protein aggregates in Lewy bodies has been found to interact directly with the C terminus of DAT. The consequences of this interaction are not clearly established showing both up- and down-regulation of DAT as a result of the interaction [62, 120]. Parkinson's disease synuclein mutants also appear to affect the DAT differently from the WT, but the evidence is not firmly established [88].

4 DAT Dysregulation and Expression in Disease

Because DAT is a target of psychostimulants and regulates DA signaling, it is not surprising that regulation of function and expression of DAT have been found to be associated with mental and neurological health issues including substance use disorders, attention deficit hyperactive disorder (ADHD), autism, and bipolar affective disorder [92].

Substance use disorders are common disorders that involve aberrant DA signaling. As mentioned above, DAT is the major target of several drugs of abuse including the psychostimulants cocaine and amphetamine. As described above, these psychostimulants produce an increase of DA concentration in the synapse. This increase in extracellular DA stimulates neurons involved in reward and motivation and leads to the euphoric high and rewarding properties of the drugs. However, following repeated use of psychostimulants, it has been found that DA signaling is decreased most likely through homeostatic mechanisms [60] and these could involve the regulation of DAT expression including surface expression. The results from clinical studies are not entirely consistent as sample sizes for these types of studies are limited and could be a result of study participants being in different stages of the addiction cycle such as initial misuse, maintenance, withdrawal, or abstinence. Some human studies have found elevated levels and uptake activity of DAT in cocaine users [67, 70, 111], but other studies have found that effects of psychostimulants are attenuated in drug users suggesting decreased levels of DAT [69, 118]. Similarly in numerous animal studies, it has been found that repeated exposure to psychostimulant and withdrawal/abstinence similarly can lead to diverging and not always consistent effects on DAT expression and function likely as a result of differences in drug administration paradigms and models [2, 10, 13, 18, 30, 52, 71, 86, 101, 122]. Most of these studies have not specifically looked at whether DAT trafficking is involved in these changes, and this is something that future studies hopefully can address. However, our study mentioned above and other studies have found evidence that DAT surface expression could play a role in modulating cocaine-elicited behaviors [5, 82, 96, 102, 113].

Rare genetic mutations in DAT have been associated with several mental health disorders with some affecting DAT trafficking [28]. In attention-deficit/hyperactivity disorder (ADHD), a chronic condition involving attention difficulty, hyperactivity, and impulsiveness, several polymorphisms of DAT have been identified suggesting a role of dysregulation of DAT function in ADHD pathology. One such mutation, in which the arginine 615 residue is changed to a cysteine, causes DAT to only go through the constitutive endocytosis described above, and this DAT mutant does not respond to the regulated trafficking triggered by amphetamine and PKC activation as described above [45, 95].

Bipolar affective disorder (BD) is a common neuropsychiatric disorder in which there are recurrent episodes of mania, depression, or mixed states. BD is predominantly characterized by unusual shifts in behaviors, mood, activity levels, as well as energy. Studies have proposed that transporter homoeostasis such as increased striatal DAT levels will lead to reduced dopaminergic function, and this could play a role in causing depression [3]. Also, during manic episodes, clinical studies have shown a decrease in DAT density [125]. Taken together, these findings suggest that DAT trafficking could play a role in this disorder.

5 Conclusion

DAT is one of the most important regulators of dopaminergic homeostasis as the main mechanism of clearing DA from the extracellular space following release. Furthermore, it is the target of psychostimulants that are in clinical use and misused in psychostimulant use disorders. This makes it critical to understand DAT regulation in the context of both normal dopaminergic neurophysiology and in the context of neuropsychiatric disorders. Because only DAT molecules expressed on the cell surface contribute to DA neurotransmission, the regulation of DAT surface expression through delivery and removal from the surface is central to proper dopaminergic function. The field as reviewed in this chapter has identified multiple important players and processes in this regulation and clearly established that surface DAT expression is a highly dynamic and regulated process and does play a role in driving DA neurophysiology and behaviors associated with DA signaling. Several of these processes involved in regulating DAT surface expression have been clearly described, but further studies on demonstrating the relevance of these processes in regulating DAT trafficking in additional animal disease models are required to fully establish their role under normal and disease conditions. This will be critical to ultimately enable the exploration of these processes as a mechanism and future target for alleviating health issues involving aberrant DA signaling and DAT surface expression.

Acknowledgments This work was funded by the National Institutes of Health grants MH106912, MH121453, and DA057054, and PA Department of Health CURE grants.

References

1. Aggarwal S, Mortensen OV (2017) Overview of monoamine transporters. Curr Protoc Pharmacol 79:12 16 1–12 16 17. https://doi.org/10.1002/cpph.32. https://www.ncbi.nlm.nih.gov/pubmed/29261228
2. Alonso IP, O'Connor BM, Bryant KG, Mandalaywala RK, Espana RA (2022) Incubation of cocaine craving coincides with changes in dopamine terminal neurotransmission. Addict Neurosci 3:100029. https://doi.org/10.1016/j.addicn.2022.100029. https://www.ncbi.nlm.nih.gov/pubmed/36081573
3. Ashok AH, Marques TR, Jauhar S, Nour MM, Goodwin GM, Young AH, Howes OD (2017) The dopamine hypothesis of bipolar affective disorder: the state of the art and implications for treatment. Mol Psychiatry 22(5):666–679. https://doi.org/10.1038/mp.2017.16. https://www.nature.com/articles/mp201716.pdf
4. Bagalkot T, Sorkin A (2024) Amphetamine induces sex-dependent loss of the Striatal Dopamine transporter in sensitized mice. eNeuro 11(1):ENEURO.0491. https://doi.org/10.1523/ENEURO.0491-23.2023. https://www.ncbi.nlm.nih.gov/pubmed/38164591
5. Bernstein DL, Lewandowski SI, Besada C, Place D, Espana RA, Mortensen OV (2024) Inactivation of ERK1/2 signaling in dopaminergic neurons by Map Kinase phosphatase MKP3 regulates Dopamine signaling and motivation for cocaine. J Neurosci 44(5):e0727232023. https://doi.org/10.1523/JNEUROSCI.0727-23.2023. https://www.ncbi.nlm.nih.gov/pubmed/38296649
6. Bolan EA, Kivell B, Jaligam V, Oz M, Jayanthi LD, Han Y, Sen N, Urizar E, Gomes I, Devi LA, Ramamoorthy S, Javitch JA, Zapata A, Shippenberg TS (2007) D2 receptors regulate dopamine transporter function via an extracellular signal-regulated kinases 1 and 2-dependent and phosphoinositide 3 kinase-independent mechanism. Mol Pharmacol 71(5):1222–1232. https://doi.org/10.1124/mol.106.027763. https://www.ncbi.nlm.nih.gov/pubmed/17267664
7. Bonifacino JS, Glick BS (2004) The mechanisms of vesicle budding and fusion. Cell 116(2):153–166. https://doi.org/10.1016/s0092-8674(03)01079-1. https://www.ncbi.nlm.nih.gov/pubmed/14744428
8. Boudanova E, Navaroli DM, Melikian HE (2008a) Amphetamine-induced decreases in dopamine transporter surface expression are protein kinase C-independent. Neuropharmacology 54(3):605–612. https://doi.org/10.1016/j.neuropharm.2007.11.007. https://www.ncbi.nlm.nih.gov/pubmed/18164041

9. Boudanova E, Navaroli DM, Stevens Z, Melikian HE (2008b) Dopamine transporter endocytic determinants: carboxy terminal residues critical for basal and PKC-stimulated internalization. Mol Cell Neurosci 39(2):211–217. https://doi.org/10.1016/j.mcn.2008.06.011
10. Calipari ES, Ferris MJ, Zimmer BA, Roberts DC, Jones SR (2013) Temporal pattern of cocaine intake determines tolerance vs sensitization of cocaine effects at the dopamine transporter. Neuropsychopharmacology 38(12):2385–2392. https://doi.org/10.1038/npp.2013.136. https://www.ncbi.nlm.nih.gov/pubmed/23719505
11. Carneiro AM, Ingram SL, Beaulieu JM, Sweeney A, Amara SG, Thomas SM, Caron MG, Torres GE (2002) The multiple LIM domain-containing adaptor protein Hic-5 synaptically colocalizes and interacts with the dopamine transporter. J Neurosci 22(16):7045–7054. https://doi.org/10.1523/JNEUROSCI.22-16-07045.2002. https://www.ncbi.nlm.nih.gov/pubmed/12177201
12. Cartier E, Garcia-Olivares J, Janezic E, Viana J, Moore M, Lin ML, Caplan JL, Torres G, Kim YH (2019) The SUMO-Conjugase Ubc9 prevents the degradation of the Dopamine transporter, enhancing its cell surface level and Dopamine uptake. Front Cell Neurosci 13:35. https://doi.org/10.3389/fncel.2019.00035
13. Cass WA, Gerhardt GA, Gillespie K, Curella P, Mayfield RD, Zahniser NR (1993) Reduced clearance of exogenous dopamine in rat nucleus accumbens, but not in dorsal striatum, following cocaine challenge in rats withdrawn from repeated cocaine administration. J Neurochem 61(1):273–283. https://doi.org/10.1111/j.1471-4159.1993.tb03565.x. https://www.ncbi.nlm.nih.gov/pubmed/8515274
14. Cen X, Nitta A, Ibi D, Zhao Y, Niwa M, Taguchi K, Hamada M, Ito Y, Ito Y, Wang L, Nabeshima T (2008) Identification of Piccolo as a regulator of behavioral plasticity and dopamine transporter internalization. Mol Psychiatry 13(4):349, 451–349, 463. https://doi.org/10.1038/sj.mp.4002132. https://www.ncbi.nlm.nih.gov/pubmed/18195717
15. Cervinski MA, Foster JD, Vaughan RA (2005) Psychoactive substrates stimulate dopamine transporter phosphorylation and down-regulation by cocaine-sensitive and protein kinase C-dependent mechanisms. J Biol Chem 280(49):40442–40449. https://doi.org/10.1074/jbc.M501969200
16. Cervinski MA, Foster JD, Vaughan RA (2010) Syntaxin 1A regulates dopamine transporter activity, phosphorylation and surface expression. Neuroscience 170(2):408–416. https://doi.org/10.1016/j.neuroscience.2010.07.025. https://www.ncbi.nlm.nih.gov/pubmed/20643191
17. Challasivakanaka S, Zhen J, Smith ME, Reith MEA, Foster JD, Vaughan RA (2017) Dopamine transporter phosphorylation site threonine 53 is stimulated by amphetamines and regulates dopamine transport, efflux, and cocaine analog binding. J Biol Chem 292(46):19066–19075. https://doi.org/10.1074/jbc.M117.787002. https://www.ncbi.nlm.nih.gov/pubmed/28939767
18. Chefer VI, Shippenberg TS (2002) Changes in basal and cocaine-evoked extracellular dopamine uptake and release in the rat nucleus accumbens during early abstinence from cocaine: quantitative determination under transient conditions. Neuroscience 112(4):907–919. https://doi.org/10.1016/s0306-4522(02)00099-4. https://www.ncbi.nlm.nih.gov/pubmed/12088750
19. Chen R, Daining CP, Sun H, Fraser R, Stokes SL, Leitges M, Gnegy ME (2013) Protein kinase Cbeta is a modulator of the dopamine D2 autoreceptor-activated trafficking of the dopamine transporter. J Neurochem 125(5):663–672. https://doi.org/10.1111/jnc.12229. https://www.ncbi.nlm.nih.gov/pubmed/23458603
20. Chen R, Furman CA, Zhang M, Kim MN, Gereau RWT, Leitges M, Gnegy ME (2009) Protein kinase Cbeta is a critical regulator of dopamine transporter trafficking and regulates the behavioral response to amphetamine in mice. J Pharmacol Exp Ther 328(3):912–920. https://doi.org/10.1124/jpet.108.147959
21. Cowell RM, Kantor L, Hewlett GH, Frey KA, Gnegy ME (2000) Dopamine transporter antagonists block phorbol ester-induced dopamine release and dopamine transporter phosphorylation in striatal synaptosomes. Eur J Pharmacol 389(1):59–65. https://doi.org/10.1016/s0014-2999(99)00828-6

22. Cremona ML, Matthies HJ, Pau K, Bowton E, Speed N, Lute BJ, Anderson M, Sen N, Robertson SD, Vaughan RA, Rothman JE, Galli A, Javitch JA, Yamamoto A (2011) Flotillin-1 is essential for PKC-triggered endocytosis and membrane microdomain localization of DAT. Nat Neurosci 14(4):469–477. https://doi.org/10.1038/nn.2781. https://www.ncbi.nlm.nih.gov/pubmed/21399631
23. Czuba LC, Hillgren KM, Swaan PW (2018) Post-translational modifications of transporters. Pharmacol Ther 192:88–99. https://doi.org/10.1016/j.pharmthera.2018.06.013
24. Daniels GM, Amara SG (1999) Regulated trafficking of the human dopamine transporter. Clathrin-mediated internalization and lysosomal degradation in response to phorbol esters. J Biol Chem 274(50):35794–35801. https://doi.org/10.1074/jbc.274.50.35794
25. Daws LC, Callaghan PD, Morón JA, Kahlig KM, Shippenberg TS, Javitch JA, Galli A (2002) Cocaine increases dopamine uptake and cell surface expression of dopamine transporters. Biochem Biophys Res Commun 290(5):1545–1550. https://doi.org/10.1006/bbrc.2002.6384
26. Eriksen J, Bjørn-Yoshimoto WE, Jørgensen TN, Newman AH, Gether U (2010) Postendocytic sorting of constitutively internalized dopamine transporter in cell lines and dopaminergic neurons. J Biol Chem 285(35):27289–27301. https://doi.org/10.1074/jbc.M110.131003
27. Eriksen J, Rasmussen SG, Rasmussen TN, Vaegter CB, Cha JH, Zou MF, Newman AH, Gether U (2009) Visualization of dopamine transporter trafficking in live neurons by use of fluorescent cocaine analogs. J Neurosci 29(21):6794–6808. https://doi.org/10.1523/jneurosci.4177-08.2009
28. Fagan RR, Kearney PJ, Melikian HE (2020a) In situ regulated Dopamine transporter trafficking: there's no place like home. Neurochem Res 45(6):1335–1343. https://doi.org/10.1007/s11064-020-03001-6
29. Fagan RR, Kearney PJ, Sweeney CG, Luethi D, Schoot Uiterkamp FE, Schicker K, Alejandro BS, O'Connor LC, Sitte HH, Melikian HE (2020b) Dopamine transporter trafficking and Rit2 GTPase: mechanism of action and in vivo impact. J Biol Chem 295(16):5229–5244. https://doi.org/10.1074/jbc.RA120.012628. https://www.ncbi.nlm.nih.gov/pubmed/32132171
30. Ferris MJ, Calipari ES, Mateo Y, Melchior JR, Roberts DC, Jones SR (2012) Cocaine self-administration produces pharmacodynamic tolerance: differential effects on the potency of dopamine transporter blockers, releasers, and methylphenidate. Neuropsychopharmacology 37(7):1708–1716. https://doi.org/10.1038/npp.2012.17. https://www.ncbi.nlm.nih.gov/pubmed/22395730
31. Fjorback AW, Muller HK, Haase J, Raarup MK, Wiborg O (2011) Modulation of the dopamine transporter by interaction with Secretory Carrier Membrane Protein 2. Biochem Biophys Res Commun 406(2):165–170. https://doi.org/10.1016/j.bbrc.2011.01.069. https://www.ncbi.nlm.nih.gov/pubmed/21295544
32. Foster JD, Adkins SD, Lever JR, Vaughan RA (2008) Phorbol ester induced trafficking-independent regulation and enhanced phosphorylation of the dopamine transporter associated with membrane rafts and cholesterol. J Neurochem 105(5):1683–1699. https://doi.org/10.1111/j.1471-4159.2008.05262.x. https://www.ncbi.nlm.nih.gov/pubmed/18248623
33. Foster JD, Pananusorn B, Vaughan RA (2002) Dopamine transporters are phosphorylated on N-terminal serines in rat striatum. J Biol Chem 277(28):25178–25186. https://doi.org/10.1074/jbc.M200294200
34. Foster JD, Vaughan RA (2011) Palmitoylation controls dopamine transporter kinetics, degradation, and protein kinase C-dependent regulation. J Biol Chem 286(7):5175–5186. https://doi.org/10.1074/jbc.M110.187872
35. Foster JD, Yang JW, Moritz AE, Challasivakanaka S, Smith MA, Holy M, Wilebski K, Sitte HH, Vaughan RA (2012) Dopamine transporter phosphorylation site threonine 53 regulates substrate reuptake and amphetamine-stimulated efflux. J Biol Chem 287(35):29702–29712. https://doi.org/10.1074/jbc.M112.367706. https://www.ncbi.nlm.nih.gov/pubmed/22722938
36. Furman CA, Chen R, Guptaroy B, Zhang M, Holz RW, Gnegy M (2009a) Dopamine and amphetamine rapidly increase dopamine transporter trafficking to the surface: live-cell imaging using total internal reflection fluorescence microscopy. J Neurosci 29(10):3328–3336.

https://doi.org/10.1523/JNEUROSCI.5386-08.2009. https://www.ncbi.nlm.nih.gov/pubmed/19279270
37. Furman CA, Lo CB, Stokes S, Esteban JA, Gnegy ME (2009b) Rab 11 regulates constitutive dopamine transporter trafficking and function in N2A neuroblastoma cells. Neurosci Lett 463(1):78–81. https://doi.org/10.1016/j.neulet.2009.07.049. https://www.ncbi.nlm.nih.gov/pubmed/19631257
38. Gabriel LR, Wu S, Kearney P, Bellve KD, Standley C, Fogarty KE, Melikian HE (2013) Dopamine transporter endocytic trafficking in striatal dopaminergic neurons: differential dependence on dynamin and the actin cytoskeleton. J Neurosci 33(45):17836–17846. https://doi.org/10.1523/JNEUROSCI.3284-13.2013. https://www.ncbi.nlm.nih.gov/pubmed/24198373
39. Garcia BG, Wei Y, Moron JA, Lin RZ, Javitch JA, Galli A (2005) Akt is essential for insulin modulation of amphetamine-induced human dopamine transporter cell-surface redistribution. Mol Pharmacol 68(1):102–109. https://doi.org/10.1124/mol.104.009092. https://www.ncbi.nlm.nih.gov/pubmed/15795321
40. Gowrishankar R, Gresch PJ, Davis GL, Katamish RM, Riele JR, Stewart AM, Vaughan RA, Hahn MK, Blakely RD (2018) Region-specific regulation of presynaptic Dopamine homeostasis by D(2) autoreceptors shapes the in vivo impact of the neuropsychiatric disease-associated DAT variant Val559. J Neurosci 38(23):5302–5312. https://doi.org/10.1523/JNEUROSCI.0055-18.2018. https://www.ncbi.nlm.nih.gov/pubmed/29739866
41. Granas C, Ferrer J, Loland CJ, Javitch JA, Gether U (2003) N-terminal truncation of the dopamine transporter abolishes phorbol ester- and substance P receptor-stimulated phosphorylation without impairing transporter internalization. J Biol Chem 278(7):4990–5000. https://doi.org/10.1074/jbc.M205058200
42. Grigoriadis DE, Wilson AA, Lew R, Sharkey JS, Kuhar MJ (1989) Dopamine transport sites selectively labeled by a novel photoaffinity probe: 125I-DEEP. J Neurosci 9(8):2664–2670. https://doi.org/10.1523/JNEUROSCI.09-08-02664.1989. https://www.ncbi.nlm.nih.gov/pubmed/2769361
43. Gulley JM, Doolen S, Zahniser NR (2002) Brief, repeated exposure to substrates down-regulates dopamine transporter function in Xenopus oocytes in vitro and rat dorsal striatum in vivo. J Neurochem 83(2):400–411. https://doi.org/10.1046/j.1471-4159.2002.01133.x
44. Hastrup H, Karlin A, Javitch JA (2001) Symmetrical dimer of the human dopamine transporter revealed by cross-linking Cys-306 at the extracellular end of the sixth transmembrane segment. Proc Natl Acad Sci USA 98(18):10055–10060. https://doi.org/10.1073/pnas.181344298. https://www.ncbi.nlm.nih.gov/pubmed/11526230
45. Herborg F, Andreassen TF, Berlin F, Loland CJ, Gether U (2018) Neuropsychiatric disease-associated genetic variants of the dopamine transporter display heterogeneous molecular phenotypes. J Biol Chem 293(19):7250–7262. https://doi.org/10.1074/jbc.RA118.001753. https://www.ncbi.nlm.nih.gov/pubmed/29559554
46. Hersch SM, Yi H, Heilman CJ, Edwards RH, Levey AI (1997) Subcellular localization and molecular topology of the dopamine transporter in the striatum and substantia nigra. J Comp Neurol 388(2):211–227. https://www.ncbi.nlm.nih.gov/pubmed/9368838
47. Holton KL, Loder MK, Melikian HE (2005) Nonclassical, distinct endocytic signals dictate constitutive and PKC-regulated neurotransmitter transporter internalization. Nat Neurosci 8(7):881–888. https://doi.org/10.1038/nn1478
48. Hong WC, Amara SG (2010) Membrane cholesterol modulates the outward facing conformation of the dopamine transporter and alters cocaine binding. J Biol Chem 285(42):32616–32626. https://doi.org/10.1074/jbc.M110.150565. https://www.ncbi.nlm.nih.gov/pubmed/20688912
49. Hong WC, Amara SG (2013) Differential targeting of the dopamine transporter to recycling or degradative pathways during amphetamine- or PKC-regulated endocytosis in dopamine neurons. FASEB J 27(8):2995–3007. https://doi.org/10.1096/fj.12-218727. https://www.ncbi.nlm.nih.gov/pubmed/23612789

50. Hoover BR, Everett CV, Sorkin A, Zahniser NR (2007) Rapid regulation of dopamine transporters by tyrosine kinases in rat neuronal preparations. J Neurochem 101(5):1258–1271. https://doi.org/10.1111/j.1471-4159.2007.04522.x. https://www.ncbi.nlm.nih.gov/pubmed/17419806
51. Hutagalung AH, Novick PJ (2011) Role of Rab GTPases in membrane traffic and cell physiology. Physiol Rev 91(1):119–149. https://doi.org/10.1152/physrev.00059.2009
52. Izenwasser S, Cox BM (1990) Daily cocaine treatment produces a persistent reduction of [3H]dopamine uptake in vitro in rat nucleus accumbens but not in striatum. Brain Res 531(1–2):338–341. https://doi.org/10.1016/0006-8993(90)90797-f. https://www.ncbi.nlm.nih.gov/pubmed/2289133
53. Jeger JL (2020) Endosomes, lysosomes, and the role of endosomal and lysosomal biogenesis in cancer development. Mol Biol Rep 47(12):9801–9810. https://doi.org/10.1007/s11033-020-05993-4
54. Jiang H, Jiang Q, Feng J (2004) Parkin increases dopamine uptake by enhancing the cell surface expression of dopamine transporter. J Biol Chem 279(52):54380–54386. https://doi.org/10.1074/jbc.M409282200. https://www.ncbi.nlm.nih.gov/pubmed/15492001
55. Johnson LA, Furman CA, Zhang M, Guptaroy B, Gnegy ME (2005) Rapid delivery of the dopamine transporter to the plasmalemmal membrane upon amphetamine stimulation. Neuropharmacology 49(6):750–758. https://doi.org/10.1016/j.neuropharm.2005.08.018
56. Kearney PJ, Bolden NC, Kahuno E, Conklin TL, Martin GE, Lubec G, Melikian HE (2023) Presynaptic Gq-coupled receptors drive biphasic dopamine transporter trafficking that modulates dopamine clearance and motor function. J Biol Chem 299(2):102900. https://doi.org/10.1016/j.jbc.2023.102900. https://www.ncbi.nlm.nih.gov/pubmed/36640864
57. Khoshbouei H, Sen N, Guptaroy B, Johnson L, Lund D, Gnegy ME, Galli A, Javitch JA (2004) N-terminal phosphorylation of the dopamine transporter is required for amphetamine-induced efflux. PLoS Biol 2(3):E78. https://doi.org/10.1371/journal.pbio.0020078
58. Kivell B, Uzelac Z, Sundaramurthy S, Rajamanickam J, Ewald A, Chefer V, Jaligam V, Bolan E, Simonson B, Annamalai B, Mannangatti P, Prisinzano TE, Gomes I, Devi LA, Jayanthi LD, Sitte HH, Ramamoorthy S, Shippenberg TS (2014) Salvinorin A regulates dopamine transporter function via a kappa opioid receptor and ERK1/2-dependent mechanism. Neuropharmacology 86:228–240. https://doi.org/10.1016/j.neuropharm.2014.07.016. https://www.ncbi.nlm.nih.gov/pubmed/25107591
59. Kokoshka JM, Vaughan RA, Hanson GR, Fleckenstein AE (1998) Nature of methamphetamine-induced rapid and reversible changes in dopamine transporters. Eur J Pharmacol 361(2–3):269–275. https://doi.org/10.1016/s0014-2999(98)00741-9
60. Koob GF (2009) Neurobiological substrates for the dark side of compulsivity in addiction. Neuropharmacology 56(Suppl 1):18–31. https://doi.org/10.1016/j.neuropharm.2008.07.043. https://www.ncbi.nlm.nih.gov/pubmed/18725236
61. Kristensen AS, Andersen J, Jorgensen TN, Sorensen L, Eriksen J, Loland CJ, Stromgaard K, Gether U (2011) SLC6 neurotransmitter transporters: structure, function, and regulation. Pharmacol Rev 63(3):585–640. https://doi.org/10.1124/pr.108.000869. https://www.ncbi.nlm.nih.gov/pubmed/21752877
62. Lee FJ, Liu F, Pristupa ZB, Niznik HB (2001) Direct binding and functional coupling of alpha-synuclein to the dopamine transporters accelerate dopamine-induced apoptosis. FASEB J 15(6):916–926. https://doi.org/10.1096/fj.00-0334com. https://www.ncbi.nlm.nih.gov/pubmed/11292651
63. Lee FJ, Pei L, Moszczynska A, Vukusic B, Fletcher PJ, Liu F (2007) Dopamine transporter cell surface localization facilitated by a direct interaction with the dopamine D2 receptor. EMBO J 26(8):2127–2136. https://doi.org/10.1038/sj.emboj.7601656. https://www.ncbi.nlm.nih.gov/pubmed/17380124
64. Li LB, Chen N, Ramamoorthy S, Chi L, Cui XN, Wang LC, Reith ME (2004) The role of N-glycosylation in function and surface trafficking of the human dopamine transporter. J

Biol Chem 279(20):21012–21020. https://doi.org/10.1074/jbc.M311972200. https://www.ncbi.nlm.nih.gov/pubmed/15024013
65. Lin Z, Zhang PW, Zhu X, Melgari JM, Huff R, Spieldoch RL, Uhl GR (2003) Phosphatidylinositol 3-kinase, protein kinase C, and MEK1/2 kinase regulation of dopamine transporters (DAT) require N-terminal DAT phosphoacceptor sites. J Biol Chem 278(22):20162–20170. https://doi.org/10.1074/jbc.M209584200. https://www.ncbi.nlm.nih.gov/pubmed/12660249
66. Little KY, Elmer LW, Zhong H, Scheys JO, Zhang L (2002) Cocaine induction of dopamine transporter trafficking to the plasma membrane. Mol Pharmacol 61(2):436–445. https://doi.org/10.1124/mol.61.2.436. https://www.ncbi.nlm.nih.gov/pubmed/11809869
67. Little KY, Kirkman JA, Carroll FI, Clark TB, Duncan GE (1993) Cocaine use increases [3H]WIN 35428 binding sites in human striatum. Brain Res 628(1–2):17–25. https://doi.org/10.1016/0006-8993(93)90932-d. https://www.ncbi.nlm.nih.gov/pubmed/8313144
68. Loder MK, Melikian HE (2003) The dopamine transporter constitutively internalizes and recycles in a protein kinase C-regulated manner in stably transfected PC12 cell lines. J Biol Chem 278(24):22168–22174. https://doi.org/10.1074/jbc.M301845200
69. Martinez D, Narendran R, Foltin RW, Slifstein M, Hwang DR, Broft A, Huang Y, Cooper TB, Fischman MW, Kleber HD, Laruelle M (2007) Amphetamine-induced dopamine release: markedly blunted in cocaine dependence and predictive of the choice to self-administer cocaine. Am J Psychiatry 164(4):622–629. https://doi.org/10.1176/ajp.2007.164.4.622. https://www.ncbi.nlm.nih.gov/pubmed/17403976
70. Mash DC, Pablo J, Ouyang Q, Hearn WL, Izenwasser S (2002) Dopamine transport function is elevated in cocaine users. J Neurochem 81(2):292–300. https://doi.org/10.1046/j.1471-4159.2002.00820.x. https://www.ncbi.nlm.nih.gov/pubmed/12064476
71. Mateo Y, Lack CM, Morgan D, Roberts DC, Jones SR (2005) Reduced dopamine terminal function and insensitivity to cocaine following cocaine binge self-administration and deprivation. Neuropsychopharmacology 30(8):1455–1463. https://doi.org/10.1038/sj.npp.1300687. https://www.ncbi.nlm.nih.gov/pubmed/15702135
72. Melikian HE, Buckley KM (1999) Membrane trafficking regulates the activity of the human dopamine transporter. J Neurosci 19(18):7699–7710. https://doi.org/10.1523/JNEUROSCI.19-18-07699.1999. https://www.ncbi.nlm.nih.gov/pubmed/10479674
73. Miranda M, Dionne KR, Sorkina T, Sorkin A (2007) Three ubiquitin conjugation sites in the amino terminus of the dopamine transporter mediate protein kinase C-dependent endocytosis of the transporter. Mol Biol Cell 18(1):313–323. https://doi.org/10.1091/mbc.e06-08-0704
74. Miranda M, Wu CC, Sorkina T, Korstjens DR, Sorkin A (2005) Enhanced ubiquitylation and accelerated degradation of the dopamine transporter mediated by protein kinase C. J Biol Chem 280(42):35617–35624. https://doi.org/10.1074/jbc.M506618200
75. Moritz AE, Foster JD, Gorentla BK, Mazei-Robison MS, Yang JW, Sitte HH, Blakely RD, Vaughan RA (2013) Phosphorylation of dopamine transporter serine 7 modulates cocaine analog binding. J Biol Chem 288(1):20–32. https://doi.org/10.1074/jbc.M112.407874. https://www.ncbi.nlm.nih.gov/pubmed/23161550
76. Moritz AE, Rastedt DE, Stanislowski DJ, Shetty M, Smith MA, Vaughan RA, Foster JD (2015) Reciprocal phosphorylation and palmitoylation control Dopamine transporter kinetics. J Biol Chem 290(48):29095–29105. https://doi.org/10.1074/jbc.M115.667055
77. Moron JA, Zakharova I, Ferrer JV, Merrill GA, Hope B, Lafer EM, Lin ZC, Wang JB, Javitch JA, Galli A, Shippenberg TS (2003) Mitogen-activated protein kinase regulates dopamine transporter surface expression and dopamine transport capacity. J Neurosci 23(24):8480–8488. https://doi.org/10.1523/JNEUROSCI.23-24-08480.2003. https://www.ncbi.nlm.nih.gov/pubmed/13679416
78. Mortensen OV, Larsen MB, Prasad BM, Amara SG (2008) Genetic complementation screen identifies a mitogen-activated protein kinase phosphatase, MKP3, as a regulator of dopamine transporter trafficking. Mol Biol Cell 19(7):2818–2829. https://doi.org/10.1091/mbc.e07-09-0980

79. Moszczynska A, Saleh J, Zhang H, Vukusic B, Lee FJ, Liu F (2007) Parkin disrupts the alpha-synuclein/dopamine transporter interaction: consequences toward dopamine-induced toxicity. J Mol Neurosci 32(3):217–227. https://doi.org/10.1007/s12031-007-0037-0. https://www.ncbi.nlm.nih.gov/pubmed/17873367
80. Mukherjee S, Ghosh RN, Maxfield FR (1997) Endocytosis. Physiol Rev 77(3):759–803. https://doi.org/10.1152/physrev.1997.77.3.759. https://www.ncbi.nlm.nih.gov/pubmed/9234965
81. Navaroli DM, Stevens ZH, Uzelac Z, Gabriel L, King MJ, Lifshitz LM, Sitte HH, Melikian HE (2011) The plasma membrane-associated GTPase Rin interacts with the dopamine transporter and is required for protein kinase C-regulated dopamine transporter trafficking. J Neurosci 31(39):13758–13770. https://doi.org/10.1523/JNEUROSCI.2649-11.2011. https://www.ncbi.nlm.nih.gov/pubmed/21957239
82. Nelson AM, Larson GA, Zahniser NR (2009) Low or high cocaine responding rats differ in striatal extracellular dopamine levels and dopamine transporter number. J Pharmacol Exp Ther 331(3):985–997. https://doi.org/10.1124/jpet.109.159897. https://www.ncbi.nlm.nih.gov/pubmed/19729579
83. Nirenberg MJ, Vaughan RA, Uhl GR, Kuhar MJ, Pickel VM (1996) The dopamine transporter is localized to dendritic and axonal plasma membranes of nigrostriatal dopaminergic neurons. J Neurosci 16(2):436–447. https://doi.org/10.1523/JNEUROSCI.16-02-00436.1996. https://www.ncbi.nlm.nih.gov/pubmed/8551328
84. O'Sullivan MJ, Lindsay AJ (2020) The endosomal recycling pathway-at the crossroads of the cell. Int J Mol Sci 21(17). https://doi.org/10.3390/ijms21176074
85. Penmatsa A, Wang KH, Gouaux E (2013) X-ray structure of dopamine transporter elucidates antidepressant mechanism. Nature 503(7474):85–90. https://doi.org/10.1038/nature12533. https://www.ncbi.nlm.nih.gov/pubmed/24037379
86. Pilotte NS, Sharpe LG, Kuhar MJ (1994) Withdrawal of repeated intravenous infusions of cocaine persistently reduces binding to dopamine transporters in the nucleus accumbens of Lewis rats. J Pharmacol Exp Ther 269(3):963–969. https://www.ncbi.nlm.nih.gov/pubmed/7912283
87. Praefcke GJ, McMahon HT (2004) The dynamin superfamily: universal membrane tubulation and fission molecules? Nat Rev Mol Cell Biol 5(2):133–147. https://doi.org/10.1038/nrm1313. https://www.ncbi.nlm.nih.gov/pubmed/15040446
88. Qian JJ, Cheng YB, Yang YP, Mao CJ, Qin ZH, Li K, Liu CF (2008) Differential effects of overexpression of wild-type and mutant human alpha-synuclein on MPP+-induced neurotoxicity in PC12 cells. Neurosci Lett 435(2):142–146. https://doi.org/10.1016/j.neulet.2008.02.021. https://www.ncbi.nlm.nih.gov/pubmed/18342443
89. Ragu Varman D, Subler MA, Windle JJ, Jayanthi LD, Ramamoorthy S (2021) Novelty-induced hyperactivity and suppressed cocaine induced locomotor activation in mice lacking threonine 53 phosphorylation of dopamine transporter. Behav Brain Res 408:113267. https://doi.org/10.1016/j.bbr.2021.113267. https://www.ncbi.nlm.nih.gov/pubmed/33794225
90. Rao A, Richards TL, Simmons D, Zahniser NR, Sorkin A (2012) Epitope-tagged dopamine transporter knock-in mice reveal rapid endocytic trafficking and filopodia targeting of the transporter in dopaminergic axons. FASEB J 26(5):1921–1933. https://doi.org/10.1096/fj.11-196113. https://www.ncbi.nlm.nih.gov/pubmed/22267337
91. Rao A, Simmons D, Sorkin A (2011) Differential subcellular distribution of endosomal compartments and the dopamine transporter in dopaminergic neurons. Mol Cell Neurosci 46(1):148–158. https://doi.org/10.1016/j.mcn.2010.08.016. https://www.ncbi.nlm.nih.gov/pubmed/20816972
92. Reith MEA, Kortagere S, Wiers CE, Sun H, Kurian MA, Galli A, Volkow ND, Lin Z (2022) The dopamine transporter gene SLC6A3: multidisease risks. Mol Psychiatry 27(2):1031–1046. https://doi.org/10.1038/s41380-021-01341-5. https://www.ncbi.nlm.nih.gov/pubmed/34650206
93. Rickhag M, Hansen FH, Sorensen G, Strandfelt KN, Andresen B, Gotfryd K, Madsen KL, Vestergaard-Klewe I, Ammendrup-Johnsen I, Eriksen J, Newman AH, Fuchtbauer EM,

Gomeza J, Woldbye DP, Wortwein G, Gether U (2013) A C-terminal PDZ domain-binding sequence is required for striatal distribution of the dopamine transporter. Nat Commun 4:1580. https://doi.org/10.1038/ncomms2568. https://www.ncbi.nlm.nih.gov/pubmed/23481388
94. Robinson MS (2015) Forty years of Clathrin-coated vesicles. Traffic 16(12):1210–1238. https://doi.org/10.1111/tra.12335. https://www.ncbi.nlm.nih.gov/pubmed/26403691
95. Sakrikar D, Mazei-Robison MS, Mergy MA, Richtand NW, Han Q, Hamilton PJ, Bowton E, Galli A, Veenstra-Vanderweele J, Gill M, Blakely RD (2012) Attention deficit/hyperactivity disorder-derived coding variation in the dopamine transporter disrupts microdomain targeting and trafficking regulation. J Neurosci 32(16):5385–5397. https://doi.org/10.1523/jneurosci.6033-11.2012
96. Samuvel DJ, Jayanthi LD, Manohar S, Kaliyaperumal K, See RE, Ramamoorthy S (2008) Dysregulation of dopamine transporter trafficking and function after abstinence from cocaine self-administration in rats: evidence for differential regulation in caudate putamen and nucleus accumbens. J Pharmacol Exp Ther 325(1):293–301. https://doi.org/10.1124/jpet.107.130534. https://www.ncbi.nlm.nih.gov/pubmed/18198344
97. Sandoval V, Riddle EL, Ugarte YV, Hanson GR, Fleckenstein AE (2001) Methamphetamine-induced rapid and reversible changes in dopamine transporter function: an in vitro model. J Neurosci 21(4):1413–1419. https://doi.org/10.1523/jneurosci.21-04-01413.2001
98. Saunders C, Ferrer JV, Shi L, Chen J, Merrill G, Lamb ME, Leeb-Lundberg LM, Carvelli L, Javitch JA, Galli A (2000) Amphetamine-induced loss of human dopamine transporter activity: an internalization-dependent and cocaine-sensitive mechanism. Proc Natl Acad Sci USA 97(12):6850–6855. https://doi.org/10.1073/pnas.110035297
99. Schorova L, Martin S (2016) Sumoylation in synaptic function and dysfunction. Front Synaptic Neurosci 8:9. https://doi.org/10.3389/fnsyn.2016.00009
100. Schultz W (2007) Multiple dopamine functions at different time courses. Annu Rev Neurosci 30:259–288. https://doi.org/10.1146/annurev.neuro.28.061604.135722. https://www.ncbi.nlm.nih.gov/pubmed/17600522
101. Sharpe LG, Pilotte NS, Mitchell WM, De Souza EB (1991) Withdrawal of repeated cocaine decreases autoradiographic [3H]mazindol-labelling of dopamine transporter in rat nucleus accumbens. Eur J Pharmacol 203(1):141–144. https://doi.org/10.1016/0014-2999(91)90804-y. https://www.ncbi.nlm.nih.gov/pubmed/1797552
102. Shaw JK, Pamela Alonso I, Lewandowski SI, Scott MO, O'Connor BM, Aggarwal S, De Biasi M, Mortensen OV, Espana RA (2021) Individual differences in dopamine uptake in the dorsomedial striatum prior to cocaine exposure predict motivation for cocaine in male rats. Neuropsychopharmacology 46(10):1757–1767. https://doi.org/10.1038/s41386-021-01009-2. https://www.ncbi.nlm.nih.gov/pubmed/33953341
103. Shetty M, Bolland DE, Morrell J, Grove BD, Foster JD, Vaughan RA (2023) Dopamine transporter membrane mobility is bidirectionally regulated by phosphorylation and palmitoylation. Curr Res Physiol 6:100106. https://doi.org/10.1016/j.crphys.2023.100106. https://www.ncbi.nlm.nih.gov/pubmed/38107792
104. Soh JW, Lee EH, Prywes R, Weinstein IB (1999) Novel roles of specific isoforms of protein kinase C in activation of the c-fos serum response element. Mol Cell Biol 19(2):1313–1324. https://doi.org/10.1128/mcb.19.2.1313
105. Sorensen G, Rickhag M, Leo D, Lycas MD, Ridderstrom PH, Weikop P, Lilja JH, Rifes P, Herborg F, Woldbye D, Wortwein G, Gainetdinov RR, Fink-Jensen A, Gether U (2021) Disruption of the PDZ domain-binding motif of the dopamine transporter uniquely alters nanoscale distribution, dopamine homeostasis, and reward motivation. J Biol Chem 297(6):101361. https://doi.org/10.1016/j.jbc.2021.101361. https://www.ncbi.nlm.nih.gov/pubmed/34756883
106. Sorkina T, Doolen S, Galperin E, Zahniser NR, Sorkin A (2003) Oligomerization of dopamine transporters visualized in living cells by fluorescence resonance energy transfer microscopy. J Biol Chem 278(30):28274–28283. https://doi.org/10.1074/jbc.M210652200. https://www.ncbi.nlm.nih.gov/pubmed/12746456

107. Sorkina T, Hoover BR, Zahniser NR, Sorkin A (2005) Constitutive and protein kinase C-induced internalization of the dopamine transporter is mediated by a clathrin-dependent mechanism. Traffic 6(2):157–170. https://doi.org/10.1111/j.1600-0854.2005.00259.x
108. Sorkina T, Ma S, Larsen MB, Watkins SC, Sorkin A (2018) Small molecule induced oligomerization, clustering and clathrin-independent endocytosis of the dopamine transporter. elife 7. https://doi.org/10.7554/eLife.32293. https://www.ncbi.nlm.nih.gov/pubmed/29630493
109. Sorkina T, Miranda M, Dionne KR, Hoover BR, Zahniser NR, Sorkin A (2006) RNA interference screen reveals an essential role of Nedd4-2 in dopamine transporter ubiquitination and endocytosis. J Neurosci 26(31):8195–8205. https://doi.org/10.1523/jneurosci.1301-06.2006
110. Sorkina T, Richards TL, Rao A, Zahniser NR, Sorkin A (2009) Negative regulation of dopamine transporter endocytosis by membrane-proximal N-terminal residues. J Neurosci 29(5):1361–1374. https://doi.org/10.1523/jneurosci.3250-08.2009
111. Staley JK, Hearn WL, Ruttenber AJ, Wetli CV, Mash DC (1994) High affinity cocaine recognition sites on the dopamine transporter are elevated in fatal cocaine overdose victims. J Pharmacol Exp Ther 271(3):1678–1685. https://www.ncbi.nlm.nih.gov/pubmed/7996484
112. Sucic S, El-Kasaby A, Kudlacek O, Sarker S, Sitte HH, Marin P, Freissmuth M (2011) The serotonin transporter is an exclusive client of the coat protein complex II (COPII) component SEC24C. J Biol Chem 286(18):16482–16490. https://doi.org/10.1074/jbc.M111.230037. https://www.ncbi.nlm.nih.gov/pubmed/21454670
113. Sweeney CG, Kearney PJ, Fagan RR, Smith LA, Bolden NC, Zhao-Shea R, Rivera IV, Kolpakova J, Xie J, Gao G, Tapper AR, Martin GE, Melikian HE (2020) Conditional, inducible gene silencing in dopamine neurons reveals a sex-specific role for Rit2 GTPase in acute cocaine response and striatal function. Neuropsychopharmacology 45(2):384–393. https://doi.org/10.1038/s41386-019-0457-x. https://www.ncbi.nlm.nih.gov/pubmed/31277075
114. Torres GE, Carneiro A, Seamans K, Fiorentini C, Sweeney A, Yao WD, Caron MG (2003) Oligomerization and trafficking of the human dopamine transporter. Mutational analysis identifies critical domains important for the functional expression of the transporter. J Biol Chem 278(4):2731–2739. https://doi.org/10.1074/jbc.M201926200. https://www.ncbi.nlm.nih.gov/pubmed/12429746
115. Torres GE, Yao WD, Mohn AR, Quan H, Kim KM, Levey AI, Staudinger J, Caron MG (2001) Functional interaction between monoamine plasma membrane transporters and the synaptic PDZ domain-containing protein PICK1. Neuron 30(1):121–134. https://doi.org/10.1016/s0896-6273(01)00267-7. https://www.ncbi.nlm.nih.gov/pubmed/11343649
116. Underhill SM, Amara SG (2021) Acetylcholine receptor stimulation activates protein Kinase C mediated internalization of the Dopamine transporter. Front Cell Neurosci 15:662216. https://doi.org/10.3389/fncel.2021.662216. https://www.ncbi.nlm.nih.gov/pubmed/33897375
117. Underhill SM, Hullihen PD, Chen J, Fenollar-Ferrer C, Rizzo MA, Ingram SL, Amara SG (2021) Amphetamines signal through intracellular TAAR1 receptors coupled to Galpha(13) and Galpha(S) in discrete subcellular domains. Mol Psychiatry 26(4):1208–1223. https://doi.org/10.1038/s41380-019-0469-2. https://www.ncbi.nlm.nih.gov/pubmed/31399635
118. Volkow ND, Wang GJ, Fowler JS, Logan J, Gatley SJ, Hitzemann R, Chen AD, Dewey SL, Pappas N (1997) Decreased striatal dopaminergic responsiveness in detoxified cocaine-dependent subjects. Nature 386(6627):830–833. https://doi.org/10.1038/386830a0. https://www.ncbi.nlm.nih.gov/pubmed/9126741
119. Wei Y, Williams JM, Dipace C, Sung U, Javitch JA, Galli A, Saunders C (2007) Dopamine transporter activity mediates amphetamine-induced inhibition of Akt through a Ca2+/calmodulin-dependent kinase II-dependent mechanism. Mol Pharmacol 71(3):835–842. https://doi.org/10.1124/mol.106.026351. https://www.ncbi.nlm.nih.gov/pubmed/17164407
120. Wersinger C, Sidhu A (2003) Attenuation of dopamine transporter activity by alpha-synuclein. Neurosci Lett 340(3):189–192. https://doi.org/10.1016/s0304-3940(03)00097-1. https://www.ncbi.nlm.nih.gov/pubmed/12672538
121. Wheeler DS, Underhill SM, Stolz DB, Murdoch GH, Thiels E, Romero G, Amara SG (2015) Amphetamine activates Rho GTPase signaling to mediate dopamine transporter

internalization and acute behavioral effects of amphetamine. Proc Natl Acad Sci USA 112(51):E7138–E7147. https://doi.org/10.1073/pnas.1511670112. https://www.ncbi.nlm.nih.gov/pubmed/26553986

122. Wilson JM, Nobrega JN, Carroll ME, Niznik HB, Shannak K, Lac ST, Pristupa ZB, Dixon LM, Kish SJ (1994) Heterogeneous subregional binding patterns of 3H-WIN 35,428 and 3H-GBR 12,935 are differentially regulated by chronic cocaine self-administration. J Neurosci 14(5 Pt 2):2966–2979. https://doi.org/10.1523/JNEUROSCI.14-05-02966.1994. https://www.ncbi.nlm.nih.gov/pubmed/8182452
123. Wu S, Bellve KD, Fogarty KE, Melikian HE (2015) Ack1 is a dopamine transporter endocytic brake that rescues a trafficking-dysregulated ADHD coding variant. Proc Natl Acad Sci USA 112(50):15480–15485. https://doi.org/10.1073/pnas.1512957112. https://www.ncbi.nlm.nih.gov/pubmed/26621748
124. Wu S, Fagan RR, Uttamapinant C, Lifshitz LM, Fogarty KE, Ting AY, Melikian HE (2017) The Dopamine transporter recycles via a Retromer-dependent Postendocytic mechanism: tracking studies using a novel fluorophore-coupling approach. J Neurosci 37(39):9438–9452. https://doi.org/10.1523/JNEUROSCI.3885-16.2017. https://www.ncbi.nlm.nih.gov/pubmed/28847807
125. Yatham LN, Liddle PF, Gonzalez M, Saraf G, Vafai N, Lam RW, Sossi V (2022) A positron emission tomography study of Dopamine transporter density in patients with bipolar disorder with current Mania and those with recently remitted Mania. JAMA Psychiatry 79(12):1217–1224. https://doi.org/10.1001/jamapsychiatry.2022.3541
126. Zahniser NR, Sorkin A (2009) Trafficking of dopamine transporters in psychostimulant actions. Semin Cell Dev Biol 20(4):411–417. https://doi.org/10.1016/j.semcdb.2009.01.004
127. Zapata A, Kivell B, Han Y, Javitch JA, Bolan EA, Kuraguntla D, Jaligam V, Oz M, Jayanthi LD, Samuvel DJ, Ramamoorthy S, Shippenberg TS (2007) Regulation of dopamine transporter function and cell surface expression by D3 dopamine receptors. J Biol Chem 282(49):35842–35854. https://doi.org/10.1074/jbc.M611758200. https://www.ncbi.nlm.nih.gov/pubmed/17923483

Causal Mechanisms of Monoamine Transporter Phosphorylation

Sammanda Ramamoorthy, Durairaj Ragu Varman, and Lankupalle D. Jayanthi

Contents

Abstract Monoamine transporters are essential proteins located at presynaptic terminals that play a crucial role in regulating neurotransmission of serotonin, dopamine, and norepinephrine by rapid reuptake of released amines from the synapse. Clinically used antidepressants and widely abused psychostimulants exhibit a high affinity for amine transporters. Function and expression of biogenic amine transporter are altered in subjects suffering from psychiatric diseases such as depression and in psychostimulant use disorder. Therefore, proper functional regulation of monoamine transporters is critical in maintaining normal amine homeostasis.

S. Ramamoorthy (✉) · L. D. Jayanthi (✉)
Department of Pharmacology and Toxicology, Virginia Commonwealth University, Richmond, VA, USA
e-mail: sramamoorthy@vcu.edu; sammanda.ramamoorthy@vcuhealth.org; ljayanthi@vcu.edu; lankupalle.jayanthi@vcuhealth.org

D. R. Varman
Department of Pharmacology and Toxicology, Virginia Commonwealth University, Richmond, VA, USA

Centre for Integrative Omics Data Science, Yenepoya (Deemed to be University), Mangalore, India

O. V. Mortensen (ed.), *Monoamine Transporters*, Advances in Neurobiology 46,
https://doi.org/10.1007/978-3-031-96364-3_5

Monoamine transporters possess several potential phosphorylation sites/motifs and exist in a phosphorylated state. Various cellular protein kinases and phosphatases are known to regulate the phosphorylation dynamics of amine transporters, which in turn influences subcellular expression and trafficking, microdomain-specific protein–protein interactions, transporter protein degradation, and overall transport capacity. Dysfunctional amine transporter function, phosphorylation, and association with interacting proteins are evident in neuropsychiatric disease states, including psychostimulant use disorder. However, the neurobiological consequences of in vivo amine transporter phosphorylation and its regulation remain unclear. Recent studies utilizing intact animal models are beginning to connect these molecular mechanisms with observed animal behaviors. This review summarizes current knowledge on the causal role of amine transporter phosphorylation in regulating amine transport and its relevance to animal behavior. Further understanding of phosphorylation-dependent molecular mechanisms governing amine transporter regulation potentially identifies regulatory motif(s) as potential therapeutic targets for treating neuropsychiatric disorders.

Keywords Biogenic amine transporters · Dopamine transporter · Knock-in mouse model · Neuropsychiatric behaviors and/or disorders · Norepinephrine transporter · Phosphorylation · Regulation · Serotonin transporter

1 Introduction: Overview of Monoamine Transporters

Monoamine neurotransmitters including serotonin (5-HT), dopamine (DA), and norepinephrine (NE) play a crucial role in regulating behavior and physiological processes in both central and peripheral nervous systems [13, 34, 47, 52]. Serotonergic signaling is associated with regulating mood, anxiety, aggression, and appetite. Disruption of the 5-HT signaling pathway has been linked to several illnesses, including depression, obsessive-compulsive disorder (OCD), anxiety disorders, autism spectrum disorders (ASD), and alcohol drinking disorders [19, 21, 84, 85, 108]. On the other hand, dopaminergic signaling has drawn attention for its function in reward, attention, and movement circuits [5, 59, 65]. Disrupted or dysregulated DA signaling has been linked to illnesses like addiction, attention-deficit/hyperactivity disorder (ADHD), schizophrenia, and Parkinson's disease [6, 35, 60, 74, 83, 107, 120, 122–125]. NE is a crucial transmitter of arousal, attention, executive function, and stress reactions [41, 80, 121]. Disrupted NE signaling in the central nervous system has been connected to disorders like ADHD, post-traumatic stress disorder (PTSD), and depression [33, 56, 58, 102, 112]. NE is also the primary neurotransmitter in postganglionic sympathetic synapses. Compromised NE uptake sites and activity have been implicated in conditions such as cardiomyopathy, heart failure, hypertension, and ischemia [11, 24, 44, 66, 77, 106].

Monoaminergic neurotransmission is regulated by the rapid reuptake of released monoamines from synapses by presynaptic biogenic amine transporters, including transporters for 5-HT (SERT), DA (DAT), and NE (NET) [45, 46]. Clinically, monoamine transporters serve as the primary site of action for numerous therapeutic medications used in the treatment of attention deficit hyperactive disorder, depression, and other psychiatric disorders [64, 118]. Additionally, highly addictive psychostimulants such as cocaine, amphetamine, and methamphetamine are high affinity ligands for amine transporters [7, 43, 48]. Numerous behavioral and degenerative diseases have long been linked to altered expression of biogenic amine transporters and their function [28, 29, 38, 89, 104]. In order to maintain proper monoamine homeostasis, the activity of biogenic amine transporter must be appropriately regulated and maintained. DAT, SERT, and NET play complex interactive role in the mechanisms regulating amine biosynthesis, amine storage, receptor sensitivity, and transporter expression. These interconnected neuronal events have been demonstrated through studies using animal models of DAT, SERT, or NET knockout mice [8, 31, 39, 53, 89, 111, 128]. Notably, there is growing evidence that cellular protein kinases and phosphatases influence the expression, activity, trafficking, and degradation of amine transporters (for review, see [9, 25, 61, 98]). Furthermore, several rare coding variants have been implicated in psychiatric disorders which exhibit disrupted transporter regulation resulting in altered expression and functional properties [40, 62, 63, 76, 82, 86, 103, 126]. These regulatory mechanisms are crucial for normal monoamine neurotransmission, and understanding these regulatory phenomena greatly enhances our understanding of psychostimulants use disorder and other psychiatric conditions. It is also crucial for developing disease treatments because these regulatory mechanisms are regulated by endogenous signaling pathways and impacted by external transporter ligands, including psychostimulants and antidepressants [97, 130]. A complex interaction of numerous molecular pathways, including the actions of protein kinases is involved in controlling monoamine transporters. These enzymes can act directly on transporter cytoplasmic domains or indirectly by targeting transporter-associated proteins through the post-translational modification process. Post-translational modifications such as phosphorylation, glycosylation, palmitoylation, and ubiquitination play a critical role in regulating neurotransmitter transporter activity and the availability of transporters in the plasma membrane (for review, see [9, 25, 61, 98]). The availability of functional transporters and their kinetic properties are responsible for regulating amine transport and availability of amine for receptors and subsequent signaling transmission. Importantly, post-translational modifications have the ability to alter intrinsic transport activity, alter transporter turnover, regulate exocytic fusion of transporter-containing vesicles with the plasma membrane, and regulate internalization of transporter from the plasma membrane by modulating endocytic machinery pathways [9, 23, 79, 98, 109, 113]. Studies have identified important phosphorylation sites as molecular motifs involved in phosphorylation-dependent monoamine transporter regulation. However, challenges in studying the role of endogenous phosphorylated transporters in intact animals have persisted until recently. These challenges were largely due to the absence of animal models with construct validity and the lack of reagents that can specifically and directly disrupt phosphorylated transporters at designated neuron terminals within particular brain regions in vivo. Recently, there has been progress in developing knock-in mouse models that carry site-specific phosphorylation mutant transporters, with the quest to develop alternative pharmacological strategies by defining the role of transporter phosphorylation and refining our understanding of compromised signaling mechanisms [75, 94, 95].

In this chapter, we first summarize studies that have investigated the causal relationship between changes in phosphorylation states of transporters and their transporter function. We also examine the progress made in the impact of disease-associated rare human coding variants in transporter phosphorylation. Furthermore, we delve into the role of phosphorylated transporters in animal behavior, aiming to illuminate the underlying compromised neuronal mechanisms to help us better understand the etiology of mental illnesses and substance use disorders (SUD), paving the way for the development of alternative pharmacological strategies targeting monoamine-specific regulatory motifs in the treatment of mental illnesses, including SUD.

2 The Serotonin Transporter Phosphorylation and Its Significance

2.1 Regulation of SERT by Phosphorylation

Several intracellular serine (Ser), threonine (Thr), and tyrosine (Tyr) amino acid motifs exist in SERT that are potential sites for phosphorylation by kinases [101]. Indeed, SERT is a phosphoprotein [99], and its phospho-status is regulated by several kinases, receptors, substrates, and psychostimulants [2, 3, 50, 96, 97, 99, 100, 105, 116] (for review, see [9, 98]). This suggests the intricate relationship between SERT phosphorylation and 5-HT clearance, which modulates serotoninergic neurotransmission.

2.2 Causal Mechanisms of Threonine-276 Phosphorylation of SERT

Several kinases regulate SERT activity through both traffic-dependent and independent mechanisms (for review, see [9, 98]). Concurrently, these kinases also initiate SERT phosphorylation. For instance, the activation of protein kinase G (PKG) phosphorylates endogenously expressed SERT in the rat midbrain, and similar effects have been observed with human SERT expressed in CHO-1 cells [100]. It would not be surprising if transporters utilized phosphorylation as a means to regulate their activity. To prove this, researchers need to identify kinase-specific sites within SERT and demonstrate that mutating these sites to non-phosphorylatable amino acids, like alanine (Ala), results in the loss of both kinase-mediated regulation and phosphorylation of SERT. Ramamoorthy and Jayanthi's team pinpointed a specific phosphorylation site in SERT and found that phosphorylation of Thr-276 is essential for cGMP-mediated regulation of SERT [100]. Treatment of rat midbrain synaptosomes with PKG activator 8-Br-cGMP increased trafficking independent SERT activity and phosphorylation. Notably, through phosphoamino acid analyses, the study revealed that cGMP-induced phosphorylation of native SERT occurs exclusively on Thr residues. Mutating all eighteen intracellular Thr residues to nonphosphorylatable Ala in human SERT (hSERT), only the Thr-276Ala substitution completely blocked cGMP-mediated stimulation of 5-HT transport and SERT phosphorylation. The phosphorylation of Thr-276 introduces a negative charge, which might affect SERT activity. If this hypothesis holds true, replacing Thr-276 with aspartic acid (which also carries a negative charge when ionized) could mimic the effects of PKG on 5-HT uptake. Supporting this notion, changing Thr-276 to aspartate as a way to mimic phosphorylation increased 5-HT uptake to levels similar to those seen with cGMP stimulation in wild-type SERT-expressing cells, and rendering it unresponsive to cGMP. These findings identified a critical phosphorylation site on hSERT and established a direct causal link indicating that PKG phosphorylates hSERT on Thr-276, resulting in increased 5-HT uptake. Understanding how PKG activity is regulated by presynaptic receptor-mediated or other signaling pathways will enhance our understanding of the interactions of SERT-Thr-276 phosphorylation with extracellular stimuli in serotonergic neurotransmission and behavior. Reports indicate that A3 adenosine receptor (A3AR) linked pathways upregulate SERT activity through PKG [78, 134]. However, the association of A3AR-PKG mediated SERT upregulation with Thr-276 phosphorylation is currently unknown and represents the next step in investigating the role of Thr-276 phosphorylation in A3AR-PKG signaling.

2.3 *Activity-Dependent Regulation of Threonine-276 Phosphorylation of SERT*

The Thr-276 phosphorylation site in SERT is located near the cytoplasmic end of transmembrane domain 5 (TM5). Cysteine scanning accessibility studies suggest that Thr-276 site may be poorly accessible to PKG to phosphorylate Thr-276 [131, 133]. Structural evidence suggests that TM5 engages in forming the permeation pathway for 5-HT with Na+ and Cl- ions [20]. During this process, TM5 undergoes a conformational change, shifting from an outward-open to an inward-open state. This change leads to a partial unwinding of the cytoplasmic end of the TM5 helix, which exposes Thr-276 to the cytoplasm, making it more accessible for kinase activity to catalyze the phosphorylation. Research from the Rudnick laboratory has provided evidence for activity-dependent phosphorylation at Thr-276. They demonstrated that inward-open stabilizing agents, such as the SERT substrate 5-HT and inhibitor ibogaine, increase Thr-276 phosphorylation. In contrast, agents that stabilize outward-open conformations, like Na+ and cocaine, reduce hSERT Thr-276 phosphorylation [132]. This dynamic rearrangement of SERT's conformation regulates the exposure of the Thr-276 site to kinases, thereby enhancing the SERT catalytic activity in transporting 5-HT. Although the specific phosphorylation sites are not yet fully identified, a study by Ramamoorthy and Blakely reported similar activity-dependent regulation of hSERT phosphorylation by protein kinase C (PKC) along with its surface expression [97].

2.4 Endogenous Role of Threonine-276 Phosphorylation of SERT in Behavior

To investigate the in vivo role of Thr-276 phosphorylation on SERT activity, as well as its neurochemical and behavioral effects, the Blakely and Ramamoorthy laboratories employed CRISPR/Cas9 technique to create knock-in mice with a non-phosphorylatable Ala substitution at the Thr-276 phosphorylation site in SERT (SERT-Ala276) [75]. The SERT-Ala276 mice are viable, exhibit normal growth, and show deviations from Mendelian expectations, with a reduced number of SERT-Ala276 offspring. SERT-Ala276 mice show no observable changes in physical or sensorimotor behaviors. Midbrain and hippocampal SERT mRNA levels, SERT protein expression, 5-HT uptake kinetics, and levels of neurochemicals (5-HT, DA, NE, GABA, and glutamate) and metabolites (5-HIAA, DOPAC, and HVA) in the SERT-Ala276 are comparable to those in wild-type mice. SERT-Ala276 mice did not display any significant differences in open field locomotor activity and elevated zero maze behavioral performance as well as in tail suspension and forced swim tests, suggesting that Thr-276 SERT phosphorylation does not influence anxiety- and depression-like behavior. However, female SERT-Ala276, but not male heterozygous and homozygous mice, showed reduced marble burying behavior compared to sex-matched wild-type littermates. Interestingly, in the tube test of social dominance, male SERT-Ala276, but not female homozygous mice, showed a reduction in wins compared to WT controls. These behavioral outcomes indicate a possible sex-dimorphic interaction of SERT-Thr-276 phosphorylation regulating specific behaviors [75].

2.5 Causal Relationship of Serine-48 Phosphorylation of SERT and Glycogen Synthase Kinase-3ß Mediated SERT Regulation

The canonical site for phosphorylation by glycogen synthase kinase-3ß (GSK3ß) is S/T-X-X-X-S/T(P) [54]. Analysis of the amino acid sequence of hSERT revealed that Ser-44 (**S**-N-G-Y-S) and Ser-48 (**S**-A-V-P-S) are two potential consensus phosphorylation sites for GSK3α/ß, located in the cytosolic amino tail of hSERT [101]. Ramamoorthy and Jayanthi's research team [92] continued their efforts to delineate the impact of Ser-44 and Ser-48 phosphorylation on GSK3ß-mediated hSERT regulation. Pharmacological inhibition (CHIR99021) of endogenously active GSK3ß in mouse striatal synaptosomes resulted in increased 5-HT uptake, maximal velocity of 5-HT transport, and surface expression of SERT, without altering total SERT expression, along with a concomitant decrease in SERT phosphorylation. Further studies using a heterologous expression model with HEK-293 cells, through pharmacological and genetic manipulations of GSK3ß, demonstrated that inhibiting GSK3ß upregulates hSERT function, Vmax, and surface hSERT levels by reducing internalization and enhancing membrane insertion. Notably, replacing the predicted GSK3α/ß-target phosphorylation site Ser-48, but not Ser-44, with a non-phosphorylatable Ala (hSERT-Ser-48Ala, hSERT-Ser-44Ala) prevented the stimulatory effect of CHIR99021 or GSK3α/ß siRNA on SERT activity and SERT surface expression. In contrast to the stimulatory effects of GSK3ß inhibition, co-expression of constitutively active GSK3ß (GSK3ß-Ser-9Ala) with wild-type hSERT resulted in decreased SERT activity, Vmax, surface expression, and increased transporter phosphorylation. However, the inhibition of hSERT activity and stimulation of hSERT phosphorylation by active GSK3ß-Ser-9Ala was absent when co-expressed with the hSERT-Ser-48Ala mutant but not with hSERT-Ser-44Ala. Additionally, the hSERT-Ser-44Ala mutant was sensitive to Akt, CaMKII, p38 MAPK, Src-kinase inhibition, and PKC activation similarly to hSERT-wildtype [92]. Collectively, these results suggest that Ser-48 is the specific and essential target phosphorylation site for GSK3ß-mediated hSERT regulation. Future studies are needed to determine whether GSK3ß phosphorylates SERT directly and to identify the pre-synaptic receptor linked upstream signals modulating GSK3ß activity, 5-HT transmission, and its relationship to behavior.

2.6 *Causal Mechanisms of Tyrosine Phosphorylation of SERT in Transporter Stability and Activity*

While studies have shown that tyrosine kinase activators or inhibitors modulate SERT function [42, 88, 90, 129], the molecular basis for tyrosine kinase-mediated SERT regulation remains unclear. The SERT amino acid sequence includes four tyrosine residues (Tyr-47, Tyr-142, Tyr-350, and Tyr-358) that face intracellularly and could be phosphorylated by tyrosine kinases [101]. Annamalai et al. demonstrated the tyrosine phosphorylation of SERT in rat platelets, which is sensitive to the tyrosine kinase inhibitor genistein. Using a heterologous expression model of human placental trophoblasts, the authors further showed that expressing the tyrosine kinase Src increased hSERT function, tyrosine phosphorylation, and hSERT protein expression while knocking down Src via siRNA inhibited hSERT activity. Moreover, individual mutations of Tyr-47 and Tyr-142 to non-phosphorylatable phenylalanine diminished the Src-induced increase in 5-HT transport and hSERT protein expression, as well as hSERT-tyrosine phosphorylation. Notably, single mutations of Tyr-350 or Tyr-358, or mutation of all four tyrosine residues (Tyr-47, Tyr-142, Tyr-350, and Tyr-358) to phenylalanine, completely abolished hSERT expression. Additionally, inhibiting protein tyrosine kinase reduces hSERT protein expression by promoting hSERT protein degradation [2]. Collectively, these studies establish a direct causal relationship between phosphorylation of Tyr-47, Tyr-142, Tyr-350, and Tyr-358 in SERT and the regulation of hSERT stability and function.

2.7 Role of Threonine-276 Phosphorylation of SERT in Disease-Linked SERT Variant

Murphy and colleagues reported that two unrelated probands and their family members with OCD contained the Ile-425Val coding variant in hSERT [86]. Kilic et al. reported that hSERT Ile-425Val variant shows higher 5-HT uptake activity, increased 5-HT transport velocity, decreased 5-HT affinity without any change in hSERT surface expression and is insensitive to stimulation by nitric oxide donors in transfected HeLa or COS-7 cells [55]. Further studies demonstrated that while the PKG activator 8-Br-cGMP stimulates the activity of wild-type hSERT, it does not affect the Ile-425Val variant. Notably, the elevated activity observed in the Ile-425Val mutant was comparable to that of wild-type hSERT when stimulated by PKG. The elevated activity observed in the Ile-425Val variant decreased with extended treatment with inhibitors of soluble guanylyl cyclase, unlike the wild type hSERT, indicating that Ile-425Val transporters exist in high activity forms due to sustained regulation by PKG. Mutating PKG phosphorylation site Thr-276 [100] into Ala prevented PKG-triggered elevation of wild-type hSERT activity and also abolished the inhibitory effects of PKG observed in the Ile-425Val variant. The higher activity observed in the phosphomimetic form of hSERT (Thr-276Asp) [100] is similar to the level of 5-HT uptake activity found in Ile-425Val variant [133]. These observations suggest that the Ile-425 to Val mutation found in OCD may regulate hSERT conformations toward a sustained and elevated Thr-276 phosphorylation state catalyzed by PKG, facilitating higher 5-HT clearance.

3 The Dopamine Transporter Phosphorylation and Its Significance

3.1 Regulation of DAT Activity and Phosphorylation

The DAT sequence reveals potential Ser, Thr, and Tyr residues in intracellular loops that are accessible to protein kinases for the phosphorylation of DAT, thereby regulating dopamine clearance [30, 32]. The Vaughan laboratory is at the forefront of identifying phosphorylation sites in DAT protein molecules. Through phosphopeptide mapping, in vitro phosphorylation assays using DAT-fusion proteins and specific protein kinases, mass spectrometry, and heterologous cellular models, evidence supports the phosphorylation of Ser-2, Ser-4, Ser-7, Ser-12, Ser-13, and Thr-53 sites in DAT (see reviews [9, 25, 26, 98]). Specific kinase-directed DAT phosphorylation at these residues can modulate the availability of DAT to clear released DA and influence dopaminergic neurotransmission as well as behavioral responses. If DAT phosphorylation impacts the binding affinity of psychostimulants such as cocaine and amphetamine, it could have profound effects on the behavioral outcomes of these substances, ultimately contributing to SUD.

3.2 *Mechanistic Relationship of Serine-7 Phosphorylation in DAT Regulation*

Using biochemical and heterologous cell model approaches, Vaughan's group identified Ser-7 in the N-terminal of DAT as a site phosphorylated in response to PKC activation [81]. Mutating Ser-7 to the non-phosphorylatable Ala abolished ß-PMA (a PKC activator)-induced DAT phosphorylation. Studies have shown that the Ser-7Ala mutation prevented the ß-PMA-induced decrease in DAT in COS cell lines [67], but it also reported no effects on DAT activity and surface expression in LLC-PK1 cells [81], suggesting that the different host cells contribute to these contrasting results. Interestingly, Moritz et al. reported that Ser-7 DAT phosphorylation affects the binding affinity of the cocaine analog CFT. For example, increasing DAT phosphorylation following okadaic acid (phosphatase inhibitor) and oleoyl-2-acetyl-sn-glycerol (OAG—a PKC activator) in rat striatal tissue displays higher CFT binding affinity, but there is no effect of PKC activation by ß-PMA in LLC-PK1. DAT mutants Ser-7Ala and Ser-7Asp (phosphomimetic DAT) showed reduced CFT binding affinity [81]. Structural transitions of DAT occur between an outward-facing configuration that allows the substrate and co-ions to bind and an inward-facing phase that releases substrates. Binding of DAT inhibitors, including cocaine and cocaine analogs, is favored when DAT is in an outward-facing conformation. Since Ser-7 phosphorylation modulates CFT binding affinity, the authors proposed that Ser-7 phosphorylation of DAT influences DAT conformational transition phases, potentially shifting the affinity for cocaine and consequently impacting cocaine sensitivity in dopaminergic neurotransmission to elucidate behaviors. However, understanding the impact of Ser-7 phosphorylation of DAT in a native context and its effect on behaviors is crucial for drawing any conclusions.

3.3 *Threonine-53 Phosphorylation of DAT and Regulations*

Proline-directed kinases include extracellular signal-regulated kinase (ERK1/2), mitogen-activated protein kinases (MAPK), and c-Jun N-terminal kinases (JNK1/2/3), all of which require a proline immediately following the phosphorylation site [54]. The presence of Thr-53 in the DAT, located within the cytoplasmic N-terminal of rodent DAT (rDAT), represents the consensus phosphorylation motif (P-P-Q-**T(53)**-P) for proline-directed kinases. Vaughan and colleagues confirmed the presence of phospho-Thr on the N-terminal tail of rat DAT and identified Thr-53 as a phosphorylation site for ERK1/2, p38MAPK, and JNK [36]. Expression of the non-phosphorylatable Thr-53 mutant (Thr-53Ala) and the phosphomimetic Thr-53 (Thr-53Asp) in LLCPK1 cells exhibited reduced surface expression without changes in total DAT expression or DA transport velocity and completely prevented amphetamine (AMPH)-triggered substrate (MPP) release [27]. In contrast to these findings, elevated DA uptake and AMPH-induced DA efflux have also been reported with Thr-53Ala and Thr-53Asp [17]. The Vaughan laboratory successfully developed the phospho-specific antibody for detecting Thr-53 phosphorylated DAT [27] and demonstrated that Thr-53 phosphorylation is elevated by AMPH and METH using in vitro and ex vivo models [17]. These studies also demonstrated the potential contribution of Thr-53 phosphorylation in the regulation of DA clearance and efflux and the impact of AMPH and METH interactions. Translating information obtained from in vitro and ex vivo studies to endogenous in vivo settings will enhance our understanding of the impact of Thr-53 phosphorylation on DAT in dopaminergic transmission and behavior.

3.4 *Causal Relationship of Endogenous DAT-Thr-53 Phosphorylation with Cocaine Affinity and Behavioral Effects of Cocaine*

Calipari et al. reported that Thr-53 phosphorylation of DAT and its activity are higher in females, particularly during the estrus stage, which is associated with increased ERK1/2 phosphorylation, increased DAT cocaine potency for inhibiting DA clearance, and a concomitant potentiation of cocaine conditioned place preference (CPP, reward) [16]. The findings that the enhanced DA-neuron activity in the ventral tegmental area (VTA) during the estrus stage may reflect the neuronal mechanisms underlying elevated DAT-Thr-53 phosphorylation, cocaine affinity, and CPP. Utilizing optogenetic and chemogenetic methods, VTA firing acts as the primary mechanism through which DAT-Thr-53 phosphorylation and DA release enhance cocaine reward via estradiol interactions with DA neurons. These studies also emphasize sexual dimorphism in Thr-53 phosphorylation of DAT and its role in cocaine-reward learning processes [16]. Supporting this notion, Brodnik et al. demonstrated that increased DAT affinity for cocaine and elevated DAT-Thr-53 phosphorylation, along with a decrease in cocaine consumption after DREADD-induced activation of DA neurons, contrasted with reduced DAT-Thr-53 phosphorylation and cocaine potency following inhibition of DA neurons, which resulted in increased cocaine consumption [14]. Furthermore, DA-neuron specific inhibition of ERK1/2 activity through DA-neuron specific viral mediated expression of MAP kinase phosphatase resulted in decreased DAT-Thr-53 phosphorylation and cocaine reward, suggesting a critical role for ERK1/2 in regulating DA function, which results in significant neurochemical and behavioral effects via DAT-Thr-53 phosphorylation [10]. Although these studies demonstrate a correlation between changes in DAT-Thr-53 phosphorylation to behavior, the causal relationship between endogenous DAT-Thr-53 phosphorylation and cocaine effects remains unclear. Recent findings from Ramamoorthy and Jayanthi's laboratory highlight the causal link of ERK1/2-mediated DAT-Thr-53 phosphorylation in cocaine affinity and behavioral effects of cocaine using knock-in mice lacking Thr-53 phosphorylation of DAT (DAT-Ala53 mouse model) endogenously [94]. Compared to wild type, DAT-Ala53 mice exhibit an absence of phospho-Thr-53-DAT, a normal distribution of TH, DAT, and colocalization of TH and DAT, as well as DA and DA-metabolite levels. In contrast to the results obtained from heterologous cell model systems [17, 27], DAT-Ala53 mice which lack endogenous DAT-Thr-53 phosphorylation in vivo exhibit unaltered striatal DA uptake (V_{max}, K_m), and DAT expression (total and surface), but reduced cocaine's potency to inhibit striatal DA uptake. Additionally, the absence of ERK1/2 inhibitor-mediated DAT inhibition is evident in DAT-Ala53 mice, suggesting the requirement for DAT-Thr-53 phosphorylation in ERK1/2-mediated DAT regulations. Behaviorally, DAT-Ala53 mice display novelty-induced hyperactivity and reduced cocaine-induced locomotor activation, demonstrating the direct mechanistic causal link of endogenous Thr-53 phosphorylation of DAT in regulating cocaine affinity and its consequent effects on behavior.

3.5 *Causal Mechanisms of DAT-Thr-53 Phosphorylation in Kappa Opioid Receptor-Mediated DAT Regulation and Impact on Behavioral Effects of KOR Agonism*

The receptor for opioid peptide dynorphin (DYN) kappa opioid receptor (KOR) is coexpressed with DAT on DA terminals as a heteroreceptor [117], suggesting that the presynaptic DYN/KOR system modulates DA transmission. KOR agonists decrease DA release, enhance DAT activity, and lower extracellular DA concentrations by activating KOR located on DA terminals [91, 119]. Kivell et al. reported that salvinorin A, a selective KOR agonist, stimulates DAT activity through ERK1/2 activation by increasing DAT surface expression. Additionally, evidence was presented for the existence of the DAT and KOR complex, and KOR activation enhanced the formation of the DAT/KOR complex [57]. This suggests that the dual actions of KOR on DA release and DAT regulation may contribute to the behavioral effects of KOR agonism. Recently, Mayer et al. presented that KOR upregulates rDAT via Thr-53 phosphorylation of rDAT. KOR activation increased DAT activity as well as total and surface Thr-53 phosphorylation of rDAT, but it failed to regulate the non-phosphorylatable mutant Thr-53Ala DAT in the EM4 cell model, establishing the causal relationship between DAT-Thr-53 phosphorylation and KOR-mediated DAT upregulation [70]. In rodents, KOR agonists produce reduced locomotor activity, conditioned place aversion (CPA) and potentiates cocaine CPP [15, 18, 110, 127]. Ragu Varman et al. used knock-in mice lacking Thr-53 phosphorylation of DAT (DAT-Ala53) (see Sect. 3.4) to translate the impact of KOR-mediated endogenous Thr-53-DAT phosphorylation on DAT regulation and the behavioral effects of KOR agonists. These investigators demonstrated that KOR activation increases KOR antagonist-sensitive DAT activity, DAT Vmax, phosphoThr-53-DAT, and surface expression in wild-type mice but not in DAT-Ala53 mice. Behaviorally, KOR agonists caused locomotor suppression, CPA and potentiation of cocaine CPP in wild-type mice but not in DAT-Ala53 mice. Similar to wild-type mice, DAT-Ala53 mice exhibit lithium chloride-induced CPA and morphine-induced conditioned place preference [93]. These findings establish the first causal evidence that locomotor suppression, aversive responses and enhancement of cocaine reward to KOR agonists manifest through the modulation of DAT activity via DAT-Thr-53 phosphorylation.

3.6 Endogenous Role of DAT-Thr-53 Phosphorylation in a Disease-Linked DAT Variant and Behavior

Blakely's group identified a pedigree that includes two male children diagnosed with ADHD, bipolar disorder, and ASD, who share a rare hDAT coding variant, the Ala-559Val substitution [12, 72]. In a heterologous cell model, the DAT Val559 variant shows normal DAT protein expression, surface levels, and DA uptake rates, but exhibits constitutive anomalous DAT-mediated DA efflux [71, 73]. Continuous studies from Blakely's team, involving the generation of DAT-Val559 knock-in mice, reveal that the DAT-Val559 mice, consistent with their anomalous dopamine efflux, have higher basal extracellular DA levels [76]. Behaviorally, DAT-Val559 mice display attenuated locomotor responses to methylphenidate and d-amphetamine, impulsivity, higher motivation, and altered working memory [22, 114, 115]. Further studies by this group demonstrated that the DAT Val559 mutation has sex- and brain region-specific impact of DAT-Val559 mutation on presynaptic DA autoreceptor (D2AR)-mediated DAT regulation. The elevated DA efflux and increased extracellular DA found in DAT-Val559 mice lead to tonic activation of D2AR, resulting in enhanced DAT-Val559 surface expression and Thr-53 hyperphosphorylation in the dorsal striatum of males and the ventral striatum of females [37, 114]. Mayer et al. studies revealed that DAT-Val559 mice exhibit constitutively upregulated KOR-mediated DAT surface expression and phosphorylation of Thr-53, effects that can be reversed by KOR antagonism. Additionally, KOR antagonism corrected in vivo DA release and the sex-dependent behavioral abnormalities observed in DAT-Val559 mice [70]. Significant evidence supports the in vivo alterations of DAT-Thr-53 phosphorylation in the disease variant DAT-Val559 through D2AR [37, 76, 115]. Further studies that inhibit or eliminate DAT-Thr-53 phosphorylation in DAT-Val559 mice may illuminate the causative role of DAT-Thr-53 phosphorylation in ADHD and DA-linked brain disorders.

4 The Norepinephrine Transporter Phosphorylation and Its Significance

4.1 *Regulation of NET by Phosphorylation*

Several predicted protein kinase sites in intracellular cytoplasmic domains are present in NET, which undergo phosphorylation to regulate NET function, NE clearance, and noradrenergic neurotransmission that affects behavior [87]. Research has shown that various biological stimuli, including neuronal activity, peptide hormones, trophic factors, NE-presynaptic auto- and heteroreceptors, and cocaine and amphetamine, modulate NET function by influencing several post-translational modifications. Studies from Jayanthi's laboratory have demonstrated that the modulation of PKC [49, 51], p38 MAPK, substance P/NK1 receptor signaling [4], and exposure to psychostimulants such as cocaine [68] and amphetamine [1] regulate NET phosphorylation (see reviews [9, 98]).

4.2 *Role of p38 MAPK-Mediated Thr-30 Phosphorylation in Cocaine-Mediated NET Upregulation and Behavior*

Jayanthi and colleagues [68] demonstrated that administering cocaine in vivo to rats, or treating prefrontal cortex (PFC) synaptosomes in vitro, upregulated NET activity and increased NET surface expression by inhibiting constitutive NET endocytosis, phosphorylation, and active p38 MAPK; these effects were abolished by inhibiting p38 MAPK. Continued mutational analysis confirmed the requirement of Thr-30 phosphorylation for cocaine-triggered functional upregulation of NET. Further studies by Mannangatti et al. [69] showed that in vivo administration of a p38 MAPK inhibitor SB203580 eliminated cocaine-induced NET upregulation and diminished cocaine locomotor sensitization and CPP, highlighting the role of active p38 MAPK. The authors further demonstrated that intervening NET-Thr-30 phosphorylation in vivo through the administration of a membrane-permeable TAT-NET peptide containing the NET-Thr-30 phosphorylation motif, but not the Thr-30 non-phosphorylatable Ala30 mutant peptide (TAT-NET-Ala30), blocked cocaine-induced NET upregulation and phosphorylation while also reducing cocaine CPP and the reinstatement of cocaine CPP [69]. These findings suggest a role for NET-Thr-30 phosphorylation in the behavioral effects of cocaine. Several questions remain, including how cocaine binding to NET leads to Thr-30 phosphorylation and which other p38 MAPK-modulated NET post-translational modifications and brain regions are involved in eliciting cocaine's behaviors.

4.3 Role of Thr-258/Ser-259 Phosphorylation of NET in Neurokinin 1 Receptor-PKC-Mediated NET Regulation and Its Impact on Amphetamine-Mediated NET Downregulation and Behavior

Jayanthi's group provided key evidence that NET phosphorylation is linked to PKC-mediated downregulation of NET. The activation of PKC by ß-PMA [49, 51] or PKC linked to neurokinin 1 receptor (NK1R) activation by its ligand substance P, triggered NET phosphorylation and reduced NET function and surface expression by promoting the endocytosis of lipid raft-resident NET [4]. However, mutation of Thr-258/Ser-259 motif to nonphosphorylatable Ala (Thr-258Ala/Ser-259Ala) diminished the effects of ß-PMA or NK1R-PKC-modulated NET phosphorylation, inhibited PKC-mediated NET internalization and lipid raft redistribution [4, 49]. Additional studies from the same group utilizing in vitro phosphorylation assay using NET-expressing membranes and purified PKC demonstrated that, consistent with the rat placental trophoblasts cell model [49], the Ca^{2+}-independent PKCε isoform triggers NET phosphorylation of wild-type NET, but not Thr-258Ala/Ser-259Ala NET, providing evidence that PKCε directly phosphorylates NET at the Thr-258/Ser-259 sites [51]. Jayanthi's group conducted further studies examining the role of the Thr-258/Ser-259 motif in AMPH-mediated NET downregulation and any causal relationship to the behavioral effects of AMPH. AMPH treatment of human placental trophoblast cells expressing wild-type hNET decreased NET functionality and surface expression by reducing NET membrane insertion and enhancing endocytosis; however, these NET regulations were absent in cells expressing Thr-258Ala/Ser-259Ala [1]. Extending from these findings, membrane-permeable TAT-Thr-258/Ser-259 motif was utilized to interfere with the in vivo role of TAT-Thr-258/Ser-259 motif on AMPH-induced NET regulations and behaviors. Intra-accumbal infusion of TAT-Thr-258/Ser-259 peptide in rats prevented AMPH-induced NET downregulation, AMPH-induced locomotor sensitization, CPP, and reinstatement, but not mutated TAT-Thr-258Ala/Ser-259Ala or scrambled peptide infusion suggesting that phosphorylation of the NET-Thr-258/Ser-259 motif in this brain region plays a role in AMPH-induced behaviors. In more recent studies, Jayanthi's laboratory generated a knock-in (KI) mouse carrying a NET-Thr-258Ala/Ser-259Ala mutation to directly link the in vivo causal role of Thr-258/Ser-259 phosphorylation motif. NET-T258A/S259A mice exhibit brain region-specific differential expression of NET, and NET function in regions important for drug reward. Importantly, NET-T258A/S259A mutant male mice display reduced AMPH-induced locomotor activity and CPP, which was not apparent in females [95]. Summarizing these comprehensive studies reveals the significance of NET phosphorylation in regulating NE clearance and the behavioral effects of cocaine and AMPH, contributing to substance use disorder.

5 Conclusion

The studies reviewed in this chapter indicate a causal relationship between monoamine transporter phosphorylation, regulation of amine clearance and behavior. Intriguingly, monoamine transporter phosphorylation modulates interactions with psychostimulants, and the activity-dependent regulation of monoamine transporter phosphorylation reveals the significance of the conformational state of the monoamine transporter, which can influence in vivo amine neurotransmission in response to cocaine, AMPH, or therapeutic ligands targeting monoamine transporters. These changes will affect the therapeutic effects of monoamine transporter ligands as well as the addictive properties of cocaine and AMPH. The relationship of monoamine transporter phosphorylation in human disease-linked coding variants of monoamine transporter suggests that alterations in monoamine transporter phosphorylation potentially contribute to maladaptive behaviors seen in psychiatric disorders. The regulation of monoamine transporters is complex, and several phosphorylation sites exist in monoamine transporters and their role needs to be investigated. More importantly, identifying presynaptic auto- and heteroreceptor signaling cascades that regulate kinase-mediated monoamine transporter phosphorylation will enhance our understanding of the interactions between neurobiology of amine neurotransmission and behaviors while also revealing whether dysregulated monoamine transporter phosphorylation and the consequent altered monoamine neurotransmission predispose to psychiatric disorders and SUD. Understanding these signaling pathways that regulate monoamine transporter phosphorylation will aid in developing novel targets for manipulating monoamine transporter regulation and creating more effective therapies for treating depression, addiction, and other psychiatric disorders.

Acknowledgments The support from the National Institutes of Health DA054694 and DA060210 to SR and LDJ is greatly acknowledged.

References

1. Annamalai B, Mannangatti P, Arapulisamy O, Ramamoorthy S, Jayanthi LD (2010) Involvement of threonine 258 and serine 259 motif in amphetamine-induced norepinephrine transporter endocytosis. J Neurochem 115:23–35
2. Annamalai B, Mannangatti P, Arapulisamy O, Shippenberg TS, Jayanthi LD, Ramamoorthy S (2012) Tyrosine phosphorylation of the human serotonin transporter: a role in the transporter stability and function. Mol Pharmacol 81:73–85
3. Annamalai B, Ragu Varman D, Horton RE, Daws LC, Jayanthi LD, Ramamoorthy S (2020) Histamine receptors regulate the activity, surface expression, and phosphorylation of Serotonin transporters. ACS Chem Neurosci 11:466–476
4. Arapulisamy O, Mannangatti P, Jayanthi LD (2013) Regulated norepinephrine transporter interaction with the neurokinin-1 receptor establishes transporter subcellular localization. J Biol Chem 288:28599–28610

5. Ashok AH, Mizuno Y, Volkow ND, Howes OD (2017) Association of stimulant use with dopaminergic alterations in users of cocaine, amphetamine, or methamphetamine: a systematic review and meta-analysis. JAMA Psychiatry 74:511–519
6. Bannon MJ (2005) The dopamine transporter: role in neurotoxicity and human disease. Toxicol Appl Pharmacol 204:355–360
7. Barker EL, Blakely RD (1995) Norepinephrine and serotonin transporters: molecular targets of antidepressant drugs. In: Psychopharmacology: the fourth generation of progress, pp 321–333
8. Bengel D, Murphy DL, Andrews AM, Wichems CH, Feltner D, Heils A, Mossner R, Westphal H, Lesch KP (1998) Altered brain serotonin homeostasis and locomotor insensitivity to 3,4-methylenedioxymetamphetamine ("ecstasy") in serotonin transporter-deficient mice. Mol Pharmacol 53:649–655
9. Bermingham DP, Blakely RD (2016) Kinase-dependent regulation of monoamine neurotransmitter transporters. Pharmacol Rev 68:888–953
10. Bernstein DL, Lewandowski SI, Besada C, Place D, Espana RA, Mortensen OV (2024) Inactivation of ERK1/2 signaling in dopaminergic neurons by Map Kinase phosphatase MKP3 regulates dopamine signaling and motivation for cocaine. J Neurosci 44:e0727232023
11. Bohm M, La Rosse K, Schwinger RHG, Erdmann E (1995) Evidence for reduction norepinephrine uptake sites in the failing human heart. J Am Coll Cardiol 25(1):146–153
12. Bowton E, Saunders C, Reddy IA, Campbell NG, Hamilton PJ, Henry LK, Coon H, Sakrikar D, Veenstra-VanderWeele JM, Blakely RD, Sutcliffe J, Matthies HJ, Erreger K, Galli A (2014) SLC6A3 coding variant Ala559Val found in two autism probands alters dopamine transporter function and trafficking. Transl Psychiatry 4:e464
13. Boyle N, Betts S, Lu H (2024) Monoaminergic modulation of learning and cognitive function in the prefrontal cortex. Brain Sci 14:902
14. Brodnik ZD, Xu W, Batra A, Lewandowski SI, Ruiz CM, Mortensen OV, Kortagere S, Mahler SV, Espana RA (2020) Chemogenetic manipulation of Dopamine neurons dictates cocaine potency at distal dopamine transporters. J Neurosci 40:8767–8779
15. Brust TF, Morgenweck J, Kim SA, Rose JH, Locke JL, Schmid CL, Zhou L, Stahl EL, Cameron MD, Scarry SM, Aube J, Jones SR, Martin TJ, Bohn LM (2016) Biased agonists of the kappa opioid receptor suppress pain and itch without causing sedation or dysphoria. Sci Signal 9:ra117
16. Calipari ES, Juarez B, Morel C, Walker DM, Cahill ME, Ribeiro E, Roman-Ortiz C, Ramakrishnan C, Deisseroth K, Han MH, Nestler EJ (2017) Dopaminergic dynamics underlying sex-specific cocaine reward. Nat Commun 8:13877
17. Challasivakanaka S, Zhen J, Smith ME, Reith MEA, Foster JD, Vaughan RA (2017) Dopamine transporter phosphorylation site threonine 53 is stimulated by amphetamines and regulates dopamine transport, efflux, and cocaine analog binding. J Biol Chem 292:19066–19075
18. Chefer VI, Backman CM, Gigante ED, Shippenberg TS (2013) Kappa opioid receptors on dopaminergic neurons are necessary for kappa-mediated place aversion. Neuropsychopharmacology 38:2623–2631
19. Coccaro EF (1989) Central serotonin and impulsive aggression. Br J Psychiatry Suppl 8:52–62
20. Coleman JA, Green EM, Gouaux E (2016) X-ray structures and mechanism of the human serotonin transporter. Nature 532:334–339
21. Compagnon P, Ernouf D, Narcisse G, Daoust M (1993) Serotonin in animal models of alcoholism. Alcohol Alcohol Suppl 2:215–219
22. Davis GL, Stewart A, Stanwood GD, Gowrishankar R, Hahn MK, Blakely RD (2018) Functional coding variation in the presynaptic dopamine transporter associated with neuropsychiatric disorders drives enhanced motivation and context-dependent impulsivity in mice. Behav Brain Res 337:61–69
23. Eriksen J, Jorgensen TN, Gether U (2010) Regulation of dopamine transporter function by protein-protein interactions: new discoveries and methodological challenges. J Neurochem 113:27–41

24. Esler M, Jackman G, Bobik A, Leonard P, Kelleher D, Skews H, Jennings G, Korner P (1981) Norepinephrine kinetics in essential hypertension. Defective neuronal uptake of norepinephrine in some patients. Hypertension 3:149–156
25. Foster JD, Vaughan RA (2017) Phosphorylation mechanisms in dopamine transporter regulation. J Chem Neuroanat 83-84:10–18
26. Foster JD, Cervinski MA, Gorentla BK, Vaughan RA (2006) Regulation of the dopamine transporter by phosphorylation. Handb Exp Pharmacol 175:197–214
27. Foster JD, Yang JW, Moritz AE, Challasivakanaka S, Smith MA, Holy M, Wilebski K, Sitte HH, Vaughan RA (2012) Dopamine transporter phosphorylation site threonine 53 regulates substrate reuptake and amphetamine-stimulated efflux. J Biol Chem 287:29702–29712
28. Gainetdinov RR, Caron MG (2003) Monoamine transporters: from genes to behavior. Annu Rev Pharmacol Toxicol 43:261–284
29. Garland EM, Hahn MK, Ketch TP, Keller NR, Kim CH, Kim KS, Biaggioni I, Shannon JR, Blakely RD, Robertson D (2002) Genetic basis of clinical catecholamine disorders. Ann N Y Acad Sci 971:506–514
30. Giros B, Mestikawy SE, Bertrand L, Caron MG (1991) Cloning and functional characterization of a cocaine-sensitive dopamine transporter. Feder Exp Biol Sci 295:149–154
31. Giros B, Jaber M, Jones SR, Wightman RM, Caron MG (1996) Hyperlocomotion and indifference to cocaine and amphetamine in mice lacking the dopamine transporter. Nature 379:606–612
32. Giros B, El Mestikawy S, Godinot N, Zheng K, Han H, Yang-Feng T, Caron MG (1992) Cloning, pharmacological characterization, and chromosome assignment of the human dopamine transporter. Mol Pharmacol 42:383–390
33. Goddard AW, Ball SG, Martinez J, Robinson MJ, Yang CR, Russell JM, Shekhar A (2010) Current perspectives of the roles of the central norepinephrine system in anxiety and depression. Depress Anxiety 27:339–350
34. Goldstein DS (1995) Stress,catecholamines, and cardiovascular disease, 1st edn. Oxford University Press, New York
35. Goodman A (2008) Neurobiology of addiction. An integrative review. Biochem Pharmacol 75:266–322
36. Gorentla BK, Moritz AE, Foster JD, Vaughan RA (2009) Proline-directed phosphorylation of the dopamine transporter N-terminal domain. Biochemistry 48:1067–1076
37. Gowrishankar R, Gresch PJ, Davis GL, Katamish RM, Riele JR, Stewart AM, Vaughan RA, Hahn MK, Blakely RD (2018) Region-specific regulation of presynaptic Dopamine homeostasis by D2 autoreceptors shapes the in vivo impact of the neuropsychiatric disease-associated DAT variant Val559. J Neurosci 38:5302–5312
38. Hahn MK, Blakely RD (2002) Monoamine transporter gene structure and polymorphisms in relation to psychiatric and other complex disorders. Pharmacogenomics J 2:217–235
39. Hall FS, Li XF, Sora I, Xu F, Caron M, Lesch KP, Murphy DL, Uhl GR (2002) Cocaine mechanisms: enhanced cocaine, fluoxetine and nisoxetine place preferences following monoamine transporter deletions. Neuroscience 115:153–161
40. Hansen FH et al (2014) Missense dopamine transporter mutations associate with adult parkinsonism and ADHD. J Clin Invest 124:3107–3120
41. Harley CW (2004) Norepinephrine and dopamine as learning signals. Neural Plast 11:191–204
42. Helmeste DM, Tang SW (1995) Tyrosine kinase inhibitors regulate serotonin uptake in platelets. Eur J Pharmacol 280:R5–R7
43. Howell LL, Kimmel HL (2008) Monoamine transporters and psychostimulant addiction. Biochem Pharmacol 75:196–217
44. Imamura M, Lander HM, Levi R (1996) Activation of histamine H3-receptors inhibits carrier-mediated norepinephrine release during protracted myocardial ischemia. Comparison with adenosine A1-receptors and alpha2-adrenoceptors. Circ Res 78:475–481
45. Iversen LL (1971) Role of transmitter uptake mechanisms in synaptic neurotransmission. Br J Pharmacol 41:571–591

46. Iversen LL (1978) Uptake processes for biogenic amines. In: Iversen I (ed) Handbook of psychopharmacology, 3rd edn. Prenum Press, New York, pp 381–442
47. Jacobs B, Azmitia EC (1992) Structure and function of the brain serotonin system. Physiol Rev 72:165–229
48. Jayanthi LD, Ramamoorthy S (2005) Regulation of monoamine transporters: influence of psychostimulants and therapeutic antidepressants. AAPS J 7:E728–E738
49. Jayanthi LD, Samuvel DJ, Ramamoorthy S (2004) Regulated internalization and phosphorylation of the native norepinephrine transporter in response to phorbol esters. Evidence for localization in lipid rafts and lipid raft-mediated internalization. J Biol Chem 279:19315–19326
50. Jayanthi LD, Samuvel DJ, Blakely RD, Ramamoorthy S (2005) Evidence for biphasic effects of protein kinase C on serotonin transporter function, endocytosis, and phosphorylation. Mol Pharmacol 67:2077–2087
51. Jayanthi LD, Annamalai B, Samuvel DJ, Gether U, Ramamoorthy S (2006) Phosphorylation of the norepinephrine transporter at threonine 258 and serine 259 is linked to protein kinase C-mediated transporter internalization. J Biol Chem 281:23326–23340
52. Jiang Y, Zou D, Li Y, Gu S, Dong J, Ma X, Xu S, Wang F, Huang JH (2022) Monoamine neurotransmitters control basic emotions and affect major depressive disorders. Pharmaceuticals (Basel) 15:1203
53. Keller NR, Diedrich A, Appalsamy M, Miller LC, Caron MG, McDonald MP, Shelton RC, Blakely RD, Robertson D (2006) Norepinephrine transporter-deficient mice respond to anxiety producing and fearful environments with bradycardia and hypotension. Neuroscience 139:931–946
54. Kemp BE, Pearson RB (1990) Protein kinase recognition sequence motifs. Trends Biochem Sci 15:342–346
55. Kilic F, Murphy DL, Rudnick G (2003) A human serotonin transporter mutation causes constitutive activation of transport activity. Mol Pharmacol 64:440–446
56. Kim CH, Waldman ID, Blakely RD, Kim KS (2008) Functional gene variation in the human norepinephrine transporter: association with attention deficit hyperactivity disorder. Ann N Y Acad Sci 1129:256–260
57. Kivell B, Uzelac Z, Sundaramurthy S, Rajamanickam J, Ewald A, Chefer V, Jaligam V, Bolan E, Simonson B, Annamalai B, Mannangatti P, Prisinzano TE, Gomes I, Devi LA, Jayanthi LD, Sitte HH, Ramamoorthy S, Shippenberg TS (2014) Salvinorin A regulates dopamine transporter function via a kappa opioid receptor and ERK1/2-dependent mechanism. Neuropharmacology 86:228–240
58. Klimek V, Stockmeier C, Overholser J, Meltzer HY, Kalka S, Dilley G, Ordway GA (1997) Reduced levels of norepinephrine transporters in the locus coeruleus in major depression. J Neurosci 17(21):8451–8458
59. Koob GF (1998) Circuits, drugs, and drug addiction. Adv Pharmacol 42:978–982
60. Koob GF, Volkow ND (2010) Neurocircuitry of addiction. Neuropsychopharmacology 35:217–238
61. Kristensen AS, Andersen J, Jorgensen TN, Sorensen L, Eriksen J, Loland CJ, Stromgaard K, Gether U (2011) SLC6 neurotransmitter transporters: structure, function, and regulation. Pharmacol Rev 63:585–640
62. Kurian MA, Zhen J, Cheng SY, Li Y, Mordekar SR, Jardine P, Morgan NV, Meyer E, Tee L, Pasha S, Wassmer E, Heales SJ, Gissen P, Reith ME, Maher ER (2009) Homozygous loss-of-function mutations in the gene encoding the dopamine transporter are associated with infantile parkinsonism-dystonia. J Clin Invest 119:1595–1603
63. Kurian MA et al (2011) Clinical and molecular characterisation of hereditary dopamine transporter deficiency syndrome: an observational cohort and experimental study. Lancet Neurol 10:54–62
64. Langer SZ, Schoemaker H (1988) Effects of antidepressants on monoamine transporters. Prog Neuro-Psychopharmacol Biol Psychiat 12:193–216

65. Lawrence AJ, Beart PM, Kalivas PW (2008) Neuropharmacology of addiction--setting the scene. Br J Pharmacol 154:259–260
66. Liang C, Fan T-HM, Sullebarger JT, Sakamoto S (1989) Decreased adrenergic neuronal uptake activity in experimental right heart failure. J Clin Invest 84:1267–1275
67. Lin Z, Zhang PW, Zhu X, Melgari JM, Huff R, Spieldoch RL, Uhl GR (2003) Phosphatidylinositol 3-kinase, protein kinase C, and MEK1/2 kinase regulation of dopamine transporters (DAT) require N-terminal DAT phosphoacceptor sites. J Biol Chem 278:20162–20170
68. Mannangatti P, Arapulisamy O, Shippenberg TS, Ramamoorthy S, Jayanthi LD (2011) Cocaine up-regulation of the norepinephrine transporter requires threonine 30 phosphorylation by p38 mitogen-activated protein kinase. J Biol Chem 286:20239–20250
69. Mannangatti P, NarasimhaNaidu K, Damaj MI, Ramamoorthy S, Jayanthi LD (2015) A role for p38 mitogen-activated protein kinase-mediated threonine 30-dependent Norepinephrine transporter regulation in cocaine sensitization and conditioned place preference. J Biol Chem 290:10814–10827
70. Mayer FP, Stewart A, Varman DR, Moritz AE, Foster JD, Owens AW, Areal LB, Gowrishankar R, Velez M, Wickham K, Phelps H, Katamish R, Rabil M, Jayanthi LD, Vaughan RA, Daws LC, Blakely RD, Ramamoorthy S (2025) Kappa opioid receptor antagonism restores phosphorylation, trafficking and behavior induced by a disease-associated dopamine transporter variant. Mol Psychiatry. https://doi.org/10.1038/s41380-025-03055-4
71. Mazei-Robison MS, Blakely RD (2005) Expression studies of naturally occurring human dopamine transporter variants identifies a novel state of transporter inactivation associated with Val382Ala. Neuropharmacology 49:737–749
72. Mazei-Robison MS, Couch RS, Shelton RC, Stein MA, Blakely RD (2005) Sequence variation in the human dopamine transporter gene in children with attention deficit hyperactivity disorder. Neuropharmacology 49:724–736
73. Mazei-Robison MS, Bowton E, Holy M, Schmudermaier M, Freissmuth M, Sitte HH, Galli A, Blakely RD (2008) Anomalous dopamine release associated with a human dopamine transporter coding variant. J Neurosci 28:7040–7046
74. McCutcheon RA, Abi-Dargham A, Howes OD (2019) Schizophrenia, Dopamine and the Striatum: from biology to symptoms. Trends Neurosci 42:205–220
75. Meinke C, Quinlan MA, Paffenroth KC, Harrison FE, Fenollar-Ferrer C, Katamish RM, Stillman I, Ramamoorthy S, Blakely RD (2022) Serotonin transporter Ala276 Mouse: Novel Model to assess the neurochemical and behavioral impact of Thr276 phosphorylation in vivo. Neurochem Res 47:37–60
76. Mergy MA, Gowrishankar R, Gresch PJ, Gantz SC, Williams J, Davis GL, Wheeler CA, Stanwood GD, Hahn MK, Blakely RD (2014) The rare DAT coding variant Val559 perturbs DA neuron function, changes behavior, and alters in vivo responses to psychostimulants. Proc Natl Acad Sci USA 111:E4779–E4788
77. Merlet P, Dubois-Rande J-L, Adnot S, Bourguignon MH, Benvenuti C, Loisance D, Valette H, Castaigne A, Syrota A (1992) Myocardial b-adrenergic desensitization and neuronal norepinephrine uptake function in idiopathic dilated cardiomyopathy. J Cardiovasc Pharmacol 19:10–16
78. Miller KJ, Hoffman BJ (1994) Adenosine A3 receptors regulate serotonin transport via nitric oxide and cGMP. J Biol Chem 269:27351–27356
79. Miranda M, Sorkin A (2007) Regulation of receptors and transporters by ubiquitination: new insights into surprisingly similar mechanisms. Mol Interv 7:157–167
80. Morilak DA, Barrera G, Echevarria DJ, Garcia AS, Hernandez A, Ma S, Petre CO (2005) Role of brain norepinephrine in the behavioral response to stress. Prog Neuro-Psychopharmacol Biol Psychiatry 29:1214–1224
81. Moritz AE, Foster JD, Gorentla BK, Mazei-Robison MS, Yang JW, Sitte HH, Blakely RD, Vaughan RA (2013) Phosphorylation of dopamine transporter serine 7 modulates cocaine analog binding. J Biol Chem 288:20–32

82. Ng J et al (2014) Dopamine transporter deficiency syndrome: phenotypic spectrum from infancy to adulthood. Brain 137:1107–1119
83. Nutt DJ, Lingford-Hughes A, Erritzoe D, Stokes PR (2015) The dopamine theory of addiction: 40 years of highs and lows. Nat Rev Neurosci 16:305–312
84. Olivier B (2015) Serotonin: a never-ending story. Eur J Pharmacol 753:2–18
85. Owens MJ, Nemeroff CB (1994) Role of serotonin in the pathophysiology of depression: focus on the serotonin transporter. Clin Chem 40:288–295
86. Ozaki N, Goldman D, Kaye WH, Plotnicov K, Greenberg BD, Lappalainen J, Rudnick G, Murphy DL (2003) Serotonin transporter missense mutation associated with a complex neuropsychiatric phenotype. Mol Psychiatry 8(895):933–896
87. Pacholczyk T, Blakely RD, Amara SG (1991) Expression cloning of a cocaine- and antidepressant-sensitive human noradrenaline transporter. Nature 350:350–354
88. Pavanetto M, Zarpellon A, Borgo C, Donella-Deana A, Deana R (2011) Regulation of serotonin transport in human platelets by tyrosine kinase Syk. Cell Physiol Biochem 27:139–148
89. Perona MT, Waters S, Hall FS, Sora I, Lesch KP, Murphy DL, Caron M, Uhl GR (2008) Animal models of depression in dopamine, serotonin, and norepinephrine transporter knockout mice: prominent effects of dopamine transporter deletions. Behav Pharmacol 19:566–574
90. Prasad PD, Torres-Zamorano V, Kekuda R, Leibach FH, Ganapathy V (1997) Functional link between tyrosine phosphorylation and human serotonin transporter gene expression. Eur J Pharmacol 325:85–92
91. Ragu Varman D, Jayanthi LD, Ramamoorthy S (2021a) Kappa Opioid Receptor mediated differential regulation of Serotonin and Dopamine transporters in mood and substance use disorder. Handb Exp Pharmacol 271:97
92. Ragu Varman D, Jayanthi LD, Ramamoorthy S (2021b) Glycogen synthase kinase-3ss supports serotonin transporter function and trafficking in a phosphorylation-dependent manner. J Neurochem 156:445–464
93. Ragu Varman D, Ramamoorthy S, Jayanthi LD (2025)Threonine-53 Phosphorylation of Dopamine Transporter Dictates kappa-Opioid Receptor-Mediated Locomotor Suppression, Aversion, and Cocaine Reward. J Neurosci 45:e0171252025
94. Ragu Varman D, Subler MA, Windle JJ, Jayanthi LD, Ramamoorthy S (2021c) Novelty-induced hyperactivity and suppressed cocaine induced locomotor activation in mice lacking threonine 53 phosphorylation of dopamine transporter. Behav Brain Res 408:113267
95. Ragu Varman D, Mannangatti P, Subler MA, Windle JJ, Ramamoorthy S, Jayanthi LD (2022) Blunted amphetamine-induced reinforcing behaviors and transporter downregulation in Knock-in mice carrying alanine mutations at Threonine-258 and Serine-259 of norepinephrine transporter. J Mol Neurosci 72:1965–1976
96. Rajamanickam J, Annamalai B, Rahbek-Clemmensen T, Sundaramurthy S, Gether U, Jayanthi LD, Ramamoorthy S (2015) Akt-mediated regulation of antidepressant-sensitive serotonin transporter function, cell-surface expression and phosphorylation. Biochem J 468:177–190
97. Ramamoorthy S, Blakely RD (1999) Phosphorylation and sequestration of serotonin transporters differentially modulated by psychostimulants. Science 285:763–766
98. Ramamoorthy S, Shippenberg TS, Jayanthi LD (2011) Regulation of monoamine transporters: role of transporter phosphorylation. Pharmacol Ther 129:220–238
99. Ramamoorthy S, Giovanetti E, Qian Y, Blakely RD (1998) Phosphorylation and regulation of antidepressant-sensitive serotonin transporters. J Biol Chem 273:2458–2466
100. Ramamoorthy S, Samuvel DJ, Buck ER, Rudnick G, Jayanthi LD (2007) Phosphorylation of threonine residue 276 is required for acute regulation of serotonin transporter by cyclic GMP. J Biol Chem 282:11639–11647
101. Ramamoorthy S, Bauman AL, Moore KR, Han H, Yang-Feng T, Chang AS, Ganapathy V, Blakely RD (1993) Antidepressant- and cocaine-sensitive human serotonin transporter: molecular cloning, expression, and chromosomal localization. Proc Natl Acad Sci USA 90:2542–2546

102. Ressler KJ, Nemeroff CB (2000) Role of serotonergic and noradrenergic systems in the pathophysiology of depression and anxiety disorders. Depress Anxiety 1:2–19
103. Robertson D, Flattem N, Tellioglu T, Carson R, Garland E, Shannon JR, Jordan J, Jacob G, Blakely RD, Biaggioni I (2001) Familial orthostatic tachycardia due to norepinephrine transporter deficiency. Ann N Y Acad Sci 940:527–543
104. Rudnick G, Kramer R, Blakely RD, Murphy DL, Verrey F (2014) The SLC6 transporters: perspectives on structure, functions, regulation, and models for transporter dysfunction. Pflugers Arch 466:25–42
105. Samuvel DJ, Jayanthi LD, Bhat NR, Ramamoorthy S (2005) A role for p38 mitogen-activated protein kinase in the regulation of the serotonin transporter: evidence for distinct cellular mechanisms involved in transporter surface expression. J Neurosci 25:29–41
106. Schafers M, Dutka D, Rhodes CG, Lammertsma AA, Hermansen F, Schober O, Camici PG (1998) Myocardial presynaptic and postsynaptic autonomic dysfunction in hypertonic cardiomyopathy. Circ Res 82:57–62
107. Segura-Aguilar J, Paris I, Munoz P, Ferrari E, Zecca L, Zucca FA (2014) Protective and toxic roles of dopamine in Parkinson's disease. J Neurochem 129:898–915
108. Sellers EM, Higgins GA, Tompkins DM, Romach MK (1992) Serotonin and alcohol drinking. NIDA Res Monogr 119:141–145
109. Shetty M, Bolland DE, Morrell J, Grove BD, Foster JD, Vaughan RA (2023) Dopamine transporter membrane mobility is bidirectionally regulated by phosphorylation and palmitoylation. Curr Res Physiol 6:100106
110. Shippenberg TS, Herz A (1986) Differential effects of mu and kappa opioid systems on motivational processes. NIDA Res Monogr 75:563–566
111. Sora I, Hall FS, Andrews AM, Itokawa M, Li XF, Wei HB, Wichems C, Lesch KP, Murphy DL, Uhl GR (2001) Molecular mechanisms of cocaine reward: combined dopamine and serotonin transporter knockouts eliminate cocaine place preference. Proc Natl Acad Sci USA 98:5300–5305
112. Southwick SM, Bremner JD, Rasmusson A, Morgan CA 3rd, Arnsten A, Charney DS (1999) Role of norepinephrine in the pathophysiology and treatment of posttraumatic stress disorder. Biol Psychiatry 46:1192–1204
113. Steiner JA, Carneiro AM, Blakely RD (2008) Going with the flow: trafficking-dependent and -independent regulation of serotonin transport. Traffic 9:1393–1402
114. Stewart A, Davis GL, Areal LB, Rabil MJ, Tran V, Mayer FP, Blakely RD (2022a) Male DAT Val559 mice exhibit compulsive behavior under devalued reward conditions accompanied by cellular and pharmacological changes. Cells 11:4059
115. Stewart A, Mayer FP, Gowrishankar R, Davis GL, Areal LB, Gresch PJ, Katamish RM, Peart R, Stilley SE, Spiess K, Rabil MJ, Diljohn FA, Wiggins AE, Vaughan RA, Hahn MK, Blakely RD (2022b) Behaviorally penetrant, anomalous dopamine efflux exposes sex and circuit dependent regulation of dopamine transporters. Mol Psychiatry 27:4869–4880
116. Sundaramurthy S, Annamalai B, Samuvel DJ, Shippenberg TS, Jayanthi LD, Ramamoorthy S (2017) Modulation of serotonin transporter function by kappa-opioid receptor ligands. Neuropharmacology 113:281–292
117. Svingos AL, Chavkin C, Colago EE, Pickel VM (2001) Major coexpression of kappa-opioid receptors and the dopamine transporter in nucleus accumbens axonal profiles. Synapse 42:185–192
118. Tatsumi M, Groshan K, Blakely RD, Richelson E (1997) Pharmacological profile of antidepressants and related compounds at human monoamine transporters. Eur J Pharmacol 340:249–258
119. Thompson AC, Zapata A, Justice JB Jr, Vaughan RA, Sharpe LG, Shippenberg TS (2000) Kappa-opioid receptor activation modifies dopamine uptake in the nucleus accumbens and opposes the effects of cocaine. J Neurosci 20:9333–9340
120. Viggiano D, Vallone D, Sadile A (2004a) Dysfunctions in dopamine systems and ADHD: evidence from animals and modeling. Neural Plast 11:97–114

121. Viggiano D, Ruocco LA, Arcieri S, Sadile AG (2004b) Involvement of norepinephrine in the control of activity and attentive processes in animal models of attention deficit hyperactivity disorder. Neural Plast 11:133–149
122. Volkow N (2004) Drug dependence and addiction, III: expectation and brain function in drug abuse. Am J Psychiatry 161:621
123. Volkow ND, Li TK (2004) Drug addiction: the neurobiology of behaviour gone awry. Nat Rev Neurosci 5:963–970
124. Volkow ND, Wang GJ, Fowler JS (1997) Imaging studies of cocaine in the human brain and studies of the cocaine addict. Ann NY Acad Sci 820:41–55
125. Volkow ND, Wang GJ, Telang F, Fowler JS, Logan J, Childress AR, Jayne M, Ma Y, Wong C (2006) Cocaine cues and dopamine in dorsal striatum: mechanism of craving in cocaine addiction. J Neurosci 26:6583–6588
126. Wendland JR, DeGuzman TB, McMahon F, Rudnick G, Detera-Wadleigh SD, Murphy DL (2008) SERT Ileu425Val in autism, Asperger syndrome and obsessive-compulsive disorder. Psychiatr Genet 18:31–39
127. White KL, Robinson JE, Zhu H, DiBerto JF, Polepally PR, Zjawiony JK, Nichols DE, Malanga CJ, Roth BL (2015) The G protein-biased kappa-opioid receptor agonist RB-64 is analgesic with a unique spectrum of activities in vivo. J Pharmacol Exp Ther 352:98–109
128. Xu F, Gainetdinov RR, Wetsel WC, Jones SR, Bohn LM, Miller GW, Wang YM, Caron MG (2000) Mice lacking the norepinephrine transporter are supersensitive to psychostimulants. Nat Neurosci 3:465–471
129. Zarpellon A, Donella-Deana A, Folda A, Turetta L, Pavanetto M, Deana R (2008) Serotonin (5-HT) transport in human platelets is modulated by Src-catalysed Tyr-phosphorylation of the plasma membrane transporter SERT. Cell Physiol Biochem 21:87–94
130. Zhang YW, Rudnick G (2005a) Serotonin transporter mutations associated with obsessive-compulsive disorder and phosphorylation alter binding affinity for inhibitors. Neuropharmacology 49:791–797
131. Zhang YW, Rudnick G (2005b) Cysteine-scanning mutagenesis of serotonin transporter intracellular loop 2 suggests an alpha-helical conformation. J Biol Chem 280:30807–30813
132. Zhang YW, Turk BE, Rudnick G (2016) Control of serotonin transporter phosphorylation by conformational state. Proc Natl Acad Sci USA 113:E2776–E2783
133. Zhang YW, Gesmonde J, Ramamoorthy S, Rudnick G (2007) Serotonin transporter phosphorylation by cGMP-dependent protein kinase is altered by a mutation associated with obsessive compulsive disorder. J Neurosci 27:10878–10886
134. Zhu CB, Hewlett WA, Feoktistov I, Biaggioni I, Blakely RD (2004) Adenosine receptor, protein kinase G, and p38 mitogen-activated protein kinase-dependent up-regulation of serotonin transporters involves both transporter trafficking and activation. Mol Pharmacol 65:1462–1474

Electrophysiological Methods to Explore the Function of Monoamine Transporters

Danila Boytsov, Michael Freissmuth, and Walter Sandtner

Contents

Abstract In this chapter we will show how electrophysiological recordings were used to gain insights into the transport kinetics and pharmacology of monoamine transporters (MATs). We will discuss data obtained from whole cell patch clamp recordings that allow for real time monitoring of MAT function. A notable property of MATs is that they carry so-called uncoupled currents. We will begin this chapter by reviewing the experimental evidence that has led to the conclusion that the currents carried by MATs are largely uncoupled and, therefore, not directly related to substrate transport. We will discuss how this has made it difficult to understand the operation of MATs. We will also explain why the existence of these currents has led to the proposition that MATs do not operate by alternate access but rather by a single file diffusion mechanism. However, we will show that ultimately the uncoupled currents carried by MATs can be most parsimoniously explained within the framework of the alternate access mechanism. We will review the existing evidence that MATs, like most other transporters, undergo a cycle during which they visit outward and inward-facing conformations (i.e., the transport cycle). We will outline what we have learned about the transport cycle of MATs from electrophysiological recordings. Thereafter, we will describe how electrophysiological recordings can be utilized to understand how drugs that target MATs affect their operation. To this

D. Boytsov · M. Freissmuth · W. Sandtner (✉)
Institute of Pharmacology, Center of Physiology and Pharmacology, Medical University of Vienna, Vienna, Austria
e-mail: walter.sandtner@meduniwien.ac.at

O. V. Mortensen (ed.), *Monoamine Transporters*, Advances in Neurobiology 46,
https://doi.org/10.1007/978-3-031-96364-3_6

end, we will discuss the binding modes of three different MAT ligands: (i) amphetamines, (ii) ibogaine, and (iii) zinc.

Keywords Electrophysiology · Patch clamp recordings · Alternate access mechanism · Transport cycle · Coupled and uncoupled transporter currents · Amphetamines · Ibogaine · Zinc

1 Introduction

The physiological role of the three monoamine transporters (i.e., the serotonin transporter (SERT/SLC6A4), the dopamine transporter (DAT/ SLC6A3), and the norepinephrine transporter (NET/ SLC6A2)) is to clear the extracellular space from previously released monoamines (i.e., serotonin (5-HT), dopamine, and norepinephrine) and—in relay with the vesicular monoamine transporters (vMAT1/ SLC18A1 and vMAT2/SLC18A2)—to replenish the vesicular pool of neurotransmitters. By this action, the MATs shape synaptic responses.

All three MATs utilize the energy stored in the Na^+ concentration gradient to drive the inward transport of monoamines. They operate as symporters, i.e., they translocate Na^+ ions and monoamines in the same direction across the cellular membrane. The influx of Na^+ ions down their electrochemical gradient provides the energy required for the uphill movement of monoamines against their concentration gradient. Substrate transport by monoamine transporters is believed to occur according to the "alternating access" mechanism. In this model, transported solutes bind to a site in the transporter that can be exposed to one or the other side of the cellular membrane, but not to both sides simultaneously [20, 28, 44]. It follows from this model that three distinct transporter conformations exist: (i) the outward-facing, (ii) the occluded, and (iii) the inward-facing conformation. It is gratifying to note that, in the meantime, the structure of SERT was captured in all three of these conformations, thus confirming that the MATs operate according to the alternate access mechanism.

The first insights into the function of monoamine transporters were obtained from measuring the accumulation of radiolabeled substrate in cells that natively expressed these transporters or in vesicles derived from those cells [34]. In the 1990s, the cDNAs encoding numerous transporter genes, including those of the MATs, were cloned. This afforded heterologous expression of MATs in *Xenopus laevis* oocytes and in immortalized mammalian cell lines (e.g., HEK293 cells) under the control of strong promoters. The resulting high protein levels allowed for the functional characterization of MATs with electrophysiological methods as these require a large number of transporters on the surface of the cell ($> 10^6$ copies per 10 pF). Electrophysiological recordings revealed the existence of an electric current mediated by the MATs. This current reports on charge translocation by these transporters and, because it can be recorded with high temporal resolution, it tracks the function of the MATs in real time. In addition, electrophysiological techniques are well suited for studying the binding kinetics of drugs that target the MATs (e.g., exogenous substrates, typical and atypical inhibitors, and allosteric modulators of substrate transport). In the following, we will show how MAT-associated currents can be used to obtain information on their transport function. Moreover, we will present three examples where recordings of currents through MATs were instrumental in unraveling how specific molecules (i.e., para-chloroamphetamine, ibogaine, and Zn^{2+}) act on these transporters.

2 Currents Through MATs

*2.1 The Expression System Matters—*Xenopus laevis *Oocytes Versus HEK293 Cells*

The first recordings of currents mediated by SERT and DAT were obtained from *Xenopus laevis* oocytes measured with the two-electrode voltage clamp (TEV) technique. These measurements revealed the existence of three types of MAT-associated currents: (i) an inwardly directed steady current induced by the substrate, (ii) a leak current in the absence of the substrate, which was small with Na^+ in the bath solution and large when Na^+ was replaced by Li^+, and (iii) a transient current activated by voltage jumps. The latter was present in SERT but not in DAT [24, 42].

Immortalized mammalian cells (e.g., human embryonic kidney 293/HEK293 cells) represent another widely used expression system for the MATs [13, 14, 37]. Electric currents passing through the plasma membrane of these cells can be recorded with the whole-cell patch-clamp technique. Figure 1a shows a 5-HT-induced current obtained from a HEK293 cell stably expressing SERT. Like in the oocyte, the application of 5-HT gave rise to a steady current. However, in the HEK293 cell, 5-HT also induced an initial transient current (hereafter referred to as peak current), which was absent in recordings obtained from *Xenopus laevis* oocytes [12, 24]. This difference can be attributed to a technical limitation: the large size of a *Xenopus laevis* oocyte (i.e., 0.5–1 mm in diameter) makes it difficult to apply the substrate fast enough to prevent the solution exchange rate from becoming rate-limiting. This is not the case for a HEK293 cell, which has a diameter of approximately 10–30 μm. Due to the small size of this cell, the surrounding bath solution can be exchanged within milliseconds. This allows for the detection of fast electrogenic events such as the peak current. Another notable difference between the two expression systems is that the voltage-activated transient current of SERT is present in *Xenopus laevis* oocytes but not in HEK293 cells [11]. Moreover, in HEK293 cells, the leak current is detectable only with Li^+ in the bath solution, but not with Na^+ (see Fig. 1b). The reasons for these differences are currently unknown.

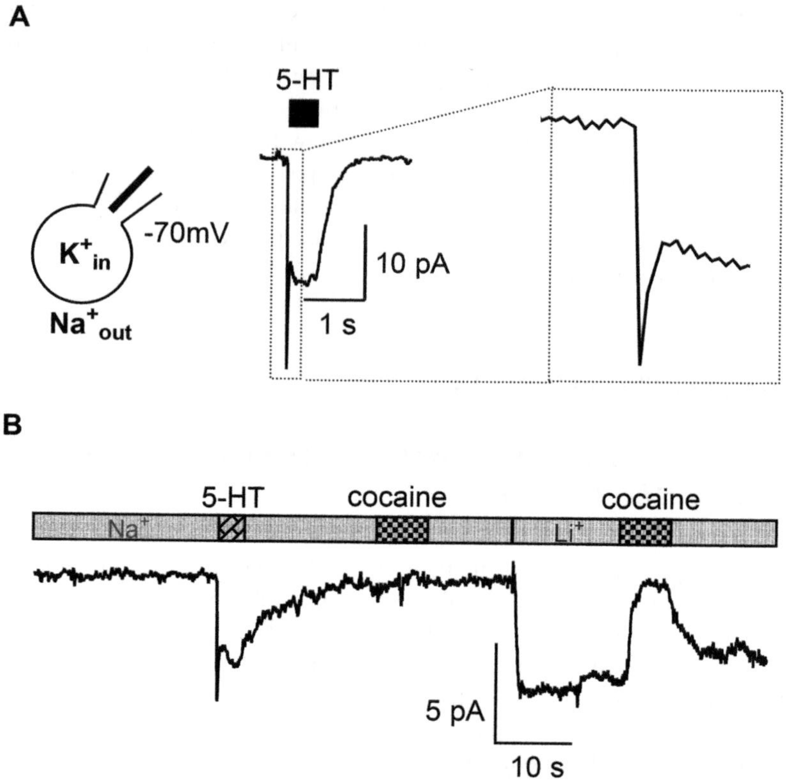

Fig. 1 Currents through SERT recorded in the whole-cell patch-clamp configuration. (**a**) Shows a current induced by 30 μM 5-HT recorded from a HEK293 cell stably expressing SERT with the membrane potential held at −70 mV. The current is comprised of two components: an initial peak (also shown at higher temporal resolution in the inset) followed by a steady current. In the experiment, the concentration of Na^+ on the extracellular side and the concentration of K^+ on the intracellular side were both set to 160 mM. (**b**) Shows a 5-HT induced current like in A. Following the removal of 5-HT from the bath solution, a saturating concentration of cocaine (30 μM) was applied to the cell. Unlike in *Xenopus laevis* oocytes, the application of cocaine in the absence of 5-HT did not result in current block suggesting that a substrate independent leak current does not exist in HEK-293 cells expressing SERT in the presence of 160 mM Na^+_{out}. However, the replacement of Na^+_{out} by Li^+_{out} (160 mM) produced an inwardly directed leak current. This current was carried by SERT because it was possible to reversibly inhibit it with 30 μM cocaine

2.2 *Substrate-Induced Currents Through MATs*

In HEK293 cells, where the substrate can be rapidly applied, all three MATs display a peak current [7]. The peak current reports on the transition that carries the substrate through the membrane. It is a "coupled current" (i.e., produced by the charge of the transported substrate and co-substrates) and, therefore, capacitive in nature. This, however, does not apply to the steady currents carried by SERT and

DAT. In the case of SERT, the following transport stoichiometry was originally proposed: $Na^+\downarrow/Cl^-\downarrow/5\text{-}HT^+\downarrow/K^+\uparrow$ (i.e., one Na^+ and one Cl^- ion are transported into the cell together with 5-HT, and one K^+ ion is counter-transported in the return step from the substrate-free inward to the substrate-free outward-facing conformation [33, 34]). According to this stoichiometry, no net charge movement through SERT ought to occur. Although this stoichiometry has since been questioned there is convincing evidence that the transport stoichiometry of SERT is indeed electroneutral. This raised the question of why SERT displayed a steady current when challenged with 5-HT [2, 24, 42]. The first insights into this enigmatic property of SERT came from experiments in which 5-HT uptake and currents were simultaneously measured in *Xenopus laevis* oocytes. These revealed that changes in transmembrane potential altered the current amplitude but did not affect the rate of 5-HT uptake. More recent studies confirmed the original finding that the amplitude of the substrate-induced steady current of SERT was voltage dependent, while substrate uptake rate was not [7, 17]. This and other observations led to the conclusion that the steady current through SERT was uncoupled (i.e., caused by a process that is not directly linked to substrate transport). Similarly, studies in DAT demonstrated the presence of an uncoupled current component: the uncoupled current was shown to account for roughly half of the current amplitude recorded in response to superfusion with dopamine [41]. Thus, with respect to these uncoupled currents, DAT is phenomenologically similar to SERT. There is also evidence that substrate-induced currents through NET comprise an uncoupled current component [14]; however, the steady current of NET is difficult to study because of its slow turnover [29].

2.3 The Origin of Uncoupled Currents Through the MATs

A model was proposed to account for the uncoupled currents through SERT [24], where the transporter was posited to occasionally adopt a channel-like conformation, when it transited through specific points in the conformational cycle. This notion was supported by the detection of single-channel currents through SERT. Occasional channel formation was explained by assuming that, in rare instances, the inner and outer gates of the transporter remained simultaneously open during transitions between inward- and outward-facing states. This model provided a good starting point for understanding the uncoupled current through SERT. However, the important issue, which was not addressed, was the question of where in the transport cycle the proposed channel-like state was formed.

Our group designed experiments to specifically address this question [37]: we analyzed the transport cycle of SERT by taking advantage of the whole cell patch clamp technique. This approach allows for reliable control of the intracellular ion composition via the patch clamp electrode. The ability to exchange intracellular ions is particularly useful when attempting to interrogate the kinetics of

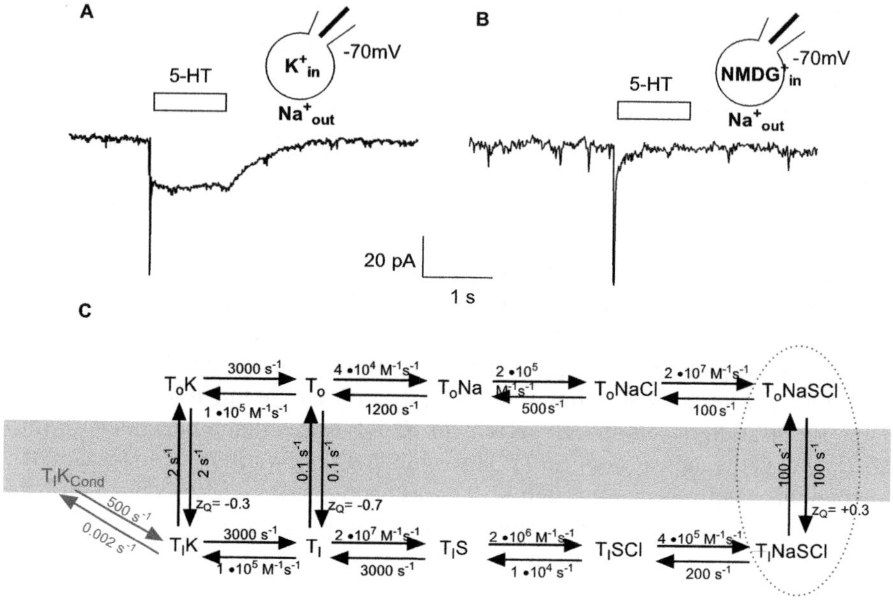

Fig. 2 The conducting state is in equilibrium with the K^+ bound inward-facing conformation of SERT. (**a**) 5-HT induced current in the presence of 160 mM K^+_{in}. (**b**) 5-HT induced current in the absence of 160 mM K^+_{in}. In this experiment K^+_{in} was replaced by the inert cation N-Methyl-D-glucamine ($NMDG^+$). (**c**) The reaction scheme of the transport cycle of SERT. T_o and T_i indicate outward and inward-facing states, respectively. Highlighted in the scheme is the reaction, which gives rise to the peak current (oval in red) and the conducting state (i.e., the channel-like state—in red). The conducting state is in rapid equilibrium with the K^+ bound inward-facing state (T_iK)

conformational transitions, which involve inward-facing states. Adams and DeFelice demonstrated that it was also possible to control the intracellular ion composition in *Xenopus laevis* oocytes by utilizing the cut-open oocyte vaseline gap (COVG) technique [2]. However, anyone who has tried this approach rapidly appreciates that the COVG method is a technical tour de force. We were grateful that Adams and DeFelice had taken on this challenge, prior to our study, because it allowed us to compare our findings with their published results. The most notable finding that we replicated in HEK293 cells was that the steady current of SERT was contingent on the presence of intracellular K^+ (i.e., the steady current was abolished when K^+ was replaced by the inert cation N-methyl-D-glucamine ($NMDG^+$). However, there was one additional piece of information, which is inaccessible when using *Xenopus laevis* oocytes as the expression system but was readily obtained from our measurements in HEK293 cells: our recordings revealed that the peak current was not affected by omission of intracellular K^+ from the electrode solution (see Fig. 2a). This observation narrowed down the possible positions in the transport cycle of SERT at which the channel-like state was formed. The only way to explain our experimental findings was to posit that the channel-like state was in equilibrium with the K^+ bound inward-facing conformation of SERT (see scheme of the kinetic cycle of SERT in Fig. 2b). The latter is an obligatory state in the transport cycle of SERT, which this transporter adopts prior to its return to the substrate-free outward-facing conformation. Hence, we were able to link the uncoupled current to the kinetic cycle of SERT. The established link allows for using the steady current as a surrogate of the transport activity of SERT.

It is worth mentioning that when faced with essentially the same data set, Adams and DeFelice asserted that their results were inconsistent with the alternating access hypothesis [20, 28] and claimed that, instead, the monoamine transporters operate by a single-file diffusion mechanism [1]. According to this mechanism, the currents of the MATs are produced by substrate and Na^+ ions permeating in a single file through a channel within the hydrophobic core of the protein. This single-file model can explain how substrate can be transported uphill against an opposing gradient, but it cannot be reconciled with the structural information on MATs, which has since become available.

For DAT, we recently provided evidence that its uncoupled current component was also linked to the K^+ bound inward-facing conformation [7]. Our observations show that K^+ can also bind to the inward-facing conformation of DAT, but—contrary to proposals by others—we do not find any evidence for antiport of K^+ by DAT: in fact, our experiments indicate that the interaction of K^+ with DAT is transient such that it unbinds from the inward-facing conformation of DAT prior to the return step [39].

3 Pharmacology of the MATs

MATs have a rich pharmacology. This includes typical inhibitors such as cocaine and selective serotonin reuptake inhibitors (SSRIs), which stabilize these transporters in the outward-facing conformation, atypical inhibitors such as ibogaine or ECSI#6 (a byproduct of 3,4-methylendioxy-N-methylamphetamin/MDMA synthesis), which inhibit transport by binding to inward-facing states, exogenous substrates, such as (S)-amphetamine or cathinone, which cause transporter-mediated release of cytosolic monoamines and compounds that bind to allosteric sites on these transporters (i.e., negative and positive modulators of substrate transport such as vilazodone and Zn^{2+}, respectively) [4, 6, 9, 19, 31, 43]. Their action on MAT function can be analyzed using recordings of transporter-associated currents. The high time resolution provided by electrophysiological recordings allows for resolving the binding kinetics of these drugs. Conversely, the effect of these drugs on the currents through MATs provide insights into how these transporters operate. In the following paragraphs, we will show how recordings of MAT-associated currents were instrumental in unraveling the mechanisms of action of exogenous substrates, ibogaine and Zn^{2+}.

3.1 Currents Induced by Exogenous Substrates of the MATs

Phenylethylamines and cathinones are chemical groups of substances that exert psychostimulatory effects by targeting the MATs [40]: prototypical representatives are (S)-amphetamine, methamphetamine, *para*-chloroamphetamine (pCA), 3,4-methylenedioxymethamphetamine (MDMA), and mephedrone: these are all MAT substrates. Their action is accounted for by their ability to increase the concentration of extracellular monoamines by a mechanism, which differs from that of MAT inhibitors: cocaine and other MAT inhibitors block the reuptake of exocytotically released monoamines and thereby raise the extracellular monoamine concentration. Exogenous substrates induce transporter-mediated release of cytosolic monoamines. Molecules which exert this type of action are, therefore, also called "releasers." From a mechanistic perspective, release of the endogenous substrate (i.e., the cognate monoamine) by an exogenous substrate is a puzzle, which has been studied for decades. The most parsimonious explanation for this phenomenon is as follows: the exogenous substrate binds to the outward-facing conformation upon which it is carried through the membrane. It then dissociates into the cytosol, which allows for binding of an intracellular monoamine. In a subsequent step, the monoamine is transported to the extracellular side, where it can dissociate from the transporter. Thus, the transporter switches from the forward cycle mode to the exchange mode, which is accomplished by a half-cycle [16]. We consider this explanation of transporter-mediated monoamine release (i.e., release by exchange) the most parsimonious because it requires the fewest assumptions: in fact, it directly follows from the alternate access mechanism.

The action of releasers on MATs was the subject of many studies some of which entailed recordings of MAT-associated currents. Initially, the recorded currents added confusion, rather than providing meaningful insights. The issue was that the currents elicited by the releasers differed phenomenologically from those evoked by the cognate substrates. Figure 3 shows representative recordings of currents elicited by 5-HT (A) and by pCA (B) obtained from a HEK293 cell expressing SERT: unlike the 5-HT induced current, the current triggered by the application of pCA was transient and reemerged upon removal of pCA from the bath solution. The peculiar profile of releaser-induced currents fueled the idea that these compounds interact with the MATs in a way that is fundamentally different from that of the endogenous substrates. An example is the study by Rodriguez-Menchaca et al. [32], who observed a persistent leak current induced by (S)-amphetamine through DAT expressed in *Xenopus laevis* oocytes. Based on these observations, they proposed the "molecular stent hypothesis" [32]: the hypothesis posited that the persistent leak current arose due to (S)-amphetamine opening a channel through DAT. It further implied that the proposed channel constituted a path via which the monoamine left the cell interior and that this was the reason why (S)-amphetamine and related compounds acted as releasers. However, in a study by our group, we showed that the persistent leak current reported by Rodriguez-Menchaca et al. had been an artifact of the used expression system [35]. In *Xenopus laevis* oocytes, we also observed a

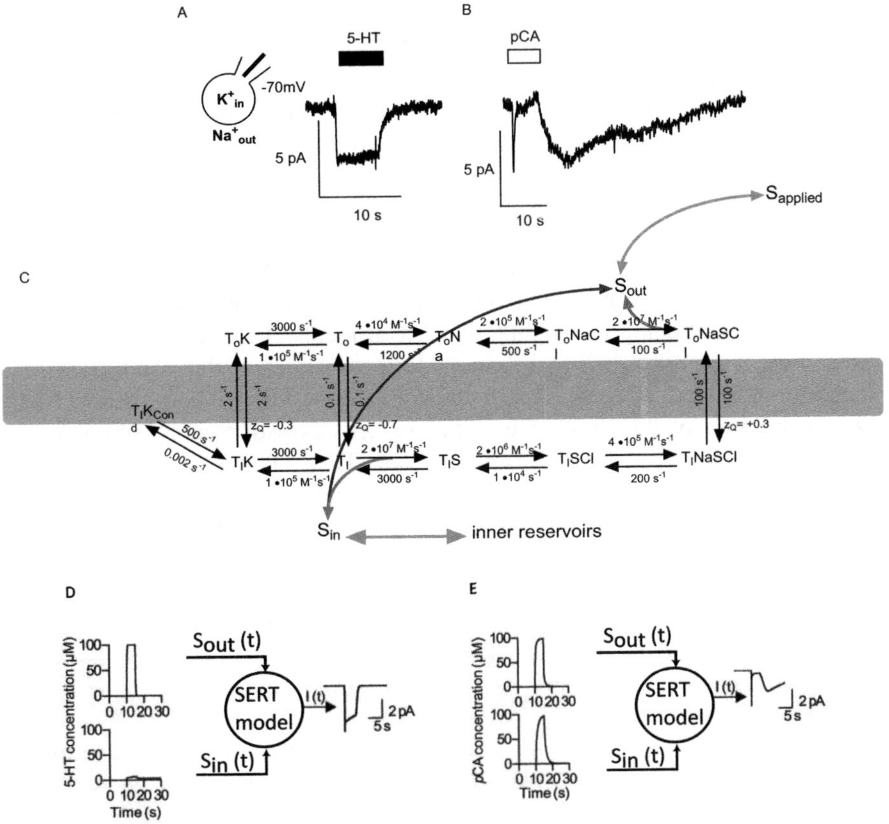

Fig. 3 Amphetamine-induced monoamine release occurs via exchange. (**a**) Current induced by 100 µM 5-HT. (**b**) Current induced by 100 µM para-chloroamphetamine (pCA). (**c**) Kinetic scheme of the transport cycle of SERT. The arrows indicate the flow of the substrate, which can pass through the membrane (blue arrow) or be transported by SERT (orange arrows). While 5-HT and pCA are both transported by SERT, only pCA can diffuse through the membrane at an appreciable rate. (**d**) Simulated concentration of 5-HT as a function of time on the outer ($S_{out}(t)$) and inner ($S_{in}(t)$) side of the membrane upon application of 100 µM 5-HT to an in silico cell. I(t) is the substrate-induced current predicted by the model shown in C. (E) Simulated concentration of pCA as a function of time on the outer ($S_{out}(t)$) and inner ($S_{in}(t)$) side of the cell upon application of 100 µM pCA to an in silico cell. I(t) induced by pCA differs from that induced by 5-HT merely because of the different concentration profiles of pCA on the outer and inner side of the membrane. The predicted currents match the currents obtained from the measurements (cf. **a** and **d**; cf. **b** and **e**)

current through SERT that persisted after the releaser pCA was removed from the bath solution. However, the same vanished within seconds when the transporter was expressed in a HEK293 cell. The molecular stent hypothesis is difficult to reconcile with the failure of the releaser pCA to produce a long-lasting leak current through SERT when expressed in a HEK293 cell. This led us to propose an alternative model, which explained the results obtained in both expression systems. We posited that the releasers are mere substrates of the MATs which, unlike the monoamines, can readily diffuse through cellular membranes. This postulate is in line with the physical-chemical properties of these molecules (i.e., logP-values and polar surface areas) and with their ability to cross the blood-brain barrier, which is evident from their psychostimulatory effects. By taking into account this property, we explained the long-lasting releaser-induced current observed in the *Xenopus laevis* oocyte as follows: because the releaser has essentially unimpeded access to the cell interior, it accumulates rapidly during its application at progressively increasing concentrations within the oocytes. Following removal of the releaser from the bath solution, the sequestered molecules start to diffuse out of the cell. This gives rise to a current because the releaser is a substrate of the MAT. The inner volume of a *Xenopus laevis* oocyte is large and comprises lipophilic compartments. Therefore, the outwardly directed flux of the releaser is long lasting. As a consequence, the resulting current is long lasting as well. The releaser also accumulates rapidly within a HEK293 cell. However, because of the small size of this cell, its cytosol cannot hold as many molecules and it therefore empties at a faster rate. This explains why in this cell the current ceases within seconds upon removal of the releaser from the bath solution. We note that this model can also account for the transient current observed upon application of pCA. Due to the rapid inflow of this compound into the cell, it accumulates to an intracellular concentration at which it more frequently rebinds to the inward-facing conformation. This hampers progression of the transport cycle and leads to the inhibition of the current. The resulting current is transient, because initially the intracellular concentration is low, but as the concentration rises the current is progressively inhibited.

Figure 3c shows the scheme of the kinetic model, which we formulated to test the described idea. There, we assumed that while 5-HT and *p*CA interact with SERT in an identical fashion, only pCA has the ability to diffuse through the membrane at a high rate. This required a plausible model to predict the diffusibility of pCA and related compounds. The details of this model are provided in Sandtner, Schmid et al. [35] and Schmid, Koenig et al. [38]. In Fig. 3d, we show an example for a synthetic (i.e., in silico) experiment, which was generated by the kinetic model: 100 μM 5-HT was applied for 5 s to a virtual HEK293 cell expressing SERT; the model allows for simulating the evolution of intra- and extracellular concentrations over time—$S_{out}(t)$ and $S_{in}(t)$—and the corresponding (simulated) current response I(t). In Fig. 3e, we assumed a 5 s application of 100 μM *p*CA. Here, the respective concentrations over time differ, because $S_{in}(t)$ accumulates more rapidly to higher concentrations. This occurs as a consequence of passive diffusion through the membrane. The time course of $S_{out}(t)$ and $S_{in}(t)$ depicted gives rise to the calculated response $I(t)$. It is evident that the simulations recapitulate the experimental findings depicted in Fig. 3a and b. This close agreement between experimental and synthetic data shows that the peculiar nature of the *p*CA-induced currents does not require any assumption that the interaction of *p*CA with SERT differs fundamentally from that of the cognate substrate (i.e., 5-HT). It also provides a circumstantial argument that the more parsimonious exchange mechanism accounts for the ability of *p*CA and other releasers to promote the efflux of monoamines from the cytosol.

Of note, many researchers in the field still assume that releasers of MATs act by opening a channel. For instance, some disease-causing mutants of DAT were recently shown to display "anomalous release": if cells expressing these mutants are preloaded with dopamine, these mutants (e.g., hDAT A559V) support a high basal substrate efflux, which is inhibited by typical releasers like amphetamine [8]. It is evident that this phenomenon of "anomalous release" can be readily explained within the framework of the alternating access mechanism [36] by positing that the mutations bias the transport cycle for the reverse transport mode. Addition of amphetamines slows the reverse mode and thus inhibits basal release, which is much larger than the release that can be achieved by amphetamine. This does not require a digression from alternating access. Nevertheless, the channel mode has been invoked to explain anomalous substrate release [3], although there is not any plausible structural equivalent and the experimental evidence is otherwise also weak.

3.2 The Effect of Ibogaine on Currents Mediated by SERT

Ibogaine is a naturally occurring psychoactive substance found in the root bark of the African shrub *Tabernanthe iboga*. It is of potential therapeutic interest, because administration of ibogaine reduces withdrawal symptoms and craving associated with opioid addiction [25]. An initial report suggested that ibogaine is a competitive inhibitor of SERT and DAT [45]. However, a more recent study revealed that ibogaine inhibits substrate transport by these transporters in a non-competitive manner ([19]. Prior to this, non-competitive inhibition of transport was observed in two other instances: the adenine nucleotide exchanger of mitochondria and the red cell glucose transporter [15, 18, 22]. In both cases, the inhibitors were found to bind to an inward-facing conformation. In agreement with these findings, ibogaine was also shown to stabilize SERT in an inward-facing state. In fact, this property of ibogaine was recently exploited to obtain a structure of this transporter captured in the inward-facing conformation. Ibogaine and 5-HT both bind to the central binding site. Originally, Jacobs et al. proposed that ibogaine is a pseudo-irreversible non-competitive inhibitor in that it forms a long-lived complex with the inward-open conformation to which 5-HT cannot bind. The formation of a long-lived complex is the most frequent cause of non-competitive inhibition in instances where the agonist (the substrate) and the antagonist (the inhibitor) share the same binding site [19].

The intriguing properties of ibogaine prompted us to examine the effects of this drug on currents mediated by SERT [10]. We found that ibogaine also inhibited the substrate-induced steady current in a non-competitive manner. However, upon removal of ibogaine from the bath solution, the current block reversed within about 1 second (see Fig. 4a). This refuted the idea that this drug stabilizes a long-lived complex and raised the question of how ibogaine can inhibit substrate transport non-competitively while binding to the same site as 5-HT. As outlined above, competitive inhibition can appear non-competitive, if a long-lived drug-protein complex is formed. A non-competitive mode of inhibition is, however, more difficult to

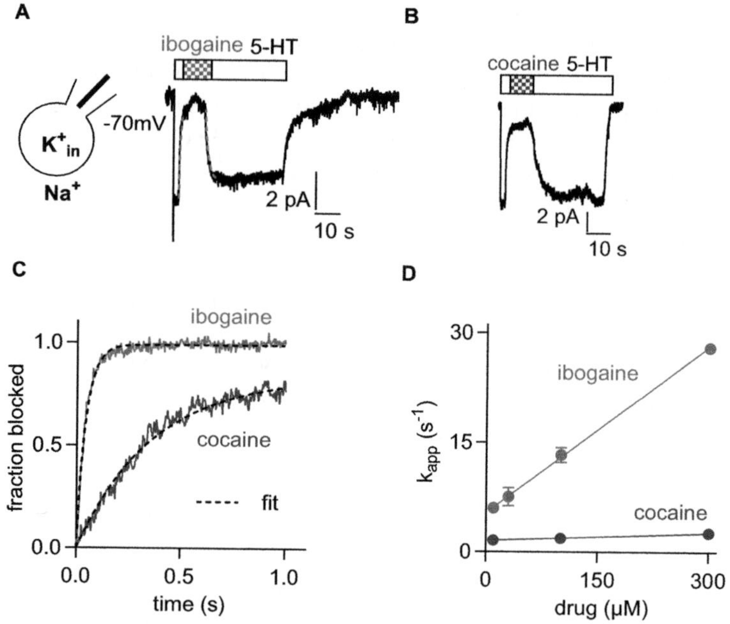

Fig. 4 The inhibitory action of ibogaine on SERT. (**a**) Representative trace of a current through SERT elicited by 10 µM 5-HT. After a steady current level was reached, 10 µM of ibogaine was coapplied and subsequently removed. The red and green line indicate fits of a monoexponential function to the association (red line) and dissociation phase (green line) of ibogaine, respectively. (**b**) The same as in A, but 10 µM cocaine was coapplied and subsequently removed instead. (**c**) Comparison of the association kinetics of 10 µM ibogaine and 10 µM cocaine. (**D**) Plotted are the k_{app} values of ibogaine and cocaine as a function of the applied concentration

explain, if the complex rapidly falls apart. To address this problem, we analyzed the association kinetics of ibogaine by measuring the rate, at which it inhibits the substrate-induced current through SERT (see Fig. 4a and c). We found that—consistent with a bimolecular reaction—the rate of current blockage increased in a linear manner, if the concentration of ibogaine was raised (Fig. 4d—red line). In contrast, the blockage rate (k_{app}) of the typical inhibitor cocaine only increased at concentrations below 10 µM but leveled out at higher concentrations (Fig. 4b, c and d—blue line). This suggested that the association kinetics of cocaine encompasses a conformational rearrangement of SERT, which becomes rate limiting at higher concentrations of cocaine. This finding was consistent with the assigned position of the channel-like state in the transport cycle. As explained in detail above, this state, which gives rise to the steady current, is in equilibrium with the K^+-bound inward-facing state of SERT. Because cocaine cannot bind to inward-facing states, it must wait until the transporter returns to the outward-facing state. It is, thus, the rate of the conversion from the substrate-free inward to the substrate-free outward-facing conformation, which limits the association kinetics of cocaine at high concentrations. Conversely, because this reaction does not become rate-limiting in the case of ibogaine, it is safe to conclude that this drug can directly bind to the inward-facing state of SERT. The combined results from our study allowed us to propose a kinetic model that explained all observed actions of this drug. According to this model, ibogaine inhibition is insurmountable because this drug binds to a state in the transport cycle (i.e., the K^+ bound inward-facing conformation), from which the substrate has already dissociated. For this reason, the substrate does not compete with ibogaine for binding to SERT. It is worth noting that the protocol shown in Fig. 4a and b can be used to determine to which conformation other atypical inhibitors bind. For instance, we recently confirmed by this approach that ECSI#6, a byproduct of the MDMA synthesis, also binds to the inward-facing conformation of SERT [5]. In addition, the above analysis lines out a way how to measure the conversion rate from the substrate-free inward to the substrate-free outward-facing conformation of the MATs. This rate can be determined by measuring the time it takes for a typical inhibitor (i.e., an inhibitor, which is selective for the outward-facing state) to block the steady current. This makes typical inhibitors useful tools for probing the kinetic cycle of MATs.

3.3 The Effect of Zinc on Currents Mediated by DAT

DAT harbors a Zn^{2+}-binding site, which is absent in the closely related transporters NET and SERT. Originally, Zn^{2+} was found to inhibit dopamine uptake through DAT [27]. However, there were other reports, which showed stimulation of dopamine uptake by Zn^{2+} or failed to detect an inhibition [26, 30]. These conflicting results prompted us to explore the action of Zn^{2+} by relying on recordings of currents carried by DAT. For the pertinent analysis, we chose HEK293 cells as the expression system [23].

Figure 5a shows an overlay of currents induced by a saturating concentration of dopamine (30 μM) recorded from a HEK293 cell stably expressing DAT in the presence and absence of 10 μM Zn^{2+}. Zn^{2+} augmented the amplitude of the steady current component, an effect, which had also been observed in a prior study, where currents through DAT were recorded in *Xenopus laevis* oocytes [26]. However, the recordings obtained from the HEK293 cell uncovered two additional effects. The first was that the dopamine-induced peak current was lower in amplitude when Zn^{2+} was present in the bath solution (see inset in Fig. 5a). The other effect of Zn^{2+} was that it accelerated current deactivation upon the washout of dopamine. The increase in the steady current amplitude and the accelerated decline of the current amplitude upon dopamine removal indicated a stimulatory effect of Zn^{2+} on dopamine uptake. This is because a larger substrate turnover rate is expected to (i) increase the number of transported ions per second (thereby augmenting the current amplitude) and (ii) to accelerate current deactivation by shortening the time required to complete a full cycle.

It is possible with electrophysiological methods to determine the substrate turnover rate by employing a two-pulse protocol (see Fig. 5b and c). In this protocol,

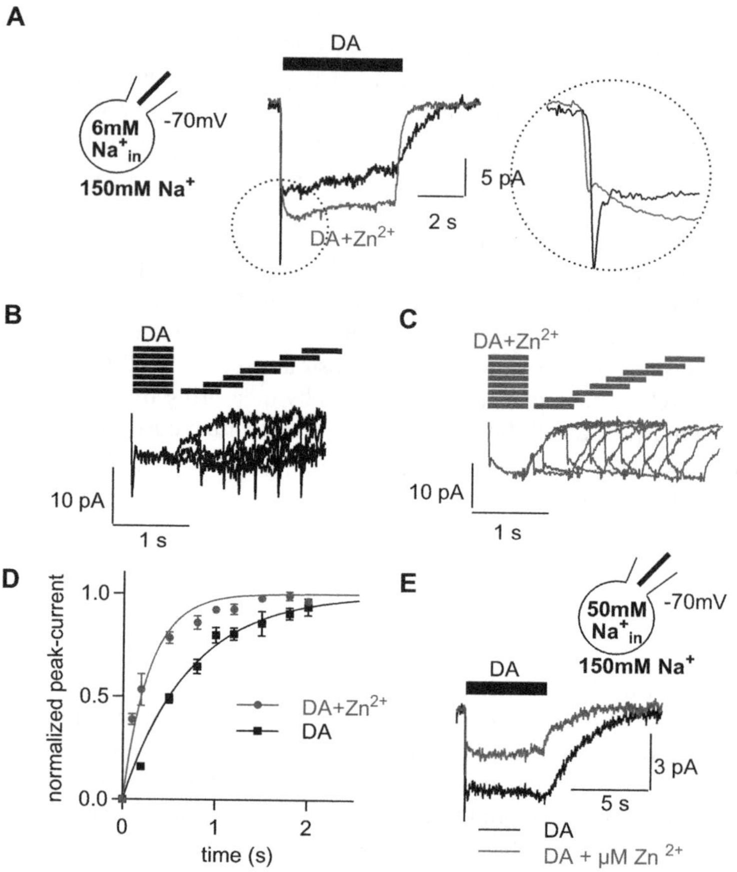

Fig. 5 Zinc is a positive allosteric modulator of DAT. (**a**) Overlay of currents induced by 30 μM dopamine in the presence and absence of 10 μM Zn^{2+}. In this experiment, the intracellular Na^+ concentration was low (i.e., 6 mM). The inset shows the peak current at a higher temporal resolution. (**b**) Two-pulse protocol conducted in the absence of Zn^{2+}. (**c**) Two-pulse protocol conducted in the presence of 10 μM Zn^{2+}. (**d**) Plotted are the normalized peak currents as a function of the time interval between the first pulse and the second pulse. In the presence of Zn^{2+}, the peak current recovery rate is larger. (**e**) 10 μM Zn^{2+} inhibits the dopamine-induced current if the intracellular Na^+ concentration is high (i.e., 50 mM)

application of a saturating dopamine concentration (i.e., the first pulse) forces all transporters into entering the transport cycle. The amplitude of the peak current elicited by this first pulse is proportional to the number of transporters residing in the resting state, which in the presence of physiological ion gradients is the Na^+ and Cl^- bound outward-facing state. A second pulse applied thereafter measures the fraction of transporters, which returned to the resting state and, thus, completed a full cycle in the time interval between the first pulse and the second pulse. This fraction is responsive to the newly applied dopamine and it can therefore be gauged from the amplitude of the peak current elicited by the second pulse. If the time interval between the two pulses is sufficiently long, all transporter will eventually complete the cycle and the peak current will recover to its initial amplitude. For this reason, the rate of peak current recovery is a surrogate of the turnover rate of the transporter.

By using this protocol we confirmed that under our experimental conditions, Zn^{2+} enhanced the transport capacity of DAT (see Fig. 5d).

To accelerate dopamine turnover, it is necessary to increase the rate of one or more partial reactions in the kinetic cycle of this transporter. However, a substantial increase of the substrate uptake rate can only be achieved by augmenting the velocity of a reaction, which is normally slow and rate-limiting. In MATs, the slowest reaction is the return of the substrate-free inward to the substrate-free outward-facing conformation. This is the reaction, which was targeted by Zn^{2+}. It is possible to increase the rate of a reaction by stabilizing its product state [9]. When DAT operates in the forward mode, the product state of the return step is the substrate-free outward-facing conformation. Prior studies showed that Zn^{2+} stabilized the outward-facing state of DAT. This property of Zn^{2+}, therefore, explains the stimulatory effect. However, while our observations were in line with the results of a study that explored the action of Zn^{2+} on DAT expressed in SK-N-MC cells, it was at odds with most other studies that used HEK293 cells as the expression system. In all of them, Zn^{2+} was found to inhibit dopamine uptake.

The first insight into the reason for this discrepancy came from the observation that Zn^{2+} decreased the peak current amplitude. The most parsimonious explanation for this effect was that Zn^{2+} forced a sizeable fraction of transporters, which converted from the outward to the inward-facing conformation with dopamine, to return immediately. The returning fraction of transporters carries the charge that normally gives rise to the inwardly directed peak current in the opposite direction. As a consequence, the net charge and the amplitude of the peak current are reduced. It is true for a reversible reaction going forward that more frequent return events will lower its overall rate. For this reason, the observed loss in the peak current amplitude also implies that Zn^{2+}slows down the reaction that carries dopamine through the membrane. This notion is in line with the reported ability of Zn^{2+} to stabilize the outward-facing conformation of DAT. It is possible to reduce the rate of a reaction by stabilizing its reactant state. When DAT operates in the forward mode, the reactant state of the aforementioned transition is the dopamine-bound outward-facing conformation.

Hence, because Zn^{2+} stabilizes both the dopamine-free and the dopamine-bound outward-facing state, it affects two partial reactions in the transport cycle of DAT: it increases the rate of the substrate-free return step and it decreases the rate of the reaction, which carries dopamine through the membrane. The former and the latter action of Zn^{2+} stimulate and inhibit dopamine uptake, respectively. It is obvious that the chosen condition in the experiment displayed in Fig. 5a, b, and c was such that it favored the stimulatory effect of Zn^{2+}. However, the inhibition of DAT, which was observed in uptake experiments with radioactive substrate, suggested that the experimental conditions favored the inhibitory effect of Zn^{2+}. The reaction that carries dopamine through the membrane is about two orders of magnitude faster than the return step of the substrate-free transporter. Accordingly, a modest decrease of its rate has little bearing on the substrate turnover rate. It is, however, possible to reduce the velocity of this reaction by increasing the intracellular concentration of Na^+. We, therefore, predicted that elevating the intracellular Na^+ concentration converts the stimulatory effect of Zn^{2+} into an inhibitory action. This was indeed the case. Figure 5e shows the overlay of dopamine-induced currents in the presence and absence of 10 μM Zn^{2+}. If the intracellular Na^+ concentration was raised from 6 mM to 50 mM, Zn^{2+} reduced the steady current amplitude in agreement with its acting as an inhibitor of dopamine uptake.

Our observations provided a plausible explanation for the contradicting reports on the Zn^{2+} action on DAT. In those studies, where Zn^{2+} exhibited an inhibitory activity, Na^+_{in} was high: Na^+ accumulation is an expected consequence of overexpression of DAT in non-excitable cells such as HEK293 cells, because in contrast to neuronal cell lines such as the SK-N-MC cells, they lack the capacity to rapidly clear the influx of Na^+. This view is supported by measurements of Na^+_{in}, which showed that upon application of dopamine to DAT-expressing HEK293 cells, Na^+_{in} accumulated up to about 50 mM within a time period of 60 s [21]. This is the concentration where we observed inhibition of the transport cycle by Zn^{2+}.

The example of DAT and Zn^{2+} shows how complex the interaction between a MAT and a given ligand can be. Arguably, it would have been difficult to fully understand how this transition metal affects the function of DAT, if it had not been explored using the patch clamp method. Both the ability to control the intracellular ion composition and the fast solution exchange rate turned out to be instrumental for unraveling the allosteric mechanism by which Zn^{2+} modulates dopamine uptake through DAT.

4 Conclusions

Recordings of currents through MATs have been employed to obtain insights into the operation of these transporters. The high temporal resolution provided by the electrophysiological measurement allows for monitoring MAT function in real-time. This is specifically true if the currents through the MATs are recorded from mammalian cells such as HEK293 cells by utilizing the whole-cell-patch-clamp technique. Due to the small size of these cells, it is possible to exchange the surrounding bath solution within milliseconds. This allows for interrogating the kinetics of individual steps in the transport cycle by recording current activation (when the substrate is applied) and current deactivation (when the substrate is removed). The kinetic information that can be extracted from the recorded currents is precise enough to justify its incorporation into a kinetic model. Such a model is useful because it can help to find the mechanism which can best explain the available experimental results. The examples shown here illustrate how the ability to rapidly apply and remove substrates or other ligands (e.g., ibogaine and Zn^{2+}) has been instrumental for the understanding of various aspects of MAT function.

The high temporal resolution that can be achieved by using the patch clamp technique is not the only advantage of this method. Other notable advantages are the ability to control (i) membrane voltage and (ii) the cytosolic ion composition (via the electrode). Both transmembrane potential and inner ion composition have a major impact on substrate translocation by MATs. Lack of control over these parameters makes it, therefore, challenging to interpret experimental observations and is arguably the most frequent cause for erroneous conclusions on functional aspects of MATs.

MATs are anomalous solute carriers, because their substrate-induced current comprises an uncoupled current component. Initially, this property of MATs made it difficult to extract from the recorded currents information on their transport activity. The then-elusive relation between currents and substrate transport not only hampered progress in understanding MAT function but also led to the idea that MATs do not operate by the alternating access mechanism. While it is now widely accepted that they do, alternative mechanisms are still—and, we argue—needlessly invoked to explain certain aspects of MAT function.

It has been established that the uncoupled currents through MATs are produced by a channel-like state. This state is occasionally formed when the substrate is present and it is in equilibrium with the K^+ bound inward-facing conformation of MATs. The inward-facing conformation is an obligatory state in the transport cycle, which the MATs can only adopt in physiological ion gradients (i.e., high external Na^+) after the substrate has forced them into entering the cycle. This explains why the uncoupled current is contingent on the presence of the substrate. More importantly, the established link allows to use substrate-induced currents as a proxy for the transport activity of the MATs.

Recordings of transporter-associated currents are well-suited for addressing open questions concerning the operation of MATs. An example of a question worth tackling is the transport stoichiometry. At this point in time, no consensus has been reached on whether the MATs transport one or two Na^+ ions along with the substrate, co-transport Cl^-, or uniformly antiport K^+. A promising approach to address these questions involves the use of site-directed mutagenesis, informed by the available structural information, and the analysis of these mutants by electrophysiological recordings.

References

1. Adams SV, DeFelice LJ (2002) Flux coupling in the human serotonin transporter. Biophys J 83(6):3268–3282. https://doi.org/10.1016/S0006-3495(02)75328-9
2. Adams SV, DeFelice LJ (2003) Ionic currents in the human serotonin transporter reveal inconsistencies in the alternating access hypothesis. Biophys J 85(3):1548–1559. https://doi.org/10.1016/s0006-3495(03)74587-1
3. Aguilar JI, Cheng MH, Font J, Schwartz AC, Ledwitch K, Duran A, Mabry SJ, Belovich AN, Zhu Y, Carter AM, Shi L, Kurian MA, Fenollar-Ferrer C, Meiler J, Ryan RM, McHaourab HS, Bahar I, Matthies HJ, Galli A (2021) Psychomotor impairments and therapeutic implica-

tions revealed by a mutation associated with infantile parkinsonism-dystonia. elife 10:e68039. https://doi.org/10.7554/eLife.68039
4. Beuming T, Kniazeff J, Bergmann ML, Shi L, Gracia L, Raniszewska K, Newman AH, Javitch JA, Weinstein H, Gether U, Loland CJ (2008) The binding sites for cocaine and dopamine in the dopamine transporter overlap. Nat Neurosci 11(7):780–789. https://doi.org/10.1038/nn.2146
5. Bhat S, El-Kasaby A, Kasture A, Boytsov D, Reichelt JB, Hummel T, Sucic S, Pifl C, Freissmuth M, Sandtner W (2022) A mechanism of uncompetitive inhibition of the serotonin transporter. bioRxiv:2022.2008.2011.503588. https://doi.org/10.1101/2022.08.11.503588
6. Bhat S, El-Kasaby A, Kasture A, Boytsov D, Reichelt JB, Hummel T, Sucic S, Pifl C, Freissmuth M, Sandtner W (2023) A mechanism of uncompetitive inhibition of the serotonin transporter. elife 12:e82641. https://doi.org/10.7554/eLife.82641
7. Bhat S, Niello M, Schicker K, Pifl C, Sitte HH, Freissmuth M, Sandtner W (2021) Handling of intracellular K(+) determines voltage dependence of plasmalemmal monoamine transporter function. elife 10:e67996. https://doi.org/10.7554/eLife.67996
8. Bowton E, Saunders C, Erreger K, Sakrikar D, Matthies HJ, Sen N, Jessen T, Colbran RJ, Caron MG, Javitch JA, Blakely RD, Galli A (2010) Dysregulation of dopamine transporters via dopamine D2 autoreceptors triggers anomalous dopamine efflux associated with attention-deficit hyperactivity disorder. J Neurosci 30(17):6048–6057. https://doi.org/10.1523/JNEUROSCI.5094-09.2010
9. Boytsov D, Schicker K, Hellsberg E, Freissmuth M, Sandtner W (2023) Allosteric modulators of solute carrier function: a theoretical framework. Front Physiol 14:1166450. https://doi.org/10.3389/fphys.2023.1166450
10. Bulling S, Schicker K, Zhang YW, Steinkellner T, Stockner T, Gruber CW, Boehm S, Freissmuth M, Rudnick G, Sitte HH, Sandtner W (2012) The mechanistic basis for noncompetitive ibogaine inhibition of serotonin and dopamine transporters. J Biol Chem 287(22):18524–18534. https://doi.org/10.1074/jbc.M112.343681
11. Burtscher V, Hotka M, Li Y, Freissmuth M, Sandtner W (2018) A label-free approach to detect ligand binding to cell surface proteins in real time. elife 7:e34944. https://doi.org/10.7554/eLife.34944
12. Cao Y, Mager S, Lester HA (1997) H+ permeation and pH regulation at a mammalian serotonin transporter. J Neurosci 17(7):2257–2266. https://doi.org/10.1523/JNEUROSCI.17-07-02257.1997
13. Erreger K, Grewer C, Javitch JA, Galli A (2008) Currents in response to rapid concentration jumps of amphetamine uncover novel aspects of human dopamine transporter function. J Neurosci 28(4):976–989. https://doi.org/10.1523/JNEUROSCI.2796-07.2008
14. Galli A, Blakely RD, DeFelice LJ (1996) Norepinephrine transporters have channel modes of conduction. Proc Natl Acad Sci USA 93(16):8671–8676. https://doi.org/10.1073/pnas.93.16.8671
15. Gorga FR, Lienhard GE (1981) Equilibria and kinetics of ligand binding to the human erythrocyte glucose transporter. Evidence for an alternating conformation model for transport. Biochemistry 20(18):5108–5113. https://doi.org/10.1021/bi00521a003
16. Hasenhuetl PS, Bhat S, Mayer FP, Sitte HH, Freissmuth M, Sandtner W (2018) A kinetic account for amphetamine-induced monoamine release. J Gen Physiol 150(3):431–451. https://doi.org/10.1085/jgp.201711915
17. Hasenhuetl PS, Freissmuth M, Sandtner W (2016) Electrogenic binding of intracellular cations defines a kinetic decision point in the transport cycle of the human serotonin transporter. J Biol Chem 291(50):25864–25876. https://doi.org/10.1074/jbc.M116.753319
18. Henderson PJ, Lardy HA, Dorschner E (1970) Factors affecting the inhibition of adenine nucleotide translocase by bongkrekic acid. Biochemistry 9(17):3453–3457. https://doi.org/10.1021/bi00819a026
19. Jacobs MT, Zhang YW, Campbell SD, Rudnick G (2007) Ibogaine, a noncompetitive inhibitor of serotonin transport, acts by stabilizing the cytoplasm-facing state of the transporter. J Biol Chem 282(40):29441–29447. https://doi.org/10.1074/jbc.M704456200

20. Jardetzky O (1966) Simple allosteric model for membrane pumps. Nature 211(5052):969–970. https://doi.org/10.1038/211969a0
21. Khoshbouei H, Wang H, Lechleiter JD, Javitch JA, Galli A (2003) Amphetamine-induced dopamine efflux. A voltage-sensitive and intracellular Na+-dependent mechanism. J Biol Chem 278(14):12070–12077. https://doi.org/10.1074/jbc.M212815200
22. Klingenberg M (2008) The ADP and ATP transport in mitochondria and its carrier. Biochim Biophys Acta 1778(10):1978–2021. https://doi.org/10.1016/j.bbamem.2008.04.011
23. Li Y, Hasenhuetl PS, Schicker K, Sitte HH, Freissmuth M, Sandtner W (2015) Dual action of Zn2+ on the transport cycle of the dopamine transporter. J Biol Chem 290(52):31069–31076. https://doi.org/10.1074/jbc.M115.688275
24. Mager S, Min C, Henry DJ, Chavkin C, Hoffman BJ, Davidson N, Lester HA (1994) Conducting states of a mammalian serotonin transporter. Neuron 12(4):845–859. https://doi.org/10.1016/0896-6273(94)90337-9
25. Mash DC, Duque L, Page B, Allen-Ferdinand K (2018) Ibogaine detoxification transitions opioid and cocaine abusers between dependence and abstinence: clinical observations and treatment outcomes. Front Pharmacol 9:529. https://doi.org/10.3389/fphar.2018.00529
26. Meinild AK, Sitte HH, Gether U (2004) Zinc potentiates an uncoupled anion conductance associated with the dopamine transporter. J Biol Chem 279(48):49671–49679. https://doi.org/10.1074/jbc.M407660200
27. Norregaard L, Frederiksen D, Nielsen EO, Gether U (1998) Delineation of an endogenous zinc-binding site in the human dopamine transporter. EMBO J 17(15):4266–4273. https://doi.org/10.1093/emboj/17.15.4266
28. Patlak CS (1957) Contributions to the theory of active transport: II. The gate type non-carrier mechanism and generalizations concerning tracer flow, efficiency, and measurement of energy expenditure. Bull Math Biophys 19(3):209–235. https://doi.org/10.1007/BF02477764
29. Pifl C, Hornykiewicz O, Giros B, Caron MG (1996) Catecholamine transporters and 1-methyl-4-phenyl-1,2,3,6-tetrahydropyridine neurotoxicity: studies comparing the cloned human noradrenaline and human dopamine transporter. J Pharmacol Exp Ther 277(3):1437–1443
30. Pifl C, Wolf A, Rebernik P, Reither H, Berger ML (2009) Zinc regulates the dopamine transporter in a membrane potential and chloride dependent manner. Neuropharmacology 56(2):531–540. https://doi.org/10.1016/j.neuropharm.2008.10.009
31. Plenge P, Yang D, Salomon K, Laursen L, Kalenderoglou IE, Newman AH, Gouaux E, Coleman JA, Loland CJ (2021) The antidepressant drug vilazodone is an allosteric inhibitor of the serotonin transporter. Nat Commun 12(1):5063. https://doi.org/10.1038/s41467-021-25363-3
32. Rodriguez-Menchaca AA, Solis E Jr, Cameron K, De Felice LJ (2012) S(+)amphetamine induces a persistent leak in the human dopamine transporter: molecular stent hypothesis. Br J Pharmacol 165(8):2749–2757. https://doi.org/10.1111/j.1476-5381.2011.01728.x
33. Rudnick G (1997) In: Reith MEA (ed) Mechanisms of biogenic amine neurotransmitter transporters. Neurotransmitter transporters: structure, function, and regulation. Humana Press, Totowa, pp 73–100. https://doi.org/10.1007/978-1-59259-470-2_3
34. Rudnick G, Nelson PJ (1978) Platelet 5-hydroxytryptamine transport, an electroneutral mechanism coupled to potassium. Biochemistry 17(22):4739–4742. https://doi.org/10.1021/bi00615a021
35. Sandtner W, Schmid D, Schicker K, Gerstbrein K, Koenig X, Mayer FP, Boehm S, Freissmuth M, Sitte HH (2014) A quantitative model of amphetamine action on the 5-HT transporter. Br J Pharmacol 171(4):1007–1018. https://doi.org/10.1111/bph.12520
36. Schicker K, Bhat S, Farr C, Burtscher V, Horner A, Freissmuth M, Sandtner W (2021) Descriptors of secondary active transporter function and how they relate to partial reactions in the transport cycle. Membranes (Basel) 11(3):178. https://doi.org/10.3390/membranes11030178
37. Schicker K, Uzelac Z, Gesmonde J, Bulling S, Stockner T, Freissmuth M, Boehm S, Rudnick G, Sitte HH, Sandtner W (2012) Unifying concept of serotonin transporter-associated currents. J Biol Chem 287(1):438–445. https://doi.org/10.1074/jbc.M111.304261

38. Schmid D, Koenig X, Bulusu S, Schicker K, Freissmuth M, Sitte HH, Sandtner W (2015) The conservative view: is it necessary to implant a stent into the dopamine transporter? Br J Pharmacol 172(19):4775–4778. https://doi.org/10.1111/bph.12766
39. Schmidt SG, Malle MG, Nielsen AK, Bohr SS, Pugh CF, Nielsen JC, Poulsen IH, Rand KD, Hatzakis NS, Loland CJ (2022) The dopamine transporter antiports potassium to increase the uptake of dopamine. Nat Commun 13(1):2446. https://doi.org/10.1038/s41467-022-30154-5
40. Sitte HH, Freissmuth M (2015) Amphetamines, new psychoactive drugs and the monoamine transporter cycle. Trends Pharmacol Sci 36(1):41–50. https://doi.org/10.1016/j.tips.2014.11.006
41. Sonders MS, Amara SG (1996) Channels in transporters. Curr Opin Neurobiol 6(3):294–302. https://doi.org/10.1016/s0959-4388(96)80111-5
42. Sonders MS, Zhu SJ, Zahniser NR, Kavanaugh MP, Amara SG (1997) Multiple ionic conductances of the human dopamine transporter: the actions of dopamine and psychostimulants. J Neurosci 17(3):960–974. https://doi.org/10.1523/jneurosci.17-03-00960.1997
43. Tavoulari S, Forrest LR, Rudnick G (2009) Fluoxetine (Prozac) binding to serotonin transporter is modulated by chloride and conformational changes. J Neurosci 29(30):9635–9643. https://doi.org/10.1523/JNEUROSCI.0440-09.2009
44. Vidaver GA (1966) Inhibition of parallel flux and augmentation of counter flux shown by transport models not involving a mobile carrier. J Theor Biol 10(2):301–306. https://doi.org/10.1016/0022-5193(66)90128-7
45. Wells GB, Lopez MC, Tanaka JC (1999) The effects of ibogaine on dopamine and serotonin transport in rat brain synaptosomes. Brain Res Bull 48(6):641–647. https://doi.org/10.1016/s0361-9230(99)00053-2

Monoamine Transporters in Drugs of Abuse: Insights from Fast Scan Cyclic Voltammetry

Shanna B. Samels, Pelin Yuksel, and Rodrigo A. España

Contents

Abstract Monoamine transmission is critical for regulating numerous physiological processes, including stress, learning, motor activity, and reward. Over the past few decades, the adoption of fast scan cyclic voltammetry has unveiled an intricate interplay between monoamine release and uptake dynamics, particularly concerning monoamine transporter involvement in reward and reinforcement processes for drugs of abuse. This review discusses how fast scan cyclic voltammetry has revolutionized our understanding of the processes that govern monoamine release and uptake, emphasizing the heterogeneity in transporter function across terminal regions, the influence of autoreceptors on monoamine transmission, and the complex interactions between drugs of abuse and monoamine transporters. While much of the review focuses on what is known about dopamine transporters—due to the wealth of evidence on dopamine transmission—we also emphasize significant gaps in knowledge regarding the serotonin and norepinephrine transporters. Finally, we highlight remaining questions about the dynamic nature of monoaminergic

S. B. Samels · P. Yuksel · R. A. España (✉)
Department of Neurobiology and Anatomy, Drexel University College of Medicine, Philadelphia, PA, USA
e-mail: rae39@drexel.edu

O. V. Mortensen (ed.), *Monoamine Transporters*, Advances in Neurobiology 46,
https://doi.org/10.1007/978-3-031-96364-3_7

transporter efficiency and suggest new areas of investigation to gain a more comprehensive understanding of the biochemical mechanisms through which monoamine transporters regulate behavior.

Keywords Monoamine transporters · Fast scan cyclic voltammetry · Monoamine uptake · Monoamine release · Heterogeneity · Substance abuse · Dopamine · Norepinephrine · Serotonin

1 Introduction

Monoamine transmitters, including dopamine, norepinephrine, and serotonin, play a pivotal role in the regulation of reward/reinforcement, stress, learning, cognition, sleep/wake activity, locomotion, and a variety of other processes. Much of what we have learned about the importance of monoaminergic regulation of behavior comes from hundreds of electrophysiological and microdialysis studies that highlight how fluctuations in monoamine activity have vast influences on behavior [38, 94, 97, 124, 126, 175, 183, 211, 220, 231]. Over the past few decades, the adoption of fast scan cyclic voltammetry (FSCV) has revealed an important interplay between monoamine release and uptake and how this relationship affects the temporal availability of monoamines in the extracellular space—particularly with respect to the regulation of terminal monoamine release and clearance rates [17, 51, 71, 93, 133, 180, 235]. Indeed, recent evidence suggests a potential disconnect between the activity of monoamine neurons and the temporal availability of monoamines in the extracellular space, which indicates the importance of local regulation of release and uptake mechanisms in terminal regions [139].

In this review, we discuss how FSCV has revolutionized our understanding of monoamine release and uptake dynamics, including what we have learned about heterogeneity in transporter function across terminal regions, the importance of autoreceptor control of release and uptake, and how drugs of abuse interact with monoamine transporters. While we focus discussion on what is known about the dopamine transporter (DAT), we also emphasize what remains unknown about the serotonin transporter (SERT) and norepinephrine transporter (NET). Finally, we discuss notable gaps in knowledge regarding processes underlying control over monoaminergic transporter efficiency, followed by suggestions for new areas of investigation that pair FSCV with biochemical approaches to examine potential mechanisms underlying changes in transporter function and the potential impact on behavior.

2 Overview of Monoamine Transmission

Monoamine transmitters include dopamine, norepinephrine, and serotonin which act on G-protein coupled receptors and exert modulatory effects on neural activity through recruitment of signaling pathways both presynaptically and postsynaptically. Although epinephrine and histamine are also monoamines, due to significant gaps in knowledge, we will not discuss these transmitters herein.

2.1 Monoamine Release

Like other neurotransmitters, monoamine release is posited to occur through canonical mechanisms including recruitment of voltage-gated calcium channels, SNARE complex interactions, and Ca^{2+}-mediated vesicle exocytosis [53, 122, 165]. However, unlike fast transmitters such as glutamate and GABA, which exert their effects within a nanometer range of a synapse [41], monoamine transmission is generally viewed as occurring primarily through "volume transmission," which is marked by release from terminal boutons—often outside of prototypical synapses [21, 31, 43, 47, 55–57, 83, 123, 195, 199, 207]. This volume transmission affords monoamines the luxury of communicating with postsynaptic neurons through diffusion that extends their actions over relatively large distances (i.e., μm beyond the release site). Although outside of the focus of this review, monoamine release is subject to robust regulation involving control of transmitter synthesis and packaging, as well as release mechanisms through heteroreceptor actions on presynaptic terminals, which can impact extracellular concentrations of monoamines over both short and long-time scales [75, 82, 157].

3 Overview of Monoamine Transporters

Dopamine, norepinephrine, and serotonin transporters (DAT, NET, SERT) are expressed primarily on presynaptic terminals and only in neurons of their respective cell-type [95, 145, 241], where they act to translocate their monoamine from the extracellular space into the intracellular compartment (Fig. 1). Once they are moved intracellularly, monoamines are recycled for future release by sequestration into synaptic vesicles by VMAT2 or they are metabolized by monoamine oxidase A and B [37, 60, 85, 184, 217]. In regions such as the prefrontal cortex, where DAT concentrations are relatively low, metabolism of dopamine can occur through catechol-O-methyltransferase [116, 230]. Notwithstanding, metabolism within the extracellular space is not considered to be the primary mode through which monoamine clearance occurs, as monoamine transporters are quite efficient at moving monoamines intracellularly [143].

Monoamine transporters are part of the SLC6 family of symporters which contain 12 alpha-helical transmembrane spanning domains where both the N- and C-termini reside in the intracellular region. Substrate translocation is thought to utilize an "alternating access" mechanism, where the transporter transitions between three sequential states following substrate interaction: outward-open, completely occluded, and inward-open [2, 120, 144, 225]. The efficiency of monoamines

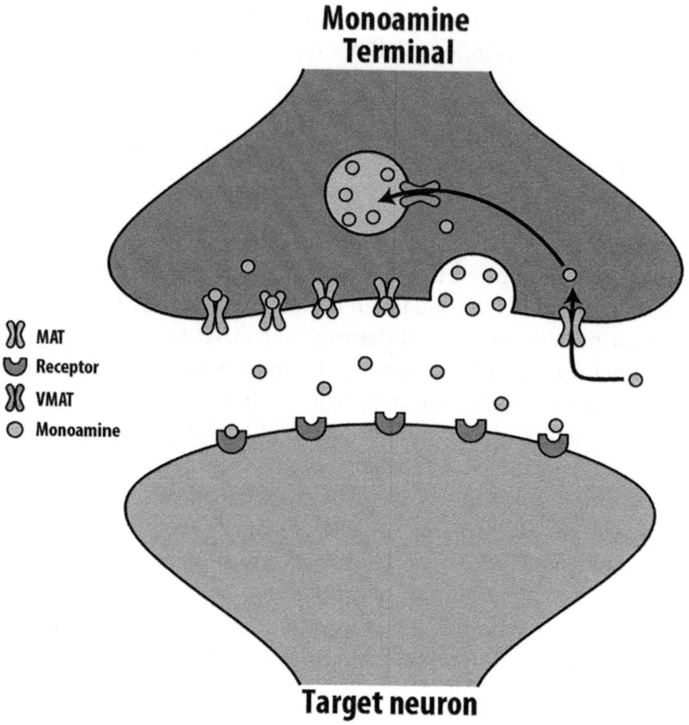

Fig. 1 Functional role of monoamine transporters. Removal of monoamines from the extracellular space aids in the termination of monoamine binding to monoamine receptors. Monoamine transporters (MAT) utilize a proposed alternating access mechanism of translocation to bring monoamines and ions back into the neuron from where they are released. Once a monoamine and ions interact with the binding site, the transporter transitions from an outward-open state facing the extracellular space to a completely occluded conformation. Then the transporter transitions to the inward-open state where it releases the monoamines into the intracellular space to be reused. The transporter can then transition back into the outward-open conformation for further removal of monoamines from the extracellular space. The monoamines that are located intracellularly are then repackaged through VMAT for future release. Disruption of transporter function with transporter inhibitors or releasers increases monoamine levels, which affects post-synaptic monoamine/receptor interactions. (See Wright and Dong [223] and Zinsmaier et al. [245] for more details)

transporters is dependent on post translational modifications, which can alter the localization, stability, and efficiency of the transporters. Beyond the monoamine transporters, there are also high-capacity, but low-affinity, transporters such as organic cation transporter 2 (OCT2), OCT3, and plasma membrane monoamine transporter—see Gasser [79] for more detail about their distribution and function.

4 Neurochemistry Reveals Basics of Monoamine Transmission

FSCV has been an instrumental tool for demonstrating the importance of monoamine transporters in the regulation of transmitter availability in the extracellular space (Fig. 2). By exploiting the oxidation and reduction potentials of monoamines, FSCV affords excellent spatial and temporal resolution for measuring both monoamine release and uptake dynamics across ex vivo and in vivo applications [40, 67, 109, 161, 170, 212]. As discussed below, FSCV experiments have led to numerous discoveries including the existence of heterogeneity of release and uptake across brain regions, the functional consequences of autoreceptor activation on release and uptake, and the contribution of monoamine transporters in the psychotropic actions of drugs of abuse.

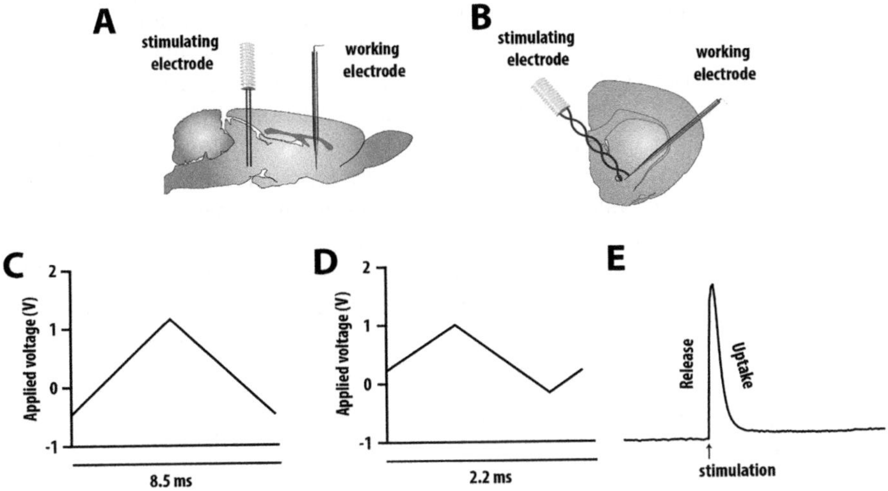

Fig. 2 FSCV for measuring monoamine release and uptake. (**a**) In vivo: a stimulating electrode is implanted into a cell body region (e.g., ventral tegmental area) and a recording electrode is implanted into a terminal region (e.g., ventral striatum). (**b**) Ex vivo: a recording electrode and stimulating electrode are both implanted into slices containing a terminal region of interest. (**a** & **b**) Using either approach, electrical stimulation will result in monoamine release, which can be detected using FSCV. (**c**) For detection of dopamine and norepinephrine, a triangle waveform scanning from −0.4 volts to 1.2 volts and back to −0.4 volts is typically used. This results in oxidation and reduction of dopamine or norepinephrine. (**d**) For detection of serotonin, an N waveform scanning from 0.2 volts to 1.0 volts, down to −0.1 volts and then back to 0.2 volts has been used [103], although an extended N waveform has also been used to avoid fouling and increase sensitivity [59]. (**e**) An example current vs. time trace depicting release and uptake of dopamine

5 FSCV Reveals Heterogeneity in Uptake Efficiency Across the Neuroaxis

5.1 *Dopamine Transporter*

Dopamine transporters are observed at varying expression levels in cell bodies of the ventral tegmental area (VTA) and substantia nigra (SN), and across terminals regions including the striatum, cortex, and amygdala [39, 72]. Within the striatum, immunohistochemical and Western blotting studies have demonstrated a dorsoventral gradient of DAT expression with the greatest expression in the dorsal striatum, moderate expression in the nucleus accumbens (NAc) core, and the lowest expression in the NAc shell [72, 174]. FSCV has confirmed the functional consequences of this differential expression, indicating that the dorsal striatum has the highest rates of dopamine uptake, followed by lower uptake rates in the NAc core, and even lower rates in the NAc shell [23].

Based on immunohistochemistry, the amygdala and the prefrontal cortex have lower DAT expression levels than the striatum [39], and FSCV has confirmed that dopamine uptake rates in the amygdala and prefrontal cortex are considerably lower than in the striatum [77, 112]. Consistent with the relationship between expression levels and uptake rate, somatodendritic expression of DAT in the VTA and SN is significantly lower than in the striatum [39] and is associated with lower dopamine uptake rates [44]. The link between DAT expression and dopamine uptake rates is further supported by studies in which DAT levels have been genetically manipulated, with overexpression of DAT resulting in faster dopamine uptake rates [24, 118], and knockdown of DAT resulting in lower uptake rates [101, 114, 185].

5.2 *Norepinephrine Transporter*

The norepinephrine transporter is located throughout the brain but is greatly expressed in the cell bodies of the locus coeruleus and terminals within the amygdala, thalamus, and cortex [194]. FSCV has been used to confirm the functional consequences of differential NET expression across brain regions [31, 92, 151, 152]. The ventral bed nucleus of the stria terminalis (BNST) is the most studied region for assessing norepinephrine uptake rate, with the dorsal BNST showing slower uptake rates than the ventral BNST [92, 150]. Notably, the dorsal BNST receives largely dopamine innervation, whereas the ventral BNST strictly receives norepinephrine [54, 73, 119]. In comparison to the ventral BNST, the anteroventral thalamic nucleus has nearly double the uptake rate [150], which matches the higher NET expression levels in the thalamus compared to BNST [194]. In contrast, the locus coeruleus has relatively slow uptake rates compared to the ventral BNST [31, 150], which is quite surprising given high levels of NET expression [194]. Consistent with what we know about DATs, NET expression determines the rate of norepinephrine uptake as NET knockout mice have slower uptake rates in the ventral BNST [138, 227]. To our knowledge, no studies have overexpressed NET in the ventral BNST or used NET knockout in alternative regions to confirm a tight relationship between NET expression and norepinephrine uptake rates.

5.3 Serotonin Transporter

SERT is expressed widely across the neuroaxis, with notable expression in cell bodies of the dorsal raphe nucleus (DRN), and in terminal regions such as the striatum, amygdala, cortex, and SN [12, 200]. Within subregions of the SN, FSCV in brain slices revealed that serotonin uptake rate was faster in the pars reticula compared to the pars compacta [110]. Further, comparing in vivo FSCV studies across brain regions suggests that the hippocampus has slightly faster uptake rates than the SN pars reticulata [87–89, 178]. The importance of SERT for serotonin uptake is further demonstrated with genetic manipulations showing that SERT knockout reduces uptake rates and is associated with increased evoked release in both subregions of the SN [48, 107, 110]. By comparison, mice over-expressing SERT display reduced evoked serotonin release but, unfortunately, no quantification of serotonin uptake rate has been reported [107]. Research comparing SERT uptake rates across brain regions will be necessary to provide a more complete determination of whether there is a linear relationship between SERT expression levels and uptake rates.

6 FSCV Reveals the Importance of Autoreceptor Function

6.1 Dopamine Transporter

FSCV has been integral in demonstrating the influence of dopamine autoreceptors on both release and uptake dynamics in terminal regions. For example, activation of autoreceptors with D_2/D_3 agonists in dorsal striatal slices using FSCV decreases dopamine uptake rate [127]. Conversely, application of the D_2 antagonist, haloperidol, increases dopamine release and reduces dopamine uptake in the dorsal striatum, NAc core, and shell in vivo [224]. Together, these results suggest that activation of D_2 autoreceptors tightly regulates dopamine transmission. Despite this evidence, there are conflicting reports on the effects of D_2 receptor knockout on baseline release within the dorsal striatum. Using FSCV in the dorsal striatum of anesthetized mice, revealed no differences in dopamine release were observed between D_2 receptor knockout and wild-type control mice [58]. However, D_2 receptor knockout reduced dopamine release in brain slices containing the dorsal striatum [181]. It is important to note that D_2 receptors are expressed on dopamine terminals as autoreceptors, but they are also expressed postsynaptically on medium spiny neurons and cholinergic interneurons [76]. Since the knockout strategy used in these experiments affected D_2 receptors both pre- and postsynaptically, follow up studies used an approach to selectively target D_2 autoreceptors. In those studies, selective knockout of D_2 autoreceptors increased dopamine release in dorsal striatal slices and application of the D_2 agonist quinpirole, which typically decreases dopamine release in a dose-dependent manner, had little effect [10].

The importance of D_2 autoreceptor regulation of DAT function is not clear as there have been multiple conflicting reports on the effects of D_2 receptor knockout on dopamine uptake. For example, FSCV in anesthetized mice with D_2 receptor knockdown showed reduced dopamine uptake in the dorsal striatum compared to control littermates [58]. However, when examining changes in dopamine uptake in dorsal striatal slices, FSCV revealed an increase in dopamine uptake in D_2 receptor knockout mice [181]. Importantly, however, selective knockout of only presynaptic D_2 receptors had no effect on dopamine uptake rates in the dorsal striatum [10]. Together, these results demonstrate that D_2 receptors may help to modulate dopamine release and uptake, but the direction of effects may depend on the experimental preparations (in vivo vs. ex vivo) and whether presynaptic or postsynaptic D_2 receptors are targeted.

6.2 Norepinephrine Transporter

Electrophysiology studies have identified α_{2A} receptors of the locus coeruleus and their presynaptic expression in terminals within the cortex and hippocampus as critical in regulating firing activity [31, 45, 187, 218, 221]. FSCV studies have confirmed the importance of these α_{2A} autoreceptors in controlling transmission wherein α_{2A} antagonists increase and α_{2A} agonists decrease norepinephrine release in the locus coeruleus [30, 31, 98]. When examining the importance of α_{2A} receptors in terminal regions, it has been shown that an a_2 antagonist increased norepinephrine release in the ventral BNST [71, 149]. To our knowledge, there are no FSCV studies indicating α_{2A} autoreceptor regulation of norepinephrine uptake.

6.3 Serotonin Transporter

The complexity of serotonin autoreceptor regulation of transmission is marked by the presence of at least three subtypes of putative serotonin autoreceptors [1, 140, 159, 196]. These receptor types are also located as heteroreceptors on postsynaptic neurons; thus, making it challenging to dissociate the effects of pre- vs. postsynaptic receptor modulation of serotonin in an intact brain. In this capacity, FSCV in slices has played a pivotal role in demonstrating differential autoreceptors actions on serotonin release and uptake across terminal regions. Within the DRN and median raphe nucleus, 5-HT_{1A} and 5-HT_{1B} antagonists have been shown to increase serotonin release [96, 169], suggesting that 5-HT and 5-HT may act as autoreceptors. While there is some evidence that 5-HT_{1D} may act as an autoreceptor, in the DRN, a selective 5-HT_{1D} antagonist only modestly reduced release and in the median raphe there appeared to be no effect [96]. Moreover, in DRN terminals in the suprachiasmatic nucleus and ventrolateral geniculate nucleus, 5-HT_{1B} and 5-HT_{1D} agonists decrease serotonin release, whereas 5-HT_{1B} and 5-HT_{1D} antagonists increased serotonin release and a selective 5-HT_{1A} agonist had little effect on release in these regions [49, 146]. There has been opposing evidence of autoreceptor regulation of serotonin release in the SN pars reticulata, with a 5-HT_{1B} antagonist having modest regulation of serotonin release [205], and 5-HT_{1B} agonists or antagonists having no effect [102]. As with norepinephrine, there are no FSCV studies that have examined autoreceptor control of serotonin uptake.

7 FSCV Reveals Acute and Lasting Effects of Drugs of Abuse on Transporter Function

7.1 Dopamine Transporter

An expansive literature suggests that exposure to drugs of abuse produces acute and lasting alterations in the mesolimbic system. In addition to changes in extracellular dopamine levels and dopamine release dynamics, there are also robust alterations in dopamine uptake rate and/or the transporter sensitivity to inhibitors such as cocaine. As we discuss below, changes in DAT function and sensitivity to DAT inhibitors appear to be dependent on the degree of drug exposure and which drugs of abuse are under investigation.

7.1.1 Psychostimulants

Psychostimulants are among the most well-studied drugs of abuse, with numerous observations indicating that cocaine and amphetamines increase dopamine levels across the neuroaxis. For example, acute delivery of cocaine increases extracellular dopamine in the dorsal striatum, NAc, prefrontal cortex, amygdala, and BNST [13, 32, 46, 78, 91, 99, 129]. Although increases in dopamine following cocaine can be partially attributed to synapsin-dependent enhancement of vesicular release [213], cocaine's inhibition of the DAT also increases extracellular dopamine by disrupting translocation of dopamine from the extracellular space into the terminals. The effects of cocaine at inhibiting the DAT have been well-characterized across species and experimental conditions, with clear evidence for the rapid onset of inhibition of dopamine uptake [4, 25–27, 111, 129, 130, 191]. For instance, using FSCV in anesthetized rats, acute intravenous cocaine increases extracellular dopamine and inhibits dopamine uptake in the NAc within 4–5 s of cocaine delivery [61, 129, 232]. The effects of cocaine on dopamine uptake are most robust after about 30 sec, with lingering effects even after 60 min [61, 129, 232]. The fast onset effects observed with cocaine may be a unique feature of cocaine itself since it is not observed with all DAT inhibitors and is not dependent on the affinity for the DAT [61, 232]. Although, the mechanisms underlying differences in the rapidity of maximal inhibition of dopamine uptake are not known, it is possible that individual DAT inhibitors possess unique chemical characteristics that bias specific DAT conformations, with cocaine potentially stabilizing the outward-facing conformation and other DAT inhibitors potentially biasing the inward-facing conformation [106, 125, 164]. In addition, other factors such as lipophilicity and speed of blood-brain barrier transport may also contribute to observed differences in the onset effects of DAT inhibitors [197, 226].

Cocaine self-administration exerts lasting changes in dopamine release, uptake rate, and the ability of cocaine to inhibit dopamine uptake (i.e., DAT sensitivity to cocaine). Recent evidence suggests that the temporal pattern of intake is a critical determinant of the directionality of effects on transporter efficiency. Following self-administration on either a short-access (2-h) or long-access (6-h) fixed ratio 1 (FR1) schedule of reinforcement (Fig. 3), there were no observed differences in either baseline dopamine release or uptake rate [25]. Nevertheless, long-access FR1 cocaine self-administration results in a decrease in DAT sensitivity to cocaine [25]. It is tempting to speculate that the effects on DAT sensitivity to cocaine are driven by differences in the amount of cocaine consumed between the short-access and long-access schedules—especially since other schedules of reinforcement (5 days of 40 injections per day; 5 × 40, see Fig. 3) that promote levels of cocaine intake similar to long access also result in decreased DAT sensitivity to cocaine [26, 62, 66]. However, unlike short- and long-access self-administration, 5 × 40 self-administration reduces baseline dopamine release and uptake in the NAc core [26, 62, 66], suggesting that there could be a reduction in DAT expression or disrupted DAT function following self-administration under this schedule.

In comparison to the sustained intake patterns associated with the short access, long access, and 5 × 40 schedules, intermittent cocaine intake engenders binge-like consumption patterns that more closely mimic how humans use cocaine [11, 244]. Consequently, investigators developed the "Intermittent Access" schedule wherein rats have 5-min access to cocaine followed by 25-min timeout periods with no cocaine access (Fig. 3). Interestingly, although self-administration on the intermittent access schedule produces similar levels of cocaine intake as short access, intermittent access to cocaine promotes increased dopamine release, uptake, and DAT sensitivity to cocaine in the NAc core compared to control, short-access, and long-access schedules [25]. These findings suggest that a binge-like pattern of cocaine intake results in faster removal of dopamine from the extracellular space, which may contribute to a hypo-dopaminergic state.

In addition to patterns and intake amounts during self-administration sessions, FSCV has also revealed that the length of abstinence from cocaine has important influences on dopamine dynamics. For example, during the first week of abstinence from long access to cocaine there is a decrease in stimulated dopamine release

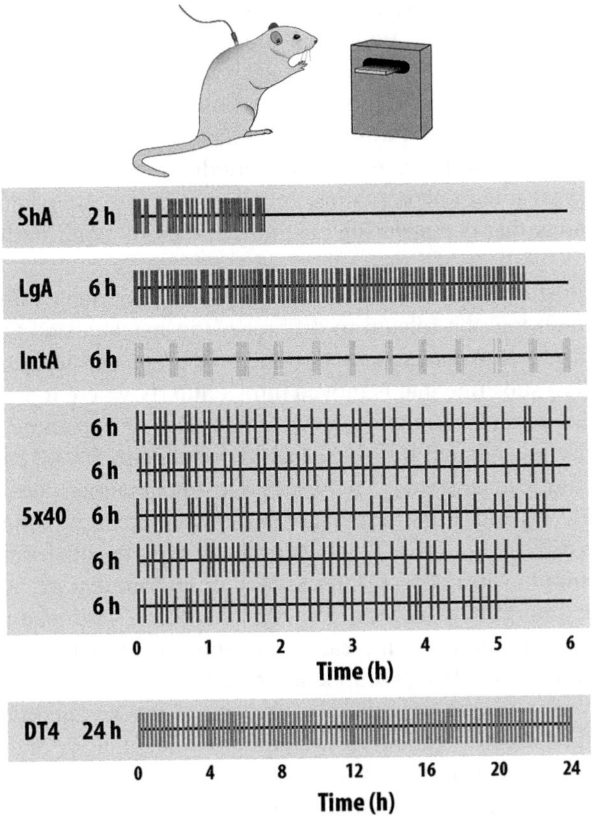

Fig. 3 Typical self-administration schedules for drugs of abuse. Short access (ShA): animals have 2-h, fixed ratio 1 (FR1) access to drug, wherein single lever presses result in single injections of drug. Long access (LgA): animals have 6-h FR1 access to drug. Intermittent access (IntA): animals have 5-min unlimited FR1 access to drug, followed by 25-min timeout periods, during which there is no access to drug. These access and time-out periods are repeated 12 times throughout a 6-h session. 5 × 40: animals undergo five, 6-h FR1 sessions during which they receive a maximum of 40, high-dose drug injections per session. Discrete trials 4 (DT4): animals are given FR1 access to drug every 15 min throughout a 24-h period. Colored vertical lines within each schedule depict lever presses that result in drug delivery

within the NAc core [65], which appears to normalize after 28 days of abstinence [4]. Abstinence from long access or 5 × 40 does not affect dopamine uptake early in abstinence or following extended abstinence [65], but abstinence from cocaine following self-administration on a Discrete-Trials 4 schedule (cocaine availability every 15 min for 24 h; see Fig. 3) is associated with an increase in dopamine uptake rates in early abstinence [130]. Similarly, after extended abstinence from intermittent access to cocaine, increases in dopamine uptake rates are still apparent [4]. These results suggest that even after drug taking has ended, robust alterations to dopamine terminals are still present.

With respect to DAT sensitivity to cocaine, there are also differences depending on patterns of cocaine intake. Following long access to cocaine, there is an initial decrease in DAT sensitivity to cocaine during the first week of abstinence, which returns to normal by day 14 [65], only to increase by abstinence day 28 [4]. Taken together, these studies suggest long-lasting adaptations to dopamine terminals that enhance transporter function even during periods of abstinence. These alterations may play a pivotal role in perpetuating drug use and may also promote the intensified drug cravings that occur during abstinence, thereby contributing to relapse to drug use.

Other psychostimulants do not necessarily produce the same effects on the DAT as cocaine, which may be related to differences in mechanisms of action. Unlike cocaine, which acts at the cocaine binding site to disrupt translocation of dopamine, amphetamine is a substrate that is moved intracellularly where it promotes accumulation of dopamine and, ultimately, reverse transport of dopamine through a non-vesicular release mechanism [68, 113, 198]. Even though 5 × 40 access to cocaine decreases dopamine uptake [65], 5 × 40 access to amphetamine does not alter uptake efficiency [29]. To our knowledge, no such assessment of amphetamine self-administration effects on DAT efficiency has been done for other schedules of reinforcement including short access, long access, or intermittent access. Nonetheless, long access or intermittent access to methylphenidate—a dopamine and norepinephrine transporter inhibitor—increases dopamine uptake at baseline and enhances the effect of methylphenidate or amphetamine when applied to NAc slices [22, 24]. These results indicate that regardless of pattern of intake, methylphenidate—but maybe not amphetamine—produces compensatory adaptations in dopamine terminals in response to elevated dopamine levels.

Despite the different mechanisms through which cocaine and amphetamines impact the DAT, there are interesting crossover effects between psychostimulants. For example, intermittent access to cocaine enhances DAT sensitivity to amphetamine or methylphenidate in NAc slices [22, 27]. However, IntA and 5 × 40 self-administration of methylphenidate does not impact DAT sensitivity to cocaine, though it does enhance sensitivity to other releasers such as amphetamine [22, 24]. Together, these observations indicate that cocaine self-administration affects DAT responses to both inhibitors and releasers, but methylphenidate may only impact the effects of other releasers on DAT function. This further suggests that DAT adaptations following cocaine exposure may be more general by affecting not only the cocaine binding site but also the ability for releasers to travel intracellularly. It has been postulated that these differences may be associated with changes in total surface expression of DAT or possibly post-translational modifications of DATs (see below). Future studies are needed to address whether the crossover features extend to other releasers and what the biochemical mechanisms may be for these interactions.

7.1.2 Opioids

Compared to stimulants, research on the effects of opioids on DAT function is limited and contains conflicting results that may be associated with the distinct experimental conditions, patterns of intake, and/or length of abstinence that vary across studies. Acutely, i.v. exposure to heroin has little effect on NAc dopamine release and uptake in vivo [155]. Similarly, oxycodone has no effect on dopamine release in drug naive NAc slices, although we have observed a robust decrease in dopamine uptake [177].

In reference to repeated exposure, FSCV reveals disagreement on the effects of opioids on dopamine release, with some studies finding decreases in release [81] and others finding no effect [80, 154, 177]. FSCV also reveals inconsistencies in dopamine uptake, with some studies reporting an increase [80], some indicating a decrease [177], and others finding no differences [81, 155]. The major distinctions between studies are the specific opioid used, the regions where recordings occurred (NAc core vs shell), and the patterns of opioid intake prior to FSCV measures. Future studies will need to disentangle which of these factors may be driving the observed changes in DAT efficiency following opioid self-administration.

7.1.3 Alcohol

Understanding the effects of alcohol on DAT function has been given greater attention than opioids but nevertheless remains understudied. Using both in vivo and ex vivo FSCV, acute high-dose ethanol inhibits dopamine release without affecting dopamine uptake in NAc slices [19, 115, 179, 233, 234]. Additionally, acute exposure of a low or high dose of ethanol to dorsal striatum slices decreases release but has no effect on dopamine uptake in either wild-type or DAT knockout mice [132]. Overall, the lack of effect on dopamine uptake throughout these experiments indicates that acute ethanol is unlikely to produce its effects through actions at the DAT.

Although acute ethanol exposure produced no changes in dopamine uptake across the striatum, chronic ethanol exposure produces unique effects depending on the region under study. For instance, chronic ethanol self-administration increases dopamine release and uptake in the NAc core but decreases release and uptake in the dorsolateral caudate of macaques [188–190]. Moreover, chronic intermittent ethanol through a two-bottle choice protocol in rats also increases dopamine uptake in the NAc core but has no effect on dopamine release [51]. However, in mouse models of chronic intermittent ethanol, there were no differences observed in dopamine release or uptake in the dorsal striatum [176]. Together, these results suggest that across species, the effects on dopamine release and uptake are conserved in the NAc core but not necessarily in the dorsal striatum.

In addition to direct consumption of ethanol, several investigators have examined the effects of chronic intermittent ethanol vapor exposure on the neurochemical effects of alcohol. Overall, most studies demonstrated that chronic intermittent ethanol vapor exposure reduced dopamine release in NAc core [117, 171, 172] (but see

[20, 134]), and that ethanol has little effect on dopamine release in the NAc shell, dorsal striatum, and dorsal BNST [20, 134, 153]. These studies also found an increase in dopamine uptake in the NAc core, dorsal striatum, and dorsal BNST, but not in the NAc shell [20, 134, 153, 171, 172]. Together, these results suggest that chronic ethanol exposure exerts a more pronounced effect on DAT efficiency than on dopamine release.

7.1.4 Nicotine

Dopamine neurons and their terminals express high levels of nicotinic acetylcholine receptors, resulting in significant impact of nicotine on mesolimbic dopamine transmission [222, 246]. As such, it is not surprising that FSCV reveals significant effects of nicotine on dopamine release and uptake. For instance, acute application of nicotine rapidly desensitizes receptors to produce the same effect as nicotinic acetylcholine receptor antagonists which reduce dopamine release in the dorsal striatum, NAc core, and NAc shell slices [168, 179, 239, 242]. Conversely, subcutaneous administration of nicotine in anesthetized rats increases dopamine uptake in the NAc, dorsal striatum, and prefrontal cortex [86, 137]. Interestingly, the effects of nicotine become more complex when considering environmental enrichment. Indeed, acute subcutaneous administration of nicotine in anesthetized rats enhanced dopamine uptake in the NAc shell in rats reared in an enriched condition [243], but in rats reared in impoverished conditions, nicotine preferentially enhanced dopamine uptake in the NAc core [243].

In general, long-term exposure to nicotine through either drinking water or repeated subcutaneous injections results in sustained reductions in dopamine release in the NAc and dorsal striatum, which is similar to the effects observed with acute nicotine exposure [63, 64, 158, 240]. However, unlike with acute exposure, long-term exposure to nicotine had no effect on dopamine uptake in either the NAc core or shell [64, 158, 240]. Future studies will need to investigate the mechanisms underlying differences in uptake between acute and chronic nicotine exposure.

7.1.5 Cannabinoids

Recent increases in marijuana use warrant inquiry into what extent cannabinoids regulate dopamine transmission. In general, both in vivo and ex vivo experiments using FSCV demonstrate that cannabinoids significantly enhance dopamine release without affecting its uptake. FSCV has revealed that a CB_1/CB_2 agonist increases dopamine release, while a CB_1 antagonist decreases dopamine release in the dorsal striatum and NAc [35, 36, 84, 131, 192]. In comparison to the involvement of cannabinoids on dopamine release, most studies either find a lack of CB_1 agonist effects on DA uptake [35, 192] or do not discuss effects on uptake at all [36, 84, 131, 147]. While the literature is limited, these findings provide a strong suggestion that cannabinoids modify dopamine release without impacting DAT efficiency. One potential avenue for future investigation is to examine to what degree chronic exposure to cannabinoids influences dopamine release or uptake. Additionally, given the common use of marijuana and related substances, questions related to how prior exposure to cannabinoids alters the effects of other drugs of abuse on dopamine transmission would be translationally important.

7.1.6 Toluene

Toluene is a major component of commonly abused inhalants, and recent studies have investigated the consequences of toluene on dopamine transmission using FSCV. Repeated toluene vapor exposure decreases dopamine release in the NAc core but not the NAc shell or dorsal striatum [6]. The effects of toluene on dopamine uptake are more complex and dependent on levels of exposure. For example, 2000 ppm exposure to toluene reduces dopamine uptake in the NAc core, but 4000 ppm exposure increases dopamine uptake in the NAc shell. Notably, there were no effects on dopamine uptake in the dorsal striatum at either toluene concentration [6]. Together, these results suggest that the effects of repeated toluene exposure produce uptake adaptations that are region and concentration specific.

7.2 Norepinephrine Transporter

An accumulating literature suggests that exposure to drugs of abuse exerts profound effects on the noradrenergic system, which manifest as shifts in neuron excitability, norepinephrine release, and norepinephrine uptake rates. Below, we discuss the importance of NETs on the effects of various drugs of abuse and to what degree changes in NET function may contribute to development of substance use disorders.

7.2.1 Psychostimulants

Significant evidence indicates that psychostimulants exert some of their effects by increasing extracellular norepinephrine concentrations through NET inhibition [5, 9, 148, 206, 216]. Comparable to its effects on the DAT, cocaine inhibits the NET to increase extracellular norepinephrine concentrations and amphetamine acts as a releaser by traveling intracellularly through the NET to promote efflux of norepinephrine into the extracellular space.

There have been several FCSV studies confirming and elaborating on the effects of psychostimulants on norepinephrine dynamics in the ventral BNST, due to its dense norepinephrine innervation and its involvement in modulating reward and stress. Using FSCV in anesthetized animals, amphetamine and both the dextrorotatory and levorotatory methamphetamine enantiomers have been shown to induce norepinephrine release and slow uptake in the ventral BNST [152, 156]. To our knowledge, there are currently no studies assessing alterations in norepinephrine transmission after chronic exposure to psychostimulants. Consequently, this area of investigation could pose as a promising research avenue.

7.2.2 Opioids, Alcohol, and Nicotine

Across studies using a variety of techniques, it is clear that opioids [3, 133, 210], alcohol [7, 214, 215], and nicotine [173, 182] alter norepinephrine neuron activity and/or norepinephrine transmission. With respect to opioids, LC norepinephrine system participates in morphine dependence and withdrawal [3, 210] through dense expression of μ-opioid receptors on norepinephrine neurons [128, 166, 208, 209]. Based on one FSCV study, it has been shown that acute administration of morphine does not affect norepinephrine release or uptake in the ventral BNST [133]. By comparison, following naloxone-precipitated morphine withdrawal, a reduction in norepinephrine uptake rate was observed, with no changes in norepinephrine release [133].

When examining the effects of acute alcohol exposure, electrophysiology studies indicate reduced norepinephrine neuron excitability [7, 214, 215]. Consequently, it is surprising that FSCV studies have generally failed to detect changes in norepinephrine release following acute or chronic ethanol exposure [52, 90]. Notably, interpretation of these findings is difficult given that these experiments examined the infralimbic cortex or central amygdala, two regions that receive both dopamine and norepinephrine innervation. Consequently, to establish the extent to which ethanol influences norepinephrine transmission, studies will need to investigate norepinephrine release and uptake dynamics in regions primarily receiving norepinephrine inputs, such as the ventral BNST.

In comparison to the lack of effect of acute opioids or alcohol on norepinephrine transmission, there is some evidence that acute administration of nicotine may have significant effects. For instance, microdialysis and autoradiography studies suggest that nicotine induces norepinephrine release in the PFC, hippocampus, striatum, and ventral BNST [104, 173, 182]. However, other than these few experiments, there is not much known of how nicotine can alter norepinephrine transmission especially in relation to norepinephrine uptake rate through NET function.

Despite evidence that opioids, alcohol, and nicotine impact norepinephrine transmission, the general lack of FSCV studies investigating how these drugs impact norepinephrine release and uptake makes it challenging to assess the importance of NET function across terminal regions.

7.3 Serotonin Transporter

Several studies have revealed that exposure to drugs of abuse can alter serotonergic system function including changes in neuronal firing rates, serotonin release, and uptake rates. Below, we consider SERT involvement in mediating these effects and how drugs of abuse impact SERT function.

7.3.1 Psychostimulants

As is evident with DAT and NET, psychostimulants also interact with SERT to elevate serotonin levels. Cocaine is a DAT, NET, and SERT inhibitor which has been shown to increase evoked serotonin release across the anterior-posterior dorsal striatum, the NAc, and the VTA through semi-derivative voltammetry [14–16]. These experiments were among the first to suggest that cocaine-induced behaviors were partially modulated by serotonin. More recently, FSCV in anesthetized animals has demonstrated that a single cocaine injection increases serotonin release and decreases serotonin uptake in the NAc core [238]. Similarly, other psychostimulants such as methamphetamine, MDMA, and amphetamine reduce serotonin uptake rates in mouse SN pars reticulata slices [108]. Overall, these experiments demonstrate that acute psychostimulant exposure inhibits SERT functionality, but future studies will need to determine if chronic exposure to psychostimulants produces adaptations or alters sensitivity to these drugs.

7.3.2 Opioids, Alcohol, Nicotine

Although there are studies indicating that opioids [74, 160, 186, 202, 203, 219], alcohol [50, 201, 204, 228, 236, 237], and nicotine [8, 167] impact serotonin neuron activity and/or serotonin levels, the evidence for these drugs to influence SERT function is lacking. Currently, there are no existing FSCV studies assessing the involvement of SERT on these drugs of abuse.

Within the DRN, serotonin neurons express μ-opioid receptors [100]. Opioids have been shown to activate serotonin neurons [74, 219] and enhance serotonin neurotransmission in the DRN [203]. However, differential pulse voltammetry in anesthetized rats revealed that acute morphine alone does not alter release of serotonin in the dorsal raphe nucleus or nucleus raphe magnus, but morphine can reduce release of serotonin resulting from presentation of a noxious stimulus [186, 202]. Furthermore, administration of the serotonin uptake inhibitor, femoxetine, has been shown to fully eliminate the effects of morphine on serotonin in the nucleus raphe magnus using differential normal pulse voltammetry in freely moving rats [160]. This suggests that alterations in SERT function or sensitivity may contribute to the interaction between morphine and serotonin, which can be addressed with future studies using FSCV at the serotonin terminals in the striatum or SN.

Similar to opioids, several microdialysis studies demonstrate that alcohol also elevates serotonin levels in the NAc, VTA, amygdala, and hippocampus [50, 201, 204, 228, 236, 237]. Although there are no studies using FSCV to study serotonin release and uptake, linear sweep voltammetry in freely moving rats revealed that low-dose ethanol decreased serotonin levels in the dorsal striatum [193]. With regard to serotonin uptake, high-speed chronoamperometry demonstrated that ethanol reduced serotonin uptake in the hippocampus [50]. However, the reduced inhibition did not appear to be due to direct actions on SERT, since SERT knockout mice have even slower uptake rates when ethanol is applied compared to control littermates [50]. These observations suggest that there may be another mechanism underlying reductions in serotonin uptake following opioids that are independent of SERT functionality.

Similar to opioids and alcohol, studies demonstrate a possible connection between nicotine and serotonin transmission. Systemic nicotine exposure increases serotonin release within the frontal cortex using microdialysis [167], and [3H]5-HT synaptosomal binding studies showed that acute and chronic nicotine exposure enhances serotonin uptake and SERT density [8, 167]. However, other than these few experiments, there is a general lack of information on how nicotine alters serotonin transmission or SERT function.

8 Biochemistry Reveals Potential Mechanisms for Changes in Transporter Function

The biochemical mechanisms underlying changes in monoamine transporter function are not known. While FSCV has demonstrated that drugs of abuse impact monoamine transporter efficiency, it remains unclear whether these effects are linked to biochemical changes in post-translational modifications that impact transporter expression, localization, stability, or efficiency. To further elucidate the potential mechanisms underlying changes in transmission, it is essential to integrate findings from different methodologies, such as combining FSCV and Western blotting approaches to assess both functional impacts and biochemical changes to transporters.

Western blotting is an excellent approach to quantify total transporter membrane surface expression, which may be among the most straightforward explanations for the putative mechanisms underlying changes in transporter efficiency. As discussed above, cocaine and methylphenidate self-administration leads to differential effects on dopamine uptake rates [26, 28, 62, 66], with long access to cocaine decreasing dopamine uptake and total DAT surface expression [28] and long access to methylphenidate increasing dopamine uptake rate and total DAT surface expression [28]. Interestingly, the alterations in total DAT surface expression for cocaine and methylphenidate were not associated with an increase in DAT mRNA in the VTA [28], suggesting that exposure to cocaine or methylphenidate changes DAT surface expression or trafficking without affecting DAT protein transcription.

Another potential mechanism for observed FSCV changes in transporter function could be post-translational modification of the transporters. One such example is phosphorylation of the DAT at the threonine 53 site (Thr53), which has previously been associated with increased cocaine analog binding [34, 69, 70]. We previously demonstrated a relationship between DAT function and phosphorylation of the DAT using chemogenetics. In these studies, we used DREADDs to activate or inhibit dopamine neurons in the VTA to assess how changes in dopamine neuron activity affect DAT sensitivity to cocaine in the NAc core. We found that activation or inhibition of dopamine neurons in the VTA did not significantly affect baseline dopamine uptake rates in the NAc [18]. However, activation of dopamine neurons increased sensitivity to cocaine (Fig. 4b) and expression of Thr53 phosphorylation of the DAT (Fig. 4d), without affecting total DAT surface expression (Fig. 4c) [18].

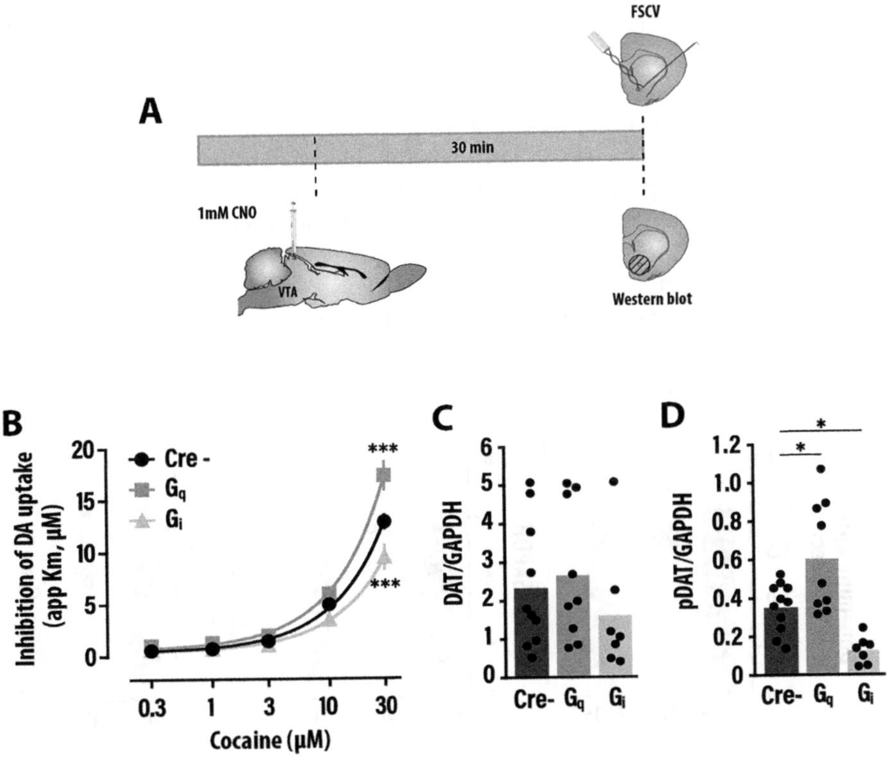

Fig. 4 Activation of dopamine neurons exerts bidirectional changes in DAT sensitivity to cocaine and changes in DAT phosphorylation at Thr53. (**a**) Experimental timeline for FSCV. (**b**) Inhibition of DA uptake following cumulative cocaine concentrations (0.3–30 μm). (**c**) Total DAT (tDAT), and (**d**) phosphorylated DAT at the Thr53 site (pDAT). Data are mean ± s.e.m. (B, C, & D) n=6 per group. Dunnett's post hoc. ***p<0.001; *p<0.05 vs. Cre⁻ control. (Modified from Brodnik et al. [18])

Next, we tested if inhibition of dopamine neurons produces the opposite effects and found that inhibition decreased sensitivity to cocaine (Fig. 4b) and reduced Thr53 phosphorylation of the DAT (Fig. 4d), again with no effects on total DAT expression (Fig. 4b) [18]. When considered together, these observations suggest that modulation of dopamine neuron activity leads to downstream alterations in Thr53 phosphorylation that modify cocaine's ability to interact with the DAT [162].

Another example of how Western blotting can elucidate mechanisms underlying changes in transporter function comes from our recent work where we examined if changes in dopamine uptake following intermittent access to oxycodone were tied to alterations in DAT expression or phosphorylation [177]. As mentioned above, using ex vivo FSCV in NAc core slices, we observed reductions in dopamine uptake rate after 2 and 15 days of abstinence from oxycodone (Fig. 5a). We performed Western blotting of NAc core tissue collected from the same rats used to conduct the FSCV experiments and found that abstinence from oxycodone did not affect total DAT expression (Fig. 5b) at either abstinence timepoint. However, we found a reduction in phosphorylated DAT at Thr53 after 15 days—but not 2 days—of abstinence (Fig. 5c). The decrease in DAT efficiency we observed on abstinence day 15 is consistent with observed reductions in Thr53 phosphorylation, which aligns with previous findings indicating a direct relationship between these two measures [34, 69, 70]. However, since we did not observe a similar reduction in phosphorylation on abstinence day 2, changes in DAT function in early abstinence may instead be attributable to other post-translational modifications, including phosphorylation of the Ser7 or Thr48 DAT sites, or decreases in palmitoylation, SUMOylation or glycosylation of the DAT [33, 121, 141, 142, 229].

Like the DAT, prior work indicates that NET and SERT can also be phosphorylated and glycosylated to alter uptake rates [42, 105, 135, 136, 163]. However, we are unaware of any studies that have paired FSCV with Western blotting to

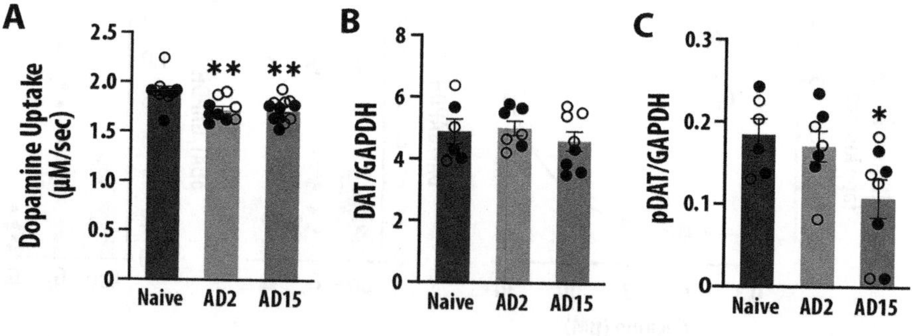

Fig. 5 Abstinence from IntA to oxycodone was associated with reduced dopamine uptake and reduced phosphorylated DAT at Threonine-53 (pDAT). (**a**) Dopamine uptake at baseline, (**b**) Total DAT (tDAT), and (**c**) phosphorylated DAT at the Thr53 site (pDAT). Data are mean ± s.e.m. (**a**) Naive, n = 10 (5 females and 5 males); AD2, n = 9 (4 females and 5 males); AD15, n = 12 (6 females and 6 males). (**b**, & **c**) Naive, n = 6 (3 females and 3 males); AD2, n = 7 (3 females and 4 males); AD15, n = 8 (4 females and 4 males). ** p<0.01; *p<0.05 vs. naive. Open circles depict females and closed circles depict males. (Modified from Samson et al. [177])

determine possible mechanisms that impact uptake efficiency for the NET and SERT. This is not surprising given there are still significant gaps in our understanding of whether acute or repeated exposure to drugs of abuse influence NET and SERT functionality. Once these gaps are addressed, future studies will surely assess to what degree potential changes in NET and SERT function are associated with shifts in transporter levels or alterations in post-translational modifications.

9 Conclusions

FSCV has been a powerful tool for demonstrating how monoamine systems are distributed, regulated, and affected by drugs of abuse. FSCV has been used to compare monoamine transmission across the neuroaxis and to assess the importance of autoreceptors in governing transmission. These foundational observations were essential to establishing release and uptake dynamics under control conditions and to determine which regions were important targets for investigating the acute and chronic effects of drugs of abuse.

Despite insights gained from FSCV on DAT functionality, significant questions remain—especially with regard to drugs of abuse. One intriguing question is why patterns of cocaine intake led to unique effects on transporter function. Another question is whether psychostimulants other than cocaine exert distinct effects across periods of abstinence or depending on drug exposure conditions. Future FSCV experiments could also determine which factors are critical for driving changes in DAT efficiency in opioid self-administration, including the effects of patterns of opioid intake, the specific opioid under investigation, and differential responses in striatal subregions. Moreover, FSCV studies will need to address if chronic alcohol exposure elicits alterations in dopamine uptake rates. Another notable inquiry could explore why nicotine administration results in acute effects on uptake, but chronic nicotine does not produce such an effect.

Aside from targeted questions for DAT research, the relative scarcity of studies focused on NET and SERT highlights the need to address many fundamental questions about their functions, which would significantly inform the field. Future questions should explore whether NET and SERT function correlates linearly with uptake rates across terminal regions, if chronic exposure to drugs of abuse leads to adaptations that alter uptake rates, and to what extent these adaptations are sustained during periods of abstinence.

Another major question that needs to be addressed for all the monoamine transporters is what molecular and biochemical mechanisms underlie changes in transporter function following exposure to drugs of abuse. For example, experiments using both FSCV and Western blotting must be used to quantify transporter uptake rate, total transporter expression, and assess which post-translational modifications may be driving alterations in uptake under various behavior contexts. Experiments that combine these approaches create a convincing and interesting story of how uptake levels are altered under various conditions. Ideally, the combination of ex vivo FSCV, Western blotting, and other biochemical methods from the same tissue would reveal the strongest link between mechanism and function, as obtaining tissue from the same animal would control for behavioral conditions and experimental timepoints. In addition, experimenters could use a similar pairing of techniques to explain differences in transmitter release by querying levels of VMAT2 or tyrosine hydroxylase. Ultimately, combining FSCV and Western blotting provides a powerful framework for elucidating the interplay between transporter function and regulation, offering profound insights into how these systems contribute to the broader understanding of neurobiological responses to drug exposure.

References

1. Adell A, Celada P, Artigas F (2001) The role of 5-HT1B receptors in the regulation of serotonin cell firing and release in the rat brain. J Neurochem 79:172–182
2. Aggarwal S, Mortensen OV (2017) Overview of monoamine transporters. Curr Protoc Pharmacol 79:12 16 11–12 16 17
3. Ahtee L, Attila LM, Carlson KR, Haikala H (1989) Changes in brain monoamine metabolism during withdrawal from chronic oral self-administration of morphine and in response to a morphine challenge in the withdrawn state. J Pharmacol Exp Ther 249:303–310
4. Alonso IP, O'Connor BM, Bryant KG, Mandalaywala RK, España RA (2022) Incubation of cocaine craving coincides with changes in dopamine terminal neurotransmission. Addict Neurosci 3:100029–100029
5. Amara SG, Sonders MS (1998) Neurotransmitter transporters as molecular targets for addictive drugs. Drug Alcohol Depend 51:87–96
6. Apawu AK, Callan SP, Mathews TA, Bowen SE (2020) Repeated toluene exposure leads to neuroadaptation in dopamine release mechanisms within the nucleus accumbens core. Toxicol Appl Pharmacol 408:115260
7. Aston-Jones G, Foote SL, Bloom FE (1982) Low doses of ethanol disrupt sensory responses of brain noradrenergic neurones. Nature 296:857–860
8. Awtry TL, Werling LL (2003) Acute and chronic effects of nicotine on serotonin uptake in prefrontal cortex and hippocampus of rats. Synapse 50:206–211
9. Axelrod J (1971) Noradrenaline: fate and control of its biosynthesis. Science 173:598–606
10. Bello EP, Mateo Y, Gelman DM, Noain D, Shin JH, Low MJ, Alvarez VA, Lovinger DM, Rubinstein M (2011) Cocaine supersensitivity and enhanced motivation for reward in mice lacking dopamine D2 autoreceptors. Nat Neurosci 14:1033–1038
11. Beveridge TJR, Wray P, Brewer A, Shapiro B, Mahoney JJ, Newton TF (2012) Analyzing human cocaine use patterns to inform animal addiction model development. Published abstract for the college on problems of drug dependence annual meeting. Pam Springs
12. Borgers AJ, Koopman KE, Bisschop PH, Serlie MJ, Swaab DF, Fliers E, la Fleur SE, Alkemade A (2014) Decreased serotonin transporter immunoreactivity in the human hypothalamic infundibular nucleus of overweight subjects. Front Neurosci 8:106
13. Bradberry CW, Nobiletti JB, Elsworth JD, Murphy B, Jatlow P, Roth RH (1993) Cocaine and cocaethylene: microdialysis comparison of brain drug levels and effects on dopamine and serotonin. J Neurochem 60:1429–1435
14. Broderick PA (1992) Cocaine's colocalized effects on synaptic serotonin and dopamine in ventral tegmentum in a reinforcement paradigm. Pharmacol Biochem Behav 42:889–898
15. Broderick PA (1993) In vivo electrochemical studies of gradient effects of (SC) cocaine on dopamine and serotonin release in dorsal striatum of conscious rats. Pharmacol Biochem Behav 46:973–984

16. Broderick PA, Kornak EP, Eng F, Wechsler R (1993) Real time detection of acute (IP) cocaine-enhanced dopamine and serotonin release in ventrolateral nucleus accumbens of the behaving Norway rat. Pharmacol Biochem Behav 46:715–722
17. Brodnik ZD, Black EM, Clark MJ, Kornsey KN, Snyder NW, España RA (2017) Susceptibility to traumatic stress sensitizes the dopaminergic response to cocaine and increases motivation for cocaine. Neuropharmacology 125:295–307
18. Brodnik ZD, Xu W, Batra A, Lewandowski SI, Ruiz CM, Mortensen OV, Kortagere S, Mahler SV, España RA (2020) Chemogenetic manipulation of dopamine neurons dictates cocaine potency at distal dopamine transporters. J Neurosci 40:8767–8779
19. Budygin EA, Mathews TA, Lapa GB, Jones SR (2005) Local effects of acute ethanol on dopamine neurotransmission in the ventral striatum in C57BL/6 mice. Eur J Pharmacol 523:40–45
20. Budygin EA, Oleson EB, Mathews TA, Läck AK, Diaz MR, McCool BA, Jones SR (2007) Effects of chronic alcohol exposure on dopamine uptake in rat nucleus accumbens and caudate putamen. Psychopharmacology 193:495–501
21. Bunin MA, Wightman RM (1998) Quantitative evaluation of 5-hydroxytryptamine (serotonin) neuronal release and uptake: an investigation of extrasynaptic transmission. J Neurosci 18:4854–4860
22. Calipari ES, Jones SR (2014) Sensitized nucleus accumbens dopamine terminal responses to methylphenidate and dopamine transporter releasers after intermittent-access self-administration. Neuropharmacology 82:1–10
23. Calipari ES, Huggins KN, Mathews TA, Jones SR (2012) Conserved dorsal-ventral gradient of dopamine release and uptake rate in mice, rats and rhesus macaques. Neurochem Int 61:986–991
24. Calipari ES, Ferris MJ, Salahpour A, Caron MG, Jones SR (2013a) Methylphenidate amplifies the potency and reinforcing effects of amphetamines by increasing dopamine transporter expression. Nat Commun 4:2720
25. Calipari ES, Ferris MJ, Zimmer BA, Roberts DC, Jones SR (2013b) Temporal pattern of cocaine intake determines tolerance vs sensitization of cocaine effects at the dopamine transporter. Neuropsychopharmacology 38:2385–2392
26. Calipari ES, Ferris MJ, Jones SR (2014a) Extended access of cocaine self-administration results in tolerance to the dopamine-elevating and locomotor-stimulating effects of cocaine. J Neurochem 128:224–232
27. Calipari ES, Ferris MJ, Siciliano CA, Zimmer BA, Jones SR (2014b) Intermittent cocaine self-administration produces sensitization of stimulant effects at the dopamine transporter. J Pharmacol Exp Ther 349:192–198
28. Calipari ES, Ferris MJ, Melchior JR, Bermejo K, Salahpour A, Roberts DC, Jones SR (2014c) Methylphenidate and cocaine self-administration produce distinct dopamine terminal alterations. Addict Biol 19:145–155
29. Calipari ES, Sun H, Eldeeb K, Luessen DJ, Feng X, Howlett AC, Jones SR, Chen R (2014d) Amphetamine self-administration attenuates dopamine D2 autoreceptor function. Neuropsychopharmacology 39:1833–1842
30. Callado LF, Stamford JA (1999) Alpha2A- but not alpha2B/C-adrenoceptors modulate noradrenaline release in rat locus coeruleus: voltammetric data. Eur J Pharmacol 366:35–39
31. Callado LF, Stamford JA (2000) Spatiotemporal interaction of alpha(2) autoreceptors and noradrenaline transporters in the rat locus coeruleus: implications for volume transmission. J Neurochem 74:2350–2358
32. Carboni E, Silvagni A, Rolando MT, Di Chiara G (2000) Stimulation of in vivo dopamine transmission in the bed nucleus of stria terminalis by reinforcing drugs. J Neurosci 20:RC102
33. Cartier E, Garcia-Olivares J, Janezic E, Viana J, Moore M, Lin ML, Caplan JL, Torres G, Kim YH (2019) The SUMO-conjugase Ubc9 prevents the degradation of the dopamine transporter, enhancing its cell surface level and dopamine uptake. Front Cell Neurosci 13:35

34. Challasivakanaka S, Zhen J, Smith ME, Reith MEA, Foster JD, Vaughan RA (2017) Dopamine transporter phosphorylation site threonine 53 is stimulated by amphetamines and regulates dopamine transport, efflux, and cocaine analog binding. J Biol Chem 292:19066–19075
35. Cheer JF, Wassum KM, Heien MLAV, Phillips PEM, Wightman RM (2004) Cannabinoids enhance subsecond dopamine release in the nucleus Accumbens of awake rats. J Neurosci 24:4393–4400
36. Cheer JF, Wassum KM, Sombers LA, Heien ML, Ariansen JL, Aragona BJ, Phillips PE, Wightman RM (2007) Phasic dopamine release evoked by abused substances requires cannabinoid receptor activation. J Neurosci 27:791–795
37. Cho HU, Kim S, Sim J, Yang S, An H, Nam MH, Jang DP, Lee CJ (2021) Redefining differential roles of MAO-A in dopamine degradation and MAO-B in tonic GABA synthesis. Exp Mol Med 53:1148–1158
38. Chu N, Bloom FE (1973) Norepinephrine-containing neurons: changes in spontaneous discharge patterns during sleeping and waking. Science 179:908–910
39. Ciliax BJ, Heilman C, Demchyshyn LL, Pristupa ZB, Ince E, Hersch SM, Niznik HB, Levey AI (1995) The dopamine transporter: immunochemical characterization and localization in brain. J Neurosci 15:1714–1723
40. Clark PJ, España RA (2023) Fast scan cyclic voltammetry to assess dopamine function: from circuits to behavior. In: Fuentealba-Evans JA, Henny P (eds) Dopaminergic system function and dysfunction: experimental approaches. Springer, New York, NY, pp 249–281
41. Clements JD, Lester RA, Tong G, Jahr CE, Westbrook GL (1992) The time course of glutamate in the synaptic cleft. Science 258:1498–1501
42. Cooper A, Woulfe D, Kilic F (2019) Post-translational modifications of serotonin transporter. Pharmacol Res 140:7–13
43. Cragg SJ, Rice ME (2004) DAncing past the DAT at a DA synapse. Trends Neurosci 27:270–277
44. Cragg S, Rice ME, Greenfield SA (1997) Heterogeneity of electrically evoked dopamine release and reuptake in substantia nigra, ventral tegmental area, and striatum. J Neurophysiol 77:863–873
45. Curet O, de Montigny C (1989) Electrophysiological characterization of adrenoceptors in the rat dorsal hippocampus. III. Evidence for the physiological role of terminal alpha 2-adrenergic autoreceptors. Brain Res 499:18–26
46. Czoty PW, Justice JB Jr, Howell LL (2000) Cocaine-induced changes in extracellular dopamine determined by microdialysis in awake squirrel monkeys. Psychopharmacology 148:299–306
47. Daniel JA, Galbraith S, Iacovitti L, Abdipranoto A, Vissel B (2009) Functional heterogeneity at dopamine release sites. J Neurosci 29:14670–14680
48. Dankoski EC, Carroll S, Wightman RM (2016) Acute selective serotonin reuptake inhibitors regulate the dorsal raphe nucleus causing amplification of terminal serotonin release. J Neurochem 136:1131–1141
49. Davidson C, Stamford JA (1996) Serotonin efflux in the rat ventral lateral geniculate nucleus assessed by fast cyclic voltammetry is modulated by 5-HT1B and 5-HT1D autoreceptors. Neuropharmacology 35:1627–1634
50. Daws LC, Montanez S, Munn JL, Owens WA, Baganz NL, Boyce-Rustay JM, Millstein RA, Wiedholz LM, Murphy DL, Holmes A (2006) Ethanol inhibits clearance of brain serotonin by a serotonin transporter-independent mechanism. J Neurosci 26:6431–6438
51. Deal AL, Konstantopoulos JK, Weiner JL, Budygin EA (2018) Exploring the consequences of social defeat stress and intermittent ethanol drinking on dopamine dynamics in the rat nucleus accumbens. Sci Rep 8:332
52. Deal AL, Mikhailova MA, Grinevich VP, Weiner JL, Gainetdinov RR, Budygin EA (2019) In vivo voltammetric evidence that locus coeruleus activation predominantly releases norepinephrine in the infralimbic cortex: effect of acute ethanol. Synapse 73:e22080

53. Del-Bel E, De-Miguel FF (2018) Extrasynaptic neurotransmission mediated by exocytosis and diffusive release of transmitter substances. Front Synaptic Neurosci 10:13
54. Delfs JM, Zhu Y, Druhan JP, Aston-Jones G (2000) Noradrenaline in the ventral forebrain is critical for opiate withdrawal-induced aversion. Nature 403:430–434
55. Descarries L, Beaudet A, Watkins KC (1975) Serotonin nerve terminals in adult rat neocortex. Brain Res 100:563–588
56. Descarries L, Watkins KC, Lapierre Y (1977) Noradrenergic axon terminals in the cerebral cortex of rat. III Topometric ultrastructural analysis. Brain Res 133:197–222
57. Descarries L, Watkins KC, Garcia S, Bosler O, Doucet G (1996) Dual character, asynaptic and synaptic, of the dopamine innervation in adult rat neostriatum: a quantitative autoradiographic and immunocytochemical analysis. J Comp Neurol 375:167–186
58. Dickinson SD, Sabeti J, Larson GA, Giardina K, Rubinstein M, Kelly MA, Grandy DK, Low MJ, Gerhardt GA, Zahniser NR (1999) Dopamine D2 receptor-deficient mice exhibit decreased dopamine transporter function but no changes in dopamine release in dorsal striatum. J Neurochem 72:148–156
59. Dunham KE, Venton BJ (2020) Improving serotonin fast-scan cyclic voltammetry detection: new waveforms to reduce electrode fouling. Analyst 145:7437–7446
60. Eiden LE, Weihe E (2011) VMAT2: a dynamic regulator of brain monoaminergic neuronal function interacting with drugs of abuse. Ann N Y Acad Sci 1216:86–98
61. España RA, Roberts DC, Jones SR (2008) Short-acting cocaine and long-acting GBR-12909 both elicit rapid dopamine uptake inhibition following intravenous delivery. Neuroscience 155:250–257
62. Estave PM, Albertson SE, Karkhanis AN, Jones SR (2024) Co-targeting the kappa opioid receptor and dopamine transporter reduces motivation to self-administer cocaine and partially reverses dopamine system dysregulation. Sci Rep 14:6509
63. Exley R, Clements MA, Hartung H, McIntosh JM, Franklin M, Bermudez I, Cragg SJ (2013) Striatal dopamine transmission is reduced after chronic nicotine with a decrease in alpha6-nicotinic receptor control in nucleus accumbens. Eur J Neurosci 38:3036–3043
64. Fennell AM, Pitts EG, Sexton LL, Ferris MJ (2020) Phasic dopamine release magnitude tracks individual differences in sensitization of locomotor response following a history of nicotine exposure. Sci Rep 10:173
65. Ferris MJ, Mateo Y, Roberts DCS, Jones SR (2011) Cocaine-insensitive dopamine transporters with intact substrate transport produced by self-administration. Biol Psychiatry 69:201–207
66. Ferris MJ, Calipari ES, Mateo Y, Melchior JR, Roberts DC, Jones SR (2012) Cocaine self-administration produces pharmacodynamic tolerance: differential effects on the potency of dopamine transporter blockers, releasers, and methylphenidate. Neuropsychopharmacology 37:1708–1716
67. Ferris MJ, Calipari ES, Yorgason JT, Jones SR (2013) Examining the complex regulation and drug-induced plasticity of dopamine release and uptake using voltammetry in brain slices. ACS Chem Neurosci 4:693–703
68. Fleckenstein AE, Volz TJ, Riddle EL, Gibb JW, Hanson GR (2007) New insights into the mechanism of action of amphetamines. Annu Rev Pharmacol Toxicol 47:681–698
69. Foster JD, Vaughan RA (2017) Phosphorylation mechanisms in dopamine transporter regulation. J Chem Neuroanat 83–84:10–18
70. Foster JD, Yang JW, Moritz AE, Challasivakanaka S, Smith MA, Holy M, Wilebski K, Sitte HH, Vaughan RA (2012) Dopamine transporter phosphorylation site threonine 53 regulates substrate reuptake and amphetamine-stimulated efflux. J Biol Chem 287:29702–29712
71. Fox ME, Isaac Studebaker R, Swofford NJ, Wightman RM (2015) Stress and drug dependence differentially modulate norepinephrine signaling in animals with varied HPA axis function. Neuropsychopharmacology 40:1752–1761
72. Freed C, Revay R, Vaughan RA, Kriek E, Grant S, Uhl GR, Kuhar MJ (1995) Dopamine transporter immunoreactivity in rat brain. J Comp Neurol 359:340–349

73. Freedman LJ, Cassell MD (1994) Distribution of dopaminergic fibers in the central division of the extended amygdala of the rat. Brain Res 633:243–252
74. Fu B, Yao JQ, Lu C, Wang B, Li Z, Huang M, Tian T, Peng H, Liu SJ (2022) Dorsal raphe nucleus serotoninergic neurons mediate morphine rewarding effect and conditioned place preference. Neuroscience 480:108–116
75. Fuxe K, Borroto-Escuela DO (2016) Volume transmission and receptor-receptor interactions in heteroreceptor complexes: understanding the role of new concepts for brain communication. Neural Regen Res 11:1220–1223
76. Gallo EF (2019) Disentangling the diverse roles of dopamine D2 receptors in striatal function and behavior. Neurochem Int 125:35–46
77. Garris PA, Wightman RM (1994) Different kinetics govern dopaminergic transmission in the amygdala, prefrontal cortex, and striatum: an in vivo voltammetric study. J Neurosci 14:442–450
78. Garris PA, Wightman RM (1995) Distinct pharmacological regulation of evoked dopamine efflux in the amygdala and striatum of the rat in vivo. Synapse 20:269–279
79. Gasser P (2021) Organic cation transporters in brain catecholamine homeostasis. Handb Exp Pharmacol 266:187–197
80. George BE, Barth SH, Kuiper LB, Holleran KM, Lacy RT, Raab-Graham KF, Jones SR (2021) Enhanced heroin self-administration and distinct dopamine adaptations in female rats. Neuropsychopharmacology 46:1724–1733
81. George BE, Dawes MH, Peck EG, Jones SR (2022) Altered Accumbal dopamine terminal dynamics following chronic heroin self-administration. Int J Mol Sci 23:8106
82. German CL, Baladi MG, McFadden LM, Hanson GR, Fleckenstein AE (2015) Regulation of the dopamine and vesicular monoamine transporters: pharmacological targets and implications for disease. Pharmacol Rev 67:1005–1024
83. Gianni G, Pasqualetti M (2023) Wiring and volume transmission: An overview of the dual modality for serotonin neurotransmission. ACS Chem Neurosci 14:4093–4104
84. Gomez DM, Everett TJ, Hamilton LR, Ranganath A, Cheer JF, Oleson EB (2021) Chronic cannabinoid exposure produces tolerance to the dopamine releasing effects of WIN 55,212-2 and heroin in adult male rats. Neuropharmacology 182:108374
85. Graves SM, Xie Z, Stout KA, Zampese E, Burbulla LF, Shih JC, Kondapalli J, Patriarchi T, Tian L, Brichta L, Greengard P, Krainc D, Schumacker PT, Surmeier DJ (2020) Dopamine metabolism by a monoamine oxidase mitochondrial shuttle activates the electron transport chain. Nat Neurosci 23:15–20
86. Hart C, Ksir C (1996) Nicotine effects on dopamine clearance in rat nucleus accumbens. J Neurochem 66:216–221
87. Hashemi P, Dankoski EC, Petrovic J, Keithley RB, Wightman RM (2009) Voltammetric detection of 5-hydroxytryptamine release in the rat brain. Anal Chem 81:9462–9471
88. Hashemi P, Dankoski EC, Wood KM, Ambrose RE, Wightman RM (2011) In vivo electrochemical evidence for simultaneous 5-HT and histamine release in the rat substantia nigra pars reticulata following medial forebrain bundle stimulation. J Neurochem 118:749–759
89. Hashemi P, Dankoski EC, Lama R, Wood KM, Takmakov P, Wightman RM (2012) Brain dopamine and serotonin differ in regulation and its consequences. Proc Natl Acad Sci USA 109:11510–11515
90. Hedges DM, Yorgason JT, Brundage JN, Wadsworth HA, Williams B, Steffensen SC, Roberto M (2020) Corticotropin releasing factor, but not alcohol, modulates norepinephrine release in the rat central nucleus of the amygdala. Neuropharmacology 179:108293
91. Hedou G, Feldon J, Heidbreder CA (1999) Effects of cocaine on dopamine in subregions of the rat prefrontal cortex and their efferents to subterritories of the nucleus accumbens. Eur J Pharmacol 372:143–155
92. Herr NR, Park J, McElligott ZA, Belle AM, Carelli RM, Wightman RM (2012) In vivo voltammetry monitoring of electrically evoked extracellular norepinephrine in subregions of the bed nucleus of the stria terminalis. J Neurophysiol 107:1731–1737

93. Hersey M, Reneaux M, Berger SN, Mena S, Buchanan AM, Ou Y, Tavakoli N, Reagan LP, Clopath C, Hashemi P (2022) A tale of two transmitters: serotonin and histamine as in vivo biomarkers of chronic stress in mice. J Neuroinflammation 19:167
94. Hoebel BG, Hernandez L, Schwartz DH, Mark GP, Hunter GA (1989) Microdialysis studies of brain norepinephrine, serotonin, and dopamine release during ingestive behavior. Theoretical and clinical implications. Ann N Y Acad Sci 575:171–191. discussion 192–173
95. Hoffman BJ, Hansson SR, Mezey E, Palkovits M (1998) Localization and dynamic regulation of biogenic amine transporters in the mammalian central nervous system. Front Neuroendocrinol 19:187–231
96. Hopwood SE, Stamford JA (2001) Multiple 5-HT(1) autoreceptor subtypes govern serotonin release in dorsal and median raphe nuclei. Neuropharmacology 40:508–519
97. Howes SR, Dalley JW, Morrison CH, Robbins TW, Everitt BJ (2000) Leftward shift in the acquisition of cocaine self-administration in isolation-reared rats: relationship to extracellular levels of dopamine, serotonin and glutamate in the nucleus accumbens and amygdala-striatal FOS expression. Psychopharmacology 151:55–63
98. Huang HP, Wang SR, Yao W, Zhang C, Zhou Y, Chen XW, Zhang B, Xiong W, Wang LY, Zheng LH, Landry M, Hokfelt T, Xu ZQ, Zhou Z (2007) Long latency of evoked quantal transmitter release from somata of locus coeruleus neurons in rat pontine slices. Proc Natl Acad Sci USA 104:1401–1406
99. Hurd YL, Weiss F, Koob GF, And NE, Ungerstedt U (1989) Cocaine reinforcement and extracellular dopamine overflow in rat nucleus accumbens: an in vivo microdialysis study. Brain Res 498:199–203
100. Hwa LS, Shimamoto A, Kayyali T, Norman KJ, Valentino RJ, DeBold JF, Miczek KA (2016) Dissociation of mu-opioid receptor and CRF-R1 antagonist effects on escalated ethanol consumption and mPFC serotonin in C57BL/6J mice. Addict Biol 21:111–124
101. Illiano P, Bass CE, Fichera L, Mus L, Budygin EA, Sotnikova TD, Leo D, Espinoza S, Gainetdinov RR (2017) Recombinant adeno-associated virus-mediated rescue of function in a mouse model of dopamine transporter deficiency syndrome. Sci Rep 7:46280
102. Iravani MM, Kruk ZL (1997) Real-time measurement of stimulated 5-hydroxytryptamine release in rat substantia nigra pars reticulata brain slices. Synapse 25:93–102
103. Jackson BP, Wightman RM (1995) Dynamics of 5-hydroxytryptamine released from dopamine neurons in the caudate putamen of the rat. Brain Res 674:163–166
104. Jadzic D, Bassareo V, Carta AR, Carboni E (2021) Nicotine, cocaine, amphetamine, morphine, and ethanol increase norepinephrine output in the bed nucleus of stria terminalis of freely moving rats. Addict Biol 26:e12864
105. Jayanthi LD, Annamalai B, Samuvel DJ, Gether U, Ramamoorthy S (2006) Phosphorylation of the norepinephrine transporter at threonine 258 and serine 259 is linked to protein kinase C-mediated transporter internalization. J Biol Chem 281:23326–23340
106. Jean B, Surratt CK, Madura JD (2017) Molecular dynamics of conformation-specific dopamine transporter-inhibitor complexes. J Mol Graph Model 76:143–151
107. Jennings KA, Lesch KP, Sharp T, Cragg SJ (2010) Non-linear relationship between 5-HT transporter gene expression and frequency sensitivity of 5-HT signals. J Neurochem 115:965–973
108. John CE, Jones SR (2007a) Voltammetric characterization of the effect of monoamine uptake inhibitors and releasers on dopamine and serotonin uptake in mouse caudate-putamen and substantia nigra slices. Neuropharmacology 52:1596–1605
109. John CE, Jones SR (2007b) Fast scan cyclic voltammetry of dopamine and serotonin in mouse brain slices. In: Michael AC, Borland LM (eds) Electrochemical methods for neuroscience, Boca Raton
110. John CE, Budygin EA, Mateo Y, Jones SR (2006) Neurochemical characterization of the release and uptake of dopamine in ventral tegmental area and serotonin in substantia nigra of the mouse. J Neurochem 96:267–282

111. Jones SR, Garris PA, Wightman RM (1995a) Different effects of cocaine and nomifensine on dopamine uptake in the caudate-putamen and nucleus accumbens. J Pharmacol Exp Ther 274:396–403
112. Jones SR, Garris PA, Kilts CD, Wightman RM (1995b) Comparison of dopamine uptake in the basolateral amygdaloid nucleus, caudate-putamen, and nucleus accumbens of the rat. J Neurochem 64:2581–2589
113. Jones SR, Gainetdinov RR, Wightman RM, Caron MG (1998a) Mechanisms of amphetamine action revealed in mice lacking the dopamine transporter. J Neurosci 18:1979–1986
114. Jones SR, Gainetdinov RR, Jaber M, Giros B, Wightman RM, Caron MG (1998b) Profound neuronal plasticity in response to inactivation of the dopamine transporter. Proc Natl Acad Sci USA 95:4029–4034
115. Jones SR, Mathews TA, Budygin EA (2006) Effect of moderate ethanol dose on dopamine uptake in rat nucleus accumbens in vivo. Synapse 60:251–255
116. Kaenmaki M, Tammimaki A, Myohanen T, Pakarinen K, Amberg C, Karayiorgou M, Gogos JA, Mannisto PT (2010) Quantitative role of COMT in dopamine clearance in the prefrontal cortex of freely moving mice. J Neurochem 114:1745–1755
117. Karkhanis AN, Rose JH, Huggins KN, Konstantopoulos JK, Jones SR (2015) Chronic intermittent ethanol exposure reduces presynaptic dopamine neurotransmission in the mouse nucleus accumbens. Drug Alcohol Depend 150:24–30
118. Kile BM, Walsh PL, McElligott ZA, Bucher ES, Guillot TS, Salahpour A, Caron MG, Wightman RM (2012) Optimizing the temporal resolution of fast-scan cyclic voltammetry. ACS Chem Neurosci 3:285–292
119. Kozicz T (2002) Met-enkephalin immunoreactive neurons recruited by acute stress are innervated by axon terminals immunopositive for tyrosine hydroxylase and dopamine-alpha-hydroxylase in the anterolateral division of bed nuclei of the stria terminalis in the rat. Eur J Neurosci 16:823–835
120. Kristensen AS, Andersen J, Jorgensen TN, Sorensen L, Eriksen J, Loland CJ, Stromgaard K, Gether U (2011) SLC6 neurotransmitter transporters: structure, function, and regulation. Pharmacol Rev 63:585–640
121. Li LB, Chen N, Ramamoorthy S, Chi L, Cui XN, Wang LC, Reith MEA (2004) The role of N-glycosylation in function and surface trafficking of the human dopamine transporter. J Biol Chem 279:21012–21020
122. Liu C, Kaeser PS (2019) Mechanisms and regulation of dopamine release. Curr Opin Neurobiol 57:46–53
123. Liu C, Kershberg L, Wang J, Schneeberger S, Kaeser PS (2018) Dopamine secretion is mediated by sparse active zone-like release sites. Cell 172:706–718 e715
124. Ljungberg T, Apicella P, Schultz W (1992) Responses of monkey dopamine neurons during learning of behavioral reactions. J Neurophysiol 67:145–163
125. Loland CJ, Desai RI, Zou MF, Cao J, Grundt P, Gerstbrein K, Sitte HH, Newman AH, Katz JL, Gether U (2008) Relationship between conformational changes in the dopamine transporter and cocaine-like subjective effects of uptake inhibitors. Mol Pharmacol 73:813–823
126. Macedo CE, Martinez RC, de Souza Silva MA, Brandao ML (2005) Increases in extracellular levels of 5-HT and dopamine in the basolateral, but not in the central, nucleus of amygdala induced by aversive stimulation of the inferior colliculus. Eur J Neurosci 21:1131–1138
127. Maina FK, Mathews TA (2010) A functional fast scan cyclic voltammetry assay to characterize dopamine D2 and D3 autoreceptors in the mouse striatum. ACS Chem Neurosci 1:450–462
128. Mansour A, Fox CA, Akil H, Watson SJ (1995) Opioid-receptor mRNA expression in the rat CNS: anatomical and functional implications. Trends Neurosci 18:22–29
129. Mateo Y, Budygin EA, Morgan D, Roberts DC, Jones SR (2004) Fast onset of dopamine uptake inhibition by intravenous cocaine. Eur J Neurosci 20:2838–2842

130. Mateo Y, Lack CM, Morgan D, Roberts DCS, Jones SR (2005) Reduced dopamine terminal function and insensitivity to cocaine following cocaine binge self-administration and deprivation. Neuropsychopharmacology 30:1455–1463
131. Mateo Y, Johnson KA, Covey DP, Atwood BK, Wang HL, Zhang S, Gildish I, Cachope R, Bellocchio L, Guzman M, Morales M, Cheer JF, Lovinger DM (2017) Endocannabinoid actions on cortical terminals orchestrate local modulation of dopamine release in the nucleus Accumbens. Neuron 96:1112–1126 e1115
132. Mathews TA, John CE, Lapa GB, Budygin EA, Jones SR (2006) No role of the dopamine transporter in acute ethanol effects on striatal dopamine dynamics. Synapse 60:288–294
133. McElligott ZA, Fox ME, Walsh PL, Urban DJ, Ferrel MS, Roth BL, Wightman RM (2013) Noradrenergic synaptic function in the bed nucleus of the Stria Terminalis varies in animal models of anxiety and addiction. Neuropsychopharmacology 38:1665–1673
134. Melchior JR, Jones SR (2017) Chronic ethanol exposure increases inhibition of optically targeted phasic dopamine release in the nucleus accumbens core and medial shell ex vivo. Mol Cell Neurosci 85:93–104
135. Melikian HE, McDonald JK, Gu H, Rudnick G, Moore KR, Blakely RD (1994) Human norepinephrine transporter. Biosynthetic studies using a site-directed polyclonal antibody. J Biol Chem 269:12290–12297
136. Melikian HE, Ramamoorthy S, Tate CG, Blakely RD (1996) Inability to N-glycosylate the human norepinephrine transporter reduces protein stability, surface trafficking, and transport activity but not ligand recognition. Mol Pharmacol 50:266–276
137. Middleton LS, Cass WA, Dwoskin LP (2004) Nicotinic receptor modulation of dopamine transporter function in rat striatum and medial prefrontal cortex. J Pharmacol Exp Ther 308:367–377
138. Miles PR, Mundorf ML, Wightman RM (2002) Release and uptake of catecholamines in the bed nucleus of the stria terminalis measured in the mouse brain slice. Synapse 44:188–197
139. Mohebi A, Pettibone JR, Hamid AA, Wong JT, Vinson LT, Patriarchi T, Tian L, Kennedy RT, Berke JD (2019) Dissociable dopamine dynamics for learning and motivation. Nature 570:65–70
140. Moret C, Briley M (1997) 5-HT autoreceptors in the regulation of 5-HT release from Guinea pig raphe nucleus and hypothalamus. Neuropharmacology 36:1713–1723
141. Moritz AE, Foster JD, Gorentla BK, Mazei-Robison MS, Yang JW, Sitte HH, Blakely RD, Vaughan RA (2013) Phosphorylation of dopamine transporter serine 7 modulates cocaine analog binding. J Biol Chem 288:20–32
142. Moritz AE, Rastedt DE, Stanislowski DJ, Shetty M, Smith MA, Vaughan RA, Foster JD (2015) Reciprocal phosphorylation and palmitoylation control dopamine transporter kinetics. J Biol Chem 290:29095–29105
143. Mundorf ML, Joseph JD, Austin CM, Caron MG, Wightman RM (2001) Catecholamine release and uptake in the mouse prefrontal cortex. J Neurochem 79:130–142
144. Nielsen AK, Moller IR, Wang Y, Rasmussen SGF, Lindorff-Larsen K, Rand KD, Loland CJ (2019) Substrate-induced conformational dynamics of the dopamine transporter. Nat Commun 10:2714
145. Nirenberg MJ, Chan J, Pohorille A, Vaughan RA, Uhl GR, Kuhar MJ, Pickel VM (1997) The dopamine transporter: comparative ultrastructure of dopaminergic axons in limbic and motor compartments of the nucleus accumbens. J Neurosci 17:6899–6907
146. O'Connor JJ, Kruk ZL (1992) Pharmacological characteristics of 5-hydroxytryptamine autoreceptors in rat brain slices incorporating the dorsal raphe or the suprachiasmatic nucleus. Br J Pharmacol 106:524–532
147. O'Neill C, Evers-Donnelly A, Nicholson D, O'Boyle KM, O'Connor JJ (2009) D2 receptor-mediated inhibition of dopamine release in the rat striatum in vitro is modulated by CB1 receptors: studies using fast cyclic voltammetry. J Neurochem 108:545–551
148. Pacholczyk T, Blakely RD, Amara SG (1991) Expression cloning of a cocaine-and antidepressant-sensitive human noradrenaline transporter. Nature 350:350–354

149. Palij P, Stamford JA (1992) Real-time monitoring of endogenous noradrenaline release in rat brain slices using fast cyclic voltammetry: 1. Characterisation of evoked noradrenaline efflux and uptake from nerve terminals in the bed nucleus of stria terminalis, pars ventralis. Brain Res 587:137–146
150. Park J, Kile BM, Wightman RM (2009) In vivo voltammetric monitoring of norepinephrine release in the rat ventral bed nucleus of the stria terminalis and anteroventral thalamic nucleus. Eur J Neurosci 30:2121–2133
151. Park J, Aragona BJ, Kile BM, Carelli RM, Wightman RM (2010) In vivo voltammetric monitoring of catecholamine release in subterritories of the nucleus accumbens shell. Neuroscience 169:132–142
152. Park J, Takmakov P, Wightman RM (2011) In vivo comparison of norepinephrine and dopamine release in rat brain by simultaneous measurements with fast-scan cyclic voltammetry. J Neurochem 119:932–944
153. Pati D, Downs AM, McElligott ZA, Kash TL (2023) Chronic ethanol exposure modulates periaqueductal gray to extended amygdala dopamine circuit. J Neurosci 43:709–721
154. Pattison LP, Bonin KD, Hemby SE, Budygin EA (2011) Speedball induced changes in electrically stimulated dopamine overflow in rat nucleus accumbens. Neuropharmacology 60:312–317
155. Pattison LP, McIntosh S, Budygin EA, Hemby SE (2012) Differential regulation of accumbal dopamine transmission in rats following cocaine, heroin and speedball self-administration. J Neurochem 122:138–146
156. Pauly RC, Bhimani RV, Li J-X, Blough BE, Landavazo A, Park J (2023) Distinct effects of methamphetamine isomers on limbic norepinephrine and dopamine transmission in the rat brain. ACS Chem Neurosci 14:1399–1411
157. Pereira DB, Sulzer D (2012) Mechanisms of dopamine quantal size regulation. Front Biosci (Landmark Ed) 17:2740–2767
158. Perez XA, Ly J, McIntosh JM, Quik M (2012) Long-term nicotine exposure depresses dopamine release in nonhuman primate nucleus accumbens. J Pharmacol Exp Ther 342:335–344
159. Pineyro G, Castanon N, Hen R, Blier P (1995) Regulation of [3H]5-HT release in raphe, frontal cortex and hippocampus of 5-HT1B knock-out mice. Neuroreport 7:353–359
160. Puig S, Rivot JP, Besson JM (1991) Femoxetine blocks the morphine-induced increase in 5-HT metabolism, as measured by in vivo voltammetry in the nucleus raphe magnus of freely-moving rats. Neuropharmacology 30:1053–1058
161. Puthongkham P, Venton BJ (2020) Recent advances in fast-scan cyclic voltammetry. Analyst 145:1087–1102
162. Ragu Varman D, Subler MA, Windle JJ, Jayanthi LD, Ramamoorthy S (2021) Novelty-induced hyperactivity and suppressed cocaine induced locomotor activation in mice lacking threonine 53 phosphorylation of dopamine transporter. Behav Brain Res 408:113267
163. Ramamoorthy S, Samuvel DJ, Buck ER, Rudnick G, Jayanthi LD (2007) Phosphorylation of threonine residue 276 is required for acute regulation of serotonin transporter by cyclic GMP. J Biol Chem 282:11639–11647
164. Reith ME, Blough BE, Hong WC, Jones KT, Schmitt KC, Baumann MH, Partilla JS, Rothman RB, Katz JL (2015) Behavioral, biological, and chemical perspectives on atypical agents targeting the dopamine transporter. Drug Alcohol Depend 147:1–19
165. Rettig J, Neher E (2002) Emerging roles of presynaptic proteins in ca++-triggered exocytosis. Science 298:781–785
166. Reyes BA, Glaser JD, Van Bockstaele EJ (2007) Ultrastructural evidence for co-localization of corticotropin-releasing factor receptor and mu-opioid receptor in the rat nucleus locus coeruleus. Neurosci Lett 413:216–221
167. Ribeiro EB, Bettiker RL, Bogdanov M, Wurtman RJ (1993) Effects of systemic nicotine on serotonin release in rat brain. Brain Res 621:311–318
168. Rice ME, Cragg SJ (2004) Nicotine amplifies reward-related dopamine signals in striatum. Nat Neurosci 7:583–584

169. Roberts C, Price GW (2001) Interaction of serotonin autoreceptor antagonists in the rat dorsal raphe nucleus: an in vitro fast cyclic voltammetry study. Neurosci Lett 300:45–48
170. Roberts JG, Sombers LA (2018) Fast-scan cyclic voltammetry: chemical sensing in the brain and beyond. Anal Chem 90:490–504
171. Rose JH, Karkhanis AN, Steiniger-Brach B, Jones SR (2016a) Distinct effects of Nalmefene on dopamine uptake rates and kappa opioid receptor activity in the nucleus Accumbens following chronic intermittent ethanol exposure. Int J Mol Sci 17:1216
172. Rose JH, Karkhanis AN, Chen R, Gioia D, Lopez MF, Becker HC, McCool BA, Jones SR (2016b) Supersensitive kappa opioid receptors promotes ethanol withdrawal-related behaviors and reduce dopamine signaling in the nucleus accumbens. Int J Neuropsychopharmacol 19:pyv127
173. Rossi S, Singer S, Shearman E, Sershen H, Lajtha A (2005) Regional heterogeneity of nicotine effects on neurotransmitters in rat brains in vivo at Low doses. Neurochem Res 30:91–103
174. Russo EE, Zovko LE, Nazari R, Steenland H, Ramsey AJ, Salahpour A (2023) Evaluation and validation of commercially available dopamine transporter antibodies. eNeuro 10:ENEURO.0341-22.2023
175. Sakai K, Crochet S (2001) Differentiation of presumed serotonergic dorsal raphe neurons in relation to behavior and wake-sleep states. Neuroscience 104:1141–1155
176. Salinas AG, Nadel JA, Mateo Y, Huynh T, Augustin SM, Pacak K, Lovinger DM (2022) Chronic ethanol consumption alters presynaptic regulation of dorsal striatal dopamine release in C57BL/6J mice. Int J Mol Sci 23:10994
177. Samson KR, Xu W, Kortagere S, España RA (2022) Intermittent access to oxycodone decreases dopamine uptake in the nucleus accumbens core during abstinence. Addict Biol 27:e13241
178. Saylor RA, Hersey M, West A, Buchanan AM, Berger SN, Nijhout HF, Reed MC, Best J, Hashemi P (2019) In vivo hippocampal serotonin dynamics in male and female mice: determining effects of acute Escitalopram using fast scan cyclic voltammetry. Front Neurosci 13:362
179. Schilaty ND, Hedges DM, Jang EY, Folsom RJ, Yorgason JT, McIntosh JM, Steffensen SC (2014) Acute ethanol inhibits dopamine release in the nucleus accumbens via alpha6 nicotinic acetylcholine receptors. J Pharmacol Exp Ther 349:559–567
180. Schmidt KT, Makhijani VH, Boyt KM, Cogan ES, Pati D, Pina MM, Bravo IM, Locke JL, Jones SR, Besheer J, McElligott ZA (2019) Stress-induced alterations of norepinephrine release in the bed nucleus of the Stria Terminalis of mice. ACS Chem Neurosci 10:1908–1914
181. Schmitz Y, Schmauss C, Sulzer D (2002) Altered dopamine release and uptake kinetics in mice lacking D2 receptors. J Neurosci 22:8002–8009
182. Scholze P, Orr-Urtreger A, Changeux JP, Mcintosh JM, Huck S (2007) Catecholamine outflow from mouse and rat brain slice preparations evoked by nicotinic acetylcholine receptor activation and electrical field stimulation. Br J Pharmacol 151:414–422
183. Schultz W (1986) Responses of midbrain dopamine neurons to behavioral trigger stimuli in the monkey. J Neurophysiol 56:1439–1461
184. Segura-Aguilar J, Paris I, Munoz P, Ferrari E, Zecca L, Zucca FA (2014) Protective and toxic roles of dopamine in Parkinson's disease. J Neurochem 129:898–915
185. Shaw JK, Pamela Alonso I, Lewandowski SI, Scott MO, O'Connor BM, Aggarwal S, De Biasi M, Mortensen OV, España RA (2021) Individual differences in dopamine uptake in the dorsomedial striatum prior to cocaine exposure predict motivation for cocaine in male rats. Neuropsychopharmacology 46:1757–1767
186. Shigenaga T, Kato M, Taguchi K (1997) In vivo voltammetric studies of the effects of intrathecal morphine on noxious heat stimuli-induced serotonin release in the nucleus raphe magnus of anesthetized rats. Neurosci Res 29:257–262
187. Shirokawa T, Ishida Y, Isobe K (2003) Age-related changes in the release and uptake activity of presynaptic axon terminals of rat locus coeruleus neurons. Neurosci Lett 344:212–214

188. Siciliano CA, Calipari ES, Cuzon Carlson VC, Helms CM, Lovinger DM, Grant KA, Jones SR (2015) Voluntary ethanol intake predicts kappa-opioid receptor supersensitivity and regionally distinct dopaminergic adaptations in macaques. J Neurosci 35:5959–5968
189. Siciliano CA, Calipari ES, Yorgason JT, Mateo Y, Helms CM, Lovinger DM, Grant KA, Jones SR (2016a) Chronic ethanol self-administration in macaques shifts dopamine feedback inhibition to predominantly D2 receptors in nucleus accumbens core. Drug Alcohol Depend 158:159–163
190. Siciliano CA, Calipari ES, Yorgason JT, Lovinger DM, Mateo Y, Jimenez VA, Helms CM, Grant KA, Jones SR (2016b) Increased presynaptic regulation of dopamine neurotransmission in the nucleus accumbens core following chronic ethanol self-administration in female macaques. Psychopharmacology 233:1435–1443
191. Siciliano CA, Mauterer MI, Fordahl SC, Jones SR (2019) Modulation of striatal dopamine dynamics by cocaine self-administration and amphetamine treatment in female rats. Eur J Neurosci 50:2740–2749
192. Sidlo Z, Reggio PH, Rice ME (2008) Inhibition of striatal dopamine release by CB1 receptor activation requires nonsynaptic communication involving GABA, H2O2, and KATP channels. Neurochem Int 52:80–88
193. Signs SA, Yamamoto BK, Schechter MD (1987) In vivo electrochemical determination of extracellular dopamine in the caudate of freely-moving rats after a low dose of ethanol. Neuropharmacology 26:1653–1656
194. Smith HR, Beveridge TJ, Porrino LJ (2006) Distribution of norepinephrine transporters in the non-human primate brain. Neuroscience 138:703–714
195. Soghomonian JJ, Descarries L, Watkins KC (1989) Serotonin innervation in adult rat neostriatum. II. Ultrastructural features: a radioautographic and immunocytochemical study. Brain Res 481:67–86
196. Sprouse JS, Aghajanian GK (1987) Electrophysiological responses of serotoninergic dorsal raphe neurons to 5-HT1A and 5-HT1B agonists. Synapse 1:3–9
197. Stathis M, Scheffel U, Lever SZ, Boja JW, Carroll FI, Kuhar MJ (1995) Rate of binding of various inhibitors at the dopamine transporter in vivo. Psychopharmacology 119:376–384
198. Sulzer D (2011) How addictive drugs disrupt presynaptic dopamine neurotransmission. Neuron 69:628–649
199. Sulzer D, Cragg SJ, Rice ME (2016) Striatal dopamine neurotransmission: regulation of release and uptake. Basal Ganglia 6:123–148
200. Sur C, Betz H, Schloss P (1996) Immunocytochemical detection of the serotonin transporter in rat brain. Neuroscience 73:217–231
201. Szumlinski KK, Diab ME, Friedman R, Henze LM, Lominac KD, Bowers MS (2007) Accumbens neurochemical adaptations produced by binge-like alcohol consumption. Psychopharmacology 190:415–431
202. Taguchi K, Suzuki Y (1991) Effects of morphine on the electrochemical signal modified by noxious stimuli in the nucleus raphe magnus of anesthetized rats. Neuropharmacology 30:1225–1231
203. Tao R, Ma Z, Auerbach SB (1998) Alteration in regulation of serotonin release in rat dorsal raphe nucleus after prolonged exposure to morphine. J Pharmacol Exp Ther 286:481–488
204. Thielen RJ, Engleman EA, Rodd ZA, Murphy JM, Lumeng L, Li T-K, Mcbride WJ (2004) Ethanol drinking and deprivation Alter dopaminergic and serotonergic function in the nucleus Accumbens of alcohol-preferring rats. J Pharmacol Exp Ther 309:216–225
205. Threlfell S, Greenfield SA, Cragg SJ (2010) 5-HT(1B) receptor regulation of serotonin (5-HT) release by endogenous 5-HT in the substantia nigra. Neuroscience 165:212–220
206. Torres GE, Gainetdinov RR, Caron MG (2003) Plasma membrane monoamine transporters: structure, regulation and function. Nat Rev Neurosci 4:13–25
207. Toyoda H, Won J, Kim W, Kim H, Davy O, Saito M, Kim D, Tanaka T, Kang Y, Oh SB (2022) The nature of noradrenergic volume transmission from locus Coeruleus to brainstem mesencephalic trigeminal sensory neurons. Front Cell Neurosci 16:841239

208. Van Bockstaele EJ, Colago EE, Moriwaki A, Uhl GR (1996a) Mu-opioid receptor is located on the plasma membrane of dendrites that receive asymmetric synapses from axon terminals containing leucine-enkephalin in the rat nucleus locus coeruleus. J Comp Neurol 376:65–74
209. Van Bockstaele EJ, Colago EE, Cheng P, Moriwaki A, Uhl GR, Pickel VM (1996b) Ultrastructural evidence for prominent distribution of the mu-opioid receptor at extrasynaptic sites on noradrenergic dendrites in the rat nucleus locus coeruleus. J Neurosci 16:5037–5048
210. Van Bockstaele EJ, Qian Y, Sterling RC, Page ME (2008) Low dose naltrexone administration in morphine dependent rats attenuates withdrawal-induced norepinephrine efflux in forebrain. Prog Neuro-Psychopharmacol Biol Psychiatry 32:1048–1056
211. van Erp AM, Miczek KA (2000) Aggressive behavior, increased accumbal dopamine, and decreased cortical serotonin in rats. J Neurosci 20:9320–9325
212. Venton BJ, Cao Q (2020) Fundamentals of fast-scan cyclic voltammetry for dopamine detection. Analyst 145:1158–1168
213. Venton BJ, Seipel AT, Phillips PE, Wetsel WC, Gitler D, Greengard P, Augustine GJ, Wightman RM (2006) Cocaine increases dopamine release by mobilization of a synapsin-dependent reserve pool. J Neurosci 26:3206–3209
214. Verbanck P, Seutin V, Dresse A, Scuvee J, Massotte L, Giesbers I, Kornreich C (1990) Electrophysiological effects of ethanol on monoaminergic neurons: an in vivo and in vitro study. Alcohol Clin Exp Res 14:728–735
215. Verbanck PM, Seutin V, Massotte L, Dresse A (1992) Differential effects of picrotoxin and RO 15-1788 on high and low ethanol concentrations on rat locus coeruleus in vitro. Eur J Pharmacol 211:15–21
216. von Euler US (1971) Adrenergic neurotransmitter functions. Science 173:202–206
217. Wang Y, Zhang P, Chao Y, Zhu Z, Yang C, Zhou Z, Li Y, Long Y, Liu Y, Li D, Wang S, Qu Q (2024) Transport and inhibition mechanism for VMAT2-mediated synaptic vesicle loading of monoamines. Cell Res 34:47–57
218. Washburn M, Moises HC (1989) Electrophysiological correlates of presynaptic alpha 2-receptor-mediated inhibition of norepinephrine release at locus coeruleus synapses in dentate gyrus. J Neurosci 9:2131–2140
219. Wei C, Han X, Weng D, Feng Q, Qi X, Li J, Luo M (2018) Response dynamics of midbrain dopamine neurons and serotonin neurons to heroin, nicotine, cocaine, and MDMA. Cell Discov 4:60
220. Weiss F, Parsons LH, Schulteis G, Hyytia P, Lorang MT, Bloom FE, Koob GF (1996) Ethanol self-administration restores withdrawal-associated deficiencies in accumbal dopamine and 5-hydroxytryptamine release in dependent rats. J Neurosci 16:3474–3485
221. Williams JT, Henderson G, North RA (1985) Characterization of alpha 2-adrenoceptors which increase potassium conductance in rat locus coeruleus neurones. Neuroscience 14:95–101
222. Wonnacott S, Kaiser S, Mogg A, Soliakov L, Jones IW (2000) Presynaptic nicotinic receptors modulating dopamine release in the rat striatum. Eur J Pharmacol 393:51–58
223. Wright WJ, Dong Y (2020) Psychostimulant-induced adaptations in nucleus Accumbens glutamatergic transmission. Cold Spring Harb Perspect Med 10:a039255
224. Wu Q, Reith ME, Walker QD, Kuhn CM, Carroll FI, Garris PA (2002) Concurrent autoreceptor-mediated control of dopamine release and uptake during neurotransmission: an in vivo voltammetric study. J Neurosci 22:6272–6281
225. Xu L, Chen LY (2022) Effects of the N-terminal dynamics on the conformational states of human dopamine transporter. Biophys Chem 283:106765
226. Xu L, Kelkar SV, Lomenzo SA, Izenwasser S, Katz JL, Kline RH, Trudell ML (1997) Synthesis, dopamine transporter affinity, dopamine uptake inhibition, and locomotor stimulant activity of 2-substituted 3 beta-phenyltropane derivatives. J Med Chem 40:858–863
227. Xu F, Gainetdinov RR, Wetsel WC, Jones SR, Bohn LM, Miller GW, Wang YM, Caron MG (2000) Mice lacking the norepinephrine transporter are supersensitive to psychostimulants. Nat Neurosci 3:465–471

228. Yan QS, Reith ME, Jobe PC, Dailey JW (1996) Focal ethanol elevates extracellular dopamine and serotonin concentrations in the rat ventral tegmental area. Eur J Pharmacol 301:49–57
229. Yang JW, Larson G, Konrad L, Shetty M, Holy M, Jäntsch K, Kastein M, Heo S, Erdem FA, Lubec G, Vaughan RA, Sitte HH, Foster JD (2019) Dephosphorylation of human dopamine transporter at threonine 48 by protein phosphatase PP1/2A up-regulates transport velocity. J Biol Chem 294:3419–3431
230. Yavich L, Forsberg MM, Karayiorgou M, Gogos JA, Mannisto PT (2007) Site-specific role of catechol-O-methyltransferase in dopamine overflow within prefrontal cortex and dorsal striatum. J Neurosci 27:10196–10209
231. Yokoo H, Tanaka M, Yoshida M, Tsuda A, Tanaka T, Mizoguchi K (1990) Direct evidence of conditioned fear-elicited enhancement of noradrenaline release in the rat hypothalamus assessed by intracranial microdialysis. Brain Res 536:305–308
232. Yorgason JT, Jones SR, España RA (2011) Low and high affinity dopamine transporter inhibitors block dopamine uptake within 5 sec of intravenous injection. Neuroscience 182:125–132
233. Yorgason JT, Ferris MJ, Steffensen SC, Jones SR (2014) Frequency-dependent effects of ethanol on dopamine release in the nucleus accumbens. Alcohol Clin Exp Res 38:438–447
234. Yorgason JT, Rose JH, McIntosh JM, Ferris MJ, Jones SR (2015) Greater ethanol inhibition of presynaptic dopamine release in C57BL/6J than DBA/2J mice: role of nicotinic acetylcholine receptors. Neuroscience 284:854–864
235. Yorgason JT, Calipari ES, Ferris MJ, Karkhanis AN, Fordahl SC, Weiner JL, Jones SR (2016) Social isolation rearing increases dopamine uptake and psychostimulant potency in the striatum. Neuropharmacology 101:471–479
236. Yoshimoto K, McBride WJ, Lumeng L, Li TK (1992) Alcohol stimulates the release of dopamine and serotonin in the nucleus accumbens. Alcohol 9:17–22
237. Yoshimoto K, Ueda S, Kato B, Takeuchi Y, Kawai Y, Noritake K, Yasuhara M (2000) Alcohol enhances characteristic releases of dopamine and serotonin in the central nucleus of the amygdala. Neurochem Int 37:369–376
238. Yuen J, Goyal A, Rusheen AE, Kouzani AZ, Berk M, Kim JH, Tye SJ, Blaha CD, Bennet KE, Lee KH, Oh Y, Shin H (2022) Cocaine increases stimulation-evoked serotonin efflux in the nucleus accumbens. J Neurophysiol 127:714–724
239. Zhang H, Sulzer D (2004) Frequency-dependent modulation of dopamine release by nicotine. Nat Neurosci 7:581–582
240. Zhang L, Dong Y, Doyon WM, Dani JA (2012) Withdrawal from chronic nicotine exposure alters dopamine signaling dynamics in the nucleus accumbens. Biol Psychiatry 71:184–191
241. Zhou FC, Tao-Cheng JH, Segu L, Patel T, Wang Y (1998) Serotonin transporters are located on the axons beyond the synaptic junctions: anatomical and functional evidence. Brain Res 805:241–254
242. Zhou FM, Liang Y, Dani JA (2001) Endogenous nicotinic cholinergic activity regulates dopamine release in the striatum. Nat Neurosci 4:1224–1229
243. Zhu J, Bardo MT, Dwoskin LP (2013) Distinct effects of enriched environment on dopamine clearance in nucleus accumbens shell and core following systemic nicotine administration. Synapse 67:57–67
244. Zimmer BA, Oleson EB, Roberts DC (2012) The motivation to self-administer is increased after a history of spiking brain levels of cocaine. Neuropsychopharmacology 37:1901–1910
245. Zinsmaier AK, Dong Y, Huang YH (2022) Cocaine-induced projection-specific and cell type-specific adaptations in the nucleus accumbens. Mol Psychiatry 27:669–686
246. Zoli M, Moretti M, Zanardi A, McIntosh JM, Clementi F, Gotti C (2002) Identification of the nicotinic receptor subtypes expressed on dopaminergic terminals in the rat striatum. J Neurosci 22:8785–8789

Dopamine–Glutamate and Dopamine–GABA Co-release

Annalisa Scimemi

Contents

Abstract Some neuronal populations in the brain have the ability to release multiple neurotransmitters, which may be packaged in the same vesicle, or released by distinct subsets of vesicles. Here, we review current knowledge on the molecular mechanisms by which multiple neurotransmitters can be stored in the same cell and the functional implications that this has for information processing throughout the brain and the control of movement execution by basal ganglia and cortical motor circuits.

Keywords Dopamine · Glutamate · GABA · Synaptic vesicle · Co-release · Transport

---

A. Scimemi (✉)
Department of Biology, SUNY Albany, Albany, NY, USA
e-mail: ascimemi@albany.edu

O. V. Mortensen (ed.), *Monoamine Transporters*, Advances in Neurobiology 46,
https://doi.org/10.1007/978-3-031-96364-3_8

1 Co-storage, Co-release, and Co-transmission

Simple principles lure the human mind, as they often provide elegant and at times easy solutions to complex problems. A simple principle in neuroscience is that the nature of the neurotransmitter(s) released by a neuron at a given terminal informs us on the neurotransmitter(s) that the same neuron can release at other terminals [1]. This current-day rephrasing of Dale's principle, read with an open mind, has served as a stakeholder in the neuroscience field since 1930 [2]. In more recent times, we have become aware that the use of plurals, in this statement, is not a coincidence and not even optional. In fact, there are numerous examples of neurons in the brain that can release more than one neurotransmitter from the same terminal and/or from different sets of terminals. The term *co-release* provides a broad description of this process, without adding any further insight on whether this happens through *co-packaging* of multiple neurotransmitters in the same vesicle, *independent packaging* of different neurotransmitters in different vesicles within the same terminal, or through spatial segregation of vesicles filled with different neurotransmitters at distinct axonal boutons (Fig. 1) [3]. The term *co-transmission*, instead, puts emphasis on the fact multiple transmitters released from the same neuron contribute to synaptic transmission on the same pre- or post-synaptic target. With this information at hand (co-release vs. co-transmission), one can imagine a scenario where co-release does not lead to co-transmission if the target compartment does not express receptors for the multiple transmitters being released from the same pre-synaptic neuron. In other instances, storage of vesicles with distinct neurotransmitters in the same terminal is not associated with co-release, because, for example, these vesicles have different functional coupling to calcium channels in the presynaptic active zone,

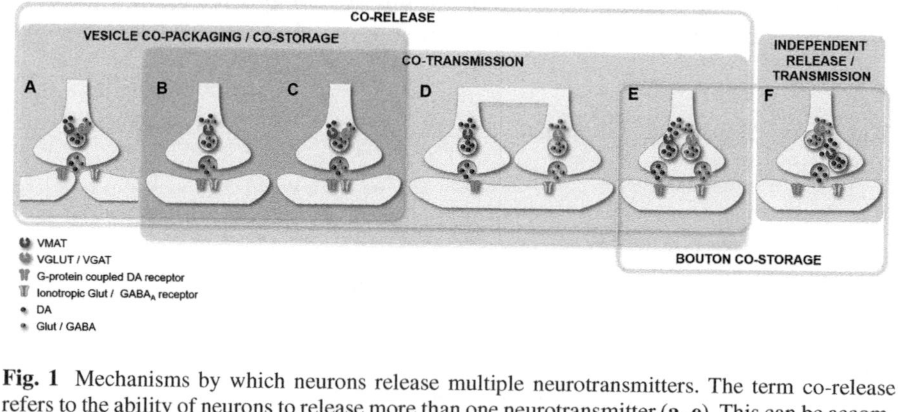

Fig. 1 Mechanisms by which neurons release multiple neurotransmitters. The term co-release refers to the ability of neurons to release more than one neurotransmitter (**a**–**e**). This can be accomplished by co-packaging multiple neurotransmitters in the same vesicle (**a**–**c**), or by storing vesicles in distinct vesicles located at different boutons (**d**) or at the same axonal bouton (**e**). In some instances, vesicles with distinct neurotransmitters can reside in the same bouton, but can be released independently, for example, because of their different coupling to calcium channels or other molecules in the release machinery (**f**). Co-transmission occurs when different neurotransmitters are released from the same neuron onto the same post-synaptic target (**b**–**e**)

pointing to the presence of specific regulatory mechanisms that control different types of information transfer from the same axonal bouton.

The neurotransmitter dopamine (DA) is interesting, in this context, because it can be co-released with other molecules, including the major excitatory and inhibitory neurotransmitters glutamate and GABA. Whether this means that there is co-storage in the same vesicle/bouton or co-transmission of DA, glutamate, and GABA from a given neuron depends on the specific cell type and brain region, the developmental stage of the organism it belongs to, and the identity of the co-released neurotransmitter. Here, we provide an overview of the molecular mechanisms that allow DA, glutamate, and GABA to be stored in synaptic vesicles, and the functional implications of this phenomenon.

2 Molecular Mechanisms for Loading DA, Glutamate, and GABA into Synaptic Vesicles

The process of filling synaptic vesicles with neurotransmitters depends on the presence of an electrochemical gradient for H^+ ($\Delta\mu_{H+}$), established by the vacuolar-type H^+-ATPase [4]. This vacuolar H^+ pump uses ATP hydrolysis to accumulate H^+ into synaptic vesicles, creating an electrical ($\Delta\psi$) and a chemical gradient for H^+ (ΔpH). The degree to which the vesicular transport of a neurotransmitter relies on ΔpH versus $\Delta\psi$ is not a given, but depends on the net charge of the neurotransmitter, and the stoichiometry of coupling to H^+ [5]. There are two variants of the vesicular monoamine transporter VMAT, the primary vesicular DA transporter known to date, namely, VMAT1 and VMAT2 [6]. VMAT1 is expressed in peripheral endocrine and paracrine cell; VMAT2 is the predominant monoamine vesicular transporter in the central nervous system, and the one we refer to in this work [6–8]. VMAT2 exchanges $2H^+$ from the vesicle lumen for each DA molecule in the cell cytoplasm. At pH 7.2 (cytoplasm) and 5.5 (vesicle lumen), DA is positively charged (pKa = 8.9) [9], so the VMAT2 stoichiometry of $2H^+$:$1DA^+$ leads to a net loss of $2H^+$ but of only 1 positive charge from the vesicle lumen [10–12]. As a result, DA storage in synaptic vesicles dissipates ΔpH more than $\Delta\psi$ and requires mechanisms other than the vacuolar H^+-ATPase to restore $\Delta\mu_{H+}$ [5]. The relationship between neurotransmitter uptake into synaptic vesicles and the dissipation of ΔpH and $\Delta\psi$ differs for the vesicular glutamate and GABA transporters VGLUT and VGAT (also known as the vesicular inhibitory amino acid transporter, VIAAT; Fig. 2).

The vesicular glutamate transporter VGLUT has three different isoforms (VGLUT1-3), which differ in their spatial distribution and in their amino-acid sequence at cytoplasmic N- and C-terminals, but not in their biophysical properties [14]. Accordingly, VGLUT1 is expressed primarily at low-release probability synapses in the cortex and hippocampus, whereas VGLUT2 is expressed at higher-release probability synapses in the diencephalon and brainstem [15]. VGLUT3 is expressed by cholinergic interneurons (CINs) in the dorsal and ventral striatum,

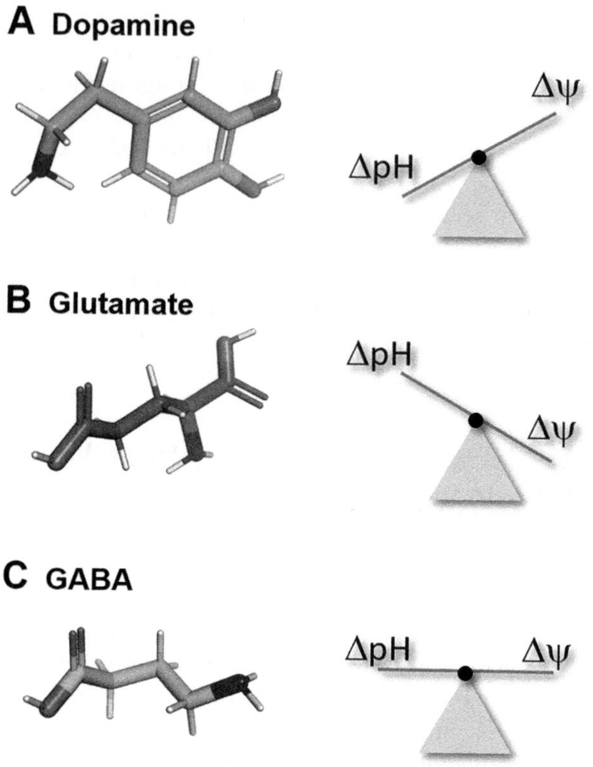

Fig. 2 Dependence on the electrochemical gradient for protons for the vesicular uptake of DA, glutamate, and GABA. Although the electrochemical H^+ gradient ($\Delta\mu_{H+}$) established by the vacuolar H^+-ATPase is required to transport neurotransmitters into synaptic vesicles, different vesicular transporters rely to different extents on the electrical ($\Delta\psi$) and on the chemical components (ΔpH) of $\Delta\mu_{H+}$. (**a**) The vesicular accumulation of DA involves the exchange of 1 cytosolic DA^+ for 2 luminal H^+, which leads to a greater dependence on ΔpH than $\Delta\psi$. (**b**) The vesicular transport of glutamate involves exchange of *n* glutamate molecules for *n* H^+, resulting in the movement of *n* + *1* charges, leading to a greater dependence on $\Delta\psi$ than on ΔpH. A Cl^- conductance associated with VGLUT may promote vesicle acidification by Cl^- [13]. VGLUT may contribute to ΔpH by dissipating $\Delta\psi$ through the transport of a negatively charged neurotransmitter. (**c**) Transport of the neutral zwitterion GABA involves the movement of an equal number of H^+ and electrical charges, leading to a similar dependence of vesicular GABA uptake on ΔpH and $\Delta\psi$ [5]

sub-populations of GABAergic basket cells in the cortex and hippocampus, and serotoninergic neurons [16]. VGLUT1-3 also differ for their endo–/exo-cytotic rates, perhaps due to differences in cell-trafficking/regulation, both of which can be controlled by the N- and C-terminal sequences of VGLUT [17–19]. There are some important differences between glutamate uptake via vesicular transporters like VGLUT and glutamate uptake by plasma membrane transporters like EAAT1-5. *First,* VGLUT has a high substrate specificity and does not recognize other substrates like aspartate, which is recognized as a substrate by EAAT1-5 [20–23]. *Second*, the K_m for glutamate uptake by VGLUT is ~1 mM, which is in the range of the cytoplasmic concentration of glutamate in presynaptic terminals, but much higher than the micromolar glutamate affinity of EAAT1-5 ([24–32] but see [33, 34]). The exact stoichiometry of glutamate uptake via VGLUT remains unknown. However, since at pH 7.2 (cytoplasm) and 5.5 (vesicle lumen), glutamate is negatively charged (pKa = 2.1), its exchange for H^+ leads to the efflux of one positive charge more than the number of exchanged H^+ (which dissipates $\Delta\psi$ more than ΔpH). In addition, a Cl^- conductance associated with the VGLUTs may also promote acidification by Cl^- [13]. High Cl^- may inhibit glutamate uptake indirectly by dissipating $\Delta\psi$, or perhaps directly by competing with glutamate for binding to VGLUT [5]. Therefore, in contrast to VMAT2, glutamate transport into synaptic vesicles depends primarily on $\Delta\psi$.

Unlike VMAT2 and VGLUT, the vesicular GABA transporter VGAT, which recognizes GABA and glycine as a substrate and is expressed in both GABAergic and glycinergic terminals [35], relies equally on ΔpH and $\Delta\psi$, although even in this case, the exact stoichiometry remains unknown [36, 37]. Despite relying on $\Delta\mu_{H+}$, VGAT displays a high affinity for Cl^- and its substrate uptake requires cotransport of $2Cl^-$, suggesting VGAT may rely on $\Delta\psi$ more than ΔpH [38].

Together, these findings show how DA, glutamate, and GABA storage into synaptic vesicles rely differently on ΔpH and $\Delta\psi$, yet as we will point out in later sections, vesicle filling with one of these neurotransmitters can influence the uptake of other molecules in the same vesicle because of its effects on ΔpH and $\Delta\psi$. Clearly, a better understanding of all the nuances of this process will benefit from future information on the stoichiometry and regulation of the uptake process through the different vesicular transporters for these neurotransmitters.

3 What Accounts for Co-release? Substrate Promiscuity, Co-expression, and Current Unknowns About Vesicular Transporters

Evidence that glutamate can be co-released with GABA has been collected at hippocampal mossy fibers, entopeduncular nucleus somatostatin projection neurons to the lateral habenula, and POMC-expressing hypothalamic neurons [39–43]. Similarly, dopaminergic (DAergic) projections from the *substantia nigra pars compacta* (SNc) and the ventral tegmental area (VTA) have been shown to release DA and glutamate in the dorsal and ventral striatum [44–46], and to inhibit action potential firing in the striatum through the release of DA and GABA [46, 47]. This could be accomplished through two non-mutually exclusive scenarios, whereby one neuron could express different types of vesicular transporters (each capable of taking up only one neurotransmitter) or vesicular transporters with poor substrate specificity (each capable of taking up multiple neurotransmitters). Whereas the vesicular glutamate transporter VGLUT does not show substrate promiscuity and can be used as a marker for glutamatergic synapses, the vesicular GABA transporter VGAT also transports glycine [48]. Consistent with this, GABA–glycine co-storage and co-release have been shown to occur in the medial nucleus of the trapezoid body [49], in spinal interneurons [50], and transiently during development in lateral superior olive neurons [51]. Likewise, the vesicular monoamine transporter VMAT2, as the name suggests, recognizes all catecholamines, serotonin and histamine, and is expressed in all monoaminergic neurons [5].

Immunolabeling experiments have shown that some neurons in the brain contain distinct sets of vesicles, each expressing a different type of vesicular transporter or expressing multiple vesicular transporter types. Both arrangements could account for co-release of multiple neurotransmitters from the same neuron [52]. Accordingly, VGLUT and VGAT are co-expressed in the same vesicles at a subset of hippocampal and cerebellar mossy fiber terminals, cerebellar basket cells, neurons in the anteroventral periventricular nucleus, and in retinal bipolar cells, which would account for glutamate–GABA co-release [53–55].

VMAT2 and VGLUT are co-expressed at a subset of synaptic vesicles in the striatum, where vesicle filling with glutamate facilitates vesicle filling with DA through its ability to increase ΔpH, accounting for DA–glutamate co-release [56, 57]. This has been suggested to be important to allow DAergic neurons to release DA during high-frequency burst firing to sustain DA release over long time windows [58].

The case of DA-GABA co-release remains puzzling. Although there is no evidence of co-expression of VMAT2 and VGAT, genetic deletion of VGAT does not affect channelrhodopsin (ChR2), optically-evoked inhibitory postsynaptic currents (oIPSCs) from DAergic neurons, suggesting that VGAT may not be the only vesicular transporter for GABA [46, 48, 59]. Because pharmacological blockade of VMAT2 eliminates inhibitory postsynaptic currents (IPSCs), and exogenous expression of VMAT2 restores GABA IPSCs, it has been suggested that VMAT2 might be the vesicular transporter responsible for vesicular GABA uptake [46]. This is surprising because GABA is not a known substrate for transport via VMAT2, and does not share structural resemblance with dopamine or other monoamines ([60]; Fig. 2). A parsimonious interpretation is therefore that VGAT is not the only route for packaging GABA into synaptic vesicles (consistent with the fact that GABA release is reduced but not absent in VGAT$^{-/-}$ mice [48]), and that VMAT2 may be able to contribute indirectly to GABA storage into synaptic vesicles through mechanisms that remain a matter of current investigations.

4 Dopamine-Glutamate Co-release: Evidence and Significance

Although DAergic neurons have been known to release peptides like cholecystokinin and neurotensin in addition to DA, evidence that they can also release small molecular neurotransmitters like glutamate and GABA has been recognized more recently [5, 61–63]. DA and glutamate can be co-released not only in cultured ventral midbrain DAergic neurons, where it was first identified, but also in vivo, in the dorsal and ventral striatum [45, 64–67] (Fig. 3). To release DA and glutamate, DAergic neurons require the expression of both VMAT2 and VGLUT. Low levels of mRNA encoding VGLUT in DAergic neurons have been collected in vitro and

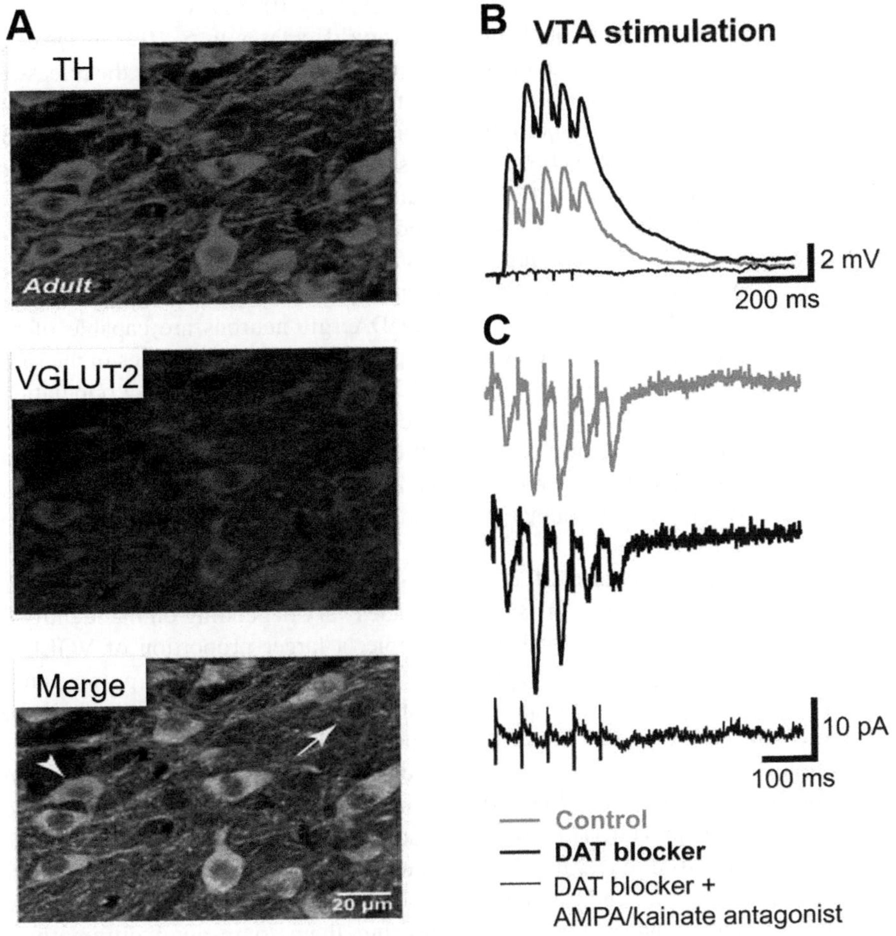

Fig. 3 Immunohistochemical and electrophysiological evidence supporting the existence of DA-glutamate co-release at VTA to *nucleus accumbens* (NAcc) synapses. *Left,* immunoreactivity for the DA biosynthetic enzyme tyrosine hydroxylase (TH; *green*), the vesicular glutamate transporter VGLUT2 (*red*) in the VTA of an adult mouse. Most neurons are TH+/VGLUT+. The arrowhead denotes a TH+/VGLUT2− neuron, and the arrow denotes a TH−/VGLUT2+ neuron. *Right:* Current clamp recordings from NAcc medium spiny neurons (MSNs) in response to VTA electrical stimulation. The evoked excitatory post-synaptic potentials, recorded in the presence of the $GABA_A$ receptor antagonist SR95531 (10 μM), are increased by the DA reuptake inhibitor nomifensine (10 μM) and blocked by the AMPA/kainate receptor antagonist CNQX (40 μM). (Modified with permission from [64])

in vivo [15, 68–71]. Although low expression of VGLUT mRNA is not a unique feature of DAergic neurons, since it also occurs in non-DAergic midbrain neurons [72], this finding is consistent with DAergic neurons being able to load multiple neurotransmitters in synaptic boutons through the activity of different vesicular transporters.

Saying that DAergic neurons *can* release glutamate and GABA does not mean that they always *do* so. For example, in the VTA, only 10–30% of all DAergic neurons also release glutamate, and this proportion decreases to 5–10% of DAergic neurons in the SNc [73–80]. An interesting question to ask is whether the release of DA and glutamate occurs from the same terminal (or even the same vesicle), from distinct terminals, or a mix of both. Initial evidence indicated that release of DA and glutamate may occur from distinct processes, pointing to a potential segregation of DA and glutamate release sites [67, 81]. Later work in cultured midbrain DAergic neurons, however, indicated that VMAT2 and VGLUT1 co-localize at ~73% of presynaptic boutons [82]. The remaining 27% of boutons show a segregated expression of VMAT2 and VGLUT1, with 20% boutons being VMAT2+, and 7% of boutons being VGLUT1+ [82]. Therefore, midbrain DAergic neurons are capable of co-releasing DA and glutamate, and even of co-packaging these molecules in the same vesicles [57], while also segregating VMAT2 and VGLUT1 to distinct populations of synaptic boutons [82]. These findings are consistent with in vivo data showing that DAergic neurons can form two distinct types of synapses: (*i*) *en passant* symmetric synapses that contact dendritic spine necks and presumably release DA; (*ii*) asymmetric synapses that contact dendritic spine heads and release only glutamate [81].

The use of optical reporters for VMAT2 or VGLUT1 allowed us to understand that the functional properties of synaptic vesicles vary depending on the identity of the neurotransmitter(s) they load. For example, a larger proportion of VGLUT1 resides on the recycling pool of synaptic vesicles, whereas a larger proportion of VMAT2 resides on the reserve pool of vesicles [82]. Optogenetic stimulation of DAergic neurons shows that DA and glutamate deplete with different kinetics and are coupled to different types of voltage-gated calcium channels and other proteins in the presynaptic active zone [83]. After stimulation, VMAT2 undergoes endocytosis more slowly than VGLUT1, likely reflecting an increased diffusion of VMAT2 along the axon before retrieval [82]. Together, these findings suggest that synaptic boutons containing DA and glutamate vesicles may be capable of regulating their release differently and independently, allowing them to respond differently to stimulation.

In the ventral striatum, where there is the highest density of DA–glutamate projections, these cells target MSNs, fast-spiking interneurons, and send the strongest projections to CINs [44]. By contrast, in the dorsal striatum, CINs receive DA–glutamate projections, but the strongest inputs are sent to the MSNs [84, 85] (Fig. 4). Further distinctions have been observed between the dorsolateral and the dorsomedial striatum with respect to DA–glutamate co-release. Accordingly, optogenetic activation of DAergic inputs drives burst firing in CINs of the dorsolateral striatum

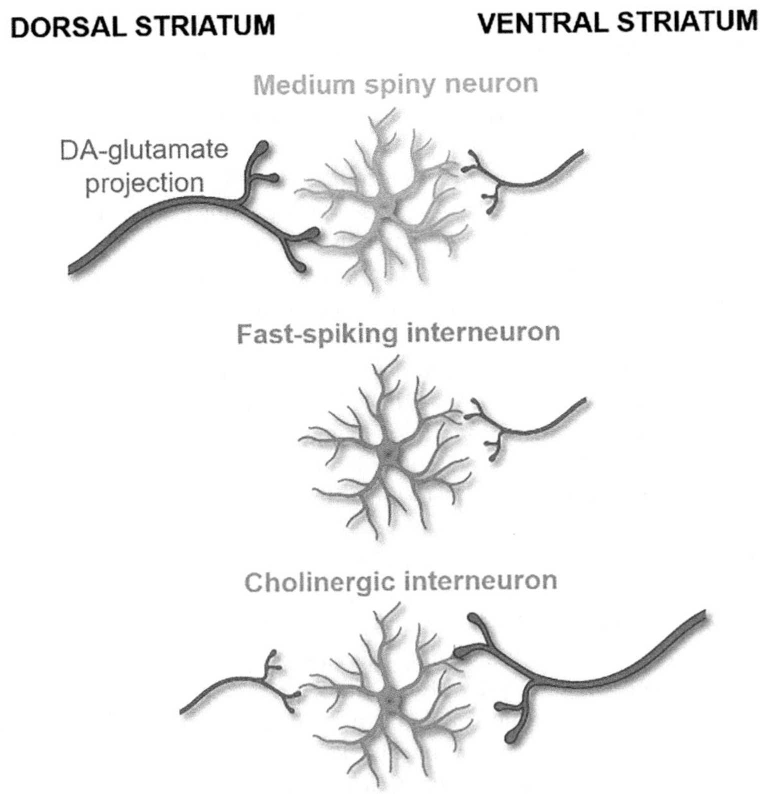

Fig. 4 Cellular targets of DA-glutamate co-release in the dorsal and ventral striatum. Schematic representation of the strength of DA-glutamate co-release in striatal MSNs, fast-spiking, and CINs. In the dorsal striatum, the strongest DA-glutamate projections target MSNs, whereas in the ventral striatum, these are formed onto CINs

by activating group I mGluRs, while it pauses action potential firing in CINs of the dorsomedial striatum by activating D2 DA receptors [84].

Other projections from which DA–glutamate are co-released have been detected, in juvenile mice, in layers II-III of the cingulate cortex, in GABAergic interneurons in the pre-frontal cortex, and in principal cells in the central amygdala and entorhinal cortex [86–89]. The functional effects of this co-release depend in part on the biophysical properties and location of the activated receptors with respect to the DA–glutamate release sites (metabotropic receptors for DA, and both ionotropic and metabotropic receptors for glutamate), in part on the wiring diagram of the DA-glutamate projections. One can imagine how glutamate released at direct synaptic connections containing AMPA and NMDA receptors could initiate fast synaptic signals, whereas DA released at non-synaptic sites would act more slowly through volume transmission and activation of different types of metabotropic DA receptors [90]. Consistent with this, in the ventral striatum, DA–glutamate release has been shown to activate CINs, triggering the release of acetylcholine that in turn potentiates DA release by activating presynaptic acetylcholine receptors on the axons of long-range projection, VTA DAergic neurons [44, 75, 91–93]. In VTA glutamatergic neurons, selective disruption of glutamate release abolishes optogenetically induced self-stimulation without altering real-time place avoidance, whereas disruption of dopamine release from these cells preserves self-stimulation and abolishes place avoidance. This suggests that the same cell can convey different types of information through the release of different neurotransmitters. Specifically, glutamate release from VTA glutamate neurons acts as a positive reinforcer, whereas DA release from these cells induces avoidance behaviors [94]. In support of the important, but still incompletely understood, physiological relevance of DA-glutamate co-release, selective loss of VGLUT in DAergic neurons leads to a compensatory increase in the AMPA/NMDA ratio in D1-MSNs and is associated with increased hippocampal excitability and impaired spatial memory function [95, 96]. This also leads to increased vulnerability to neurotoxicity, similar to that induced by MPTP and 6-OHDA [79, 97]. All these effects are likely to vary with age and disease, since the expression of VGLUT in DAergic neurons is dynamically regulated during development and in Parkinson's disease [79, 98–101]. Specifically, VGLUT-expressing DAergic neurons, whose co-localization with TH peaks at birth and declines in adulthood in mice, are more resilient to degeneration, suggesting that DA-glutamate co-transmission may exert a neuroprotective role [70, 101, 102]. The response to cues associated with reward delivery is also altered by disrupting DA–glutamate co-release, which might be of particular interest in understanding the synaptic basis of relapse in addiction [103]. Given the potential implications of DA–glutamate co-release, much attention is now devoted to understanding the dynamics and modulation of this process during development and disease.

5 Dopamine–GABA Co-release: Evidence and Significance

Although functional evidence for DA-GABA co-transmission exists, SNc DAergic neurons do not possess the GABA synthetic enzymes (GAD65/67) or the vesicular GABA transporter VGAT [104–106]. Therefore, it has been suggested that inhibitory co-transmission from SNc neurons might rely on the DA transporter VMAT2 [46]. Consistent with this hypothesis: (*i*) conditional loss of VGAT from DAergic neurons does not impair GABA release from these cells; (*ii*) blocking VMAT2 with reserpine inhibits GABA release from these cells; (*iii*) exogenous expression of VMAT2 in mice with conditional loss of VGAT mice induces a partial rescue of GABA release [46]. The dilemma, however, comes from the fact that VMAT2 is not known to transport GABA [60]. There are ways to reconcile the lack of vesicular GABA transporters and biosynthetic enzymes for GABA with the ability of DAergic neurons to release GABA. For example, one could hypothesize that (*i*) the inhibitory transmitter released by SNc neurons may not actually be GABA, but an endogenous agonist of $GABA_A$ receptors that can be transported by VMAT2; (*ii*) GABA may be synthesized in SNc neurons by other enzymes, like the mitochondrial enzyme Aldh1a1 [107], known to be expressed by a sub-population of SNc neurons [104, 108, 109]; (*iii*) GABA may be recycled from the extracellular space by membrane GABA transporters like GAT1, which is expressed at the mRNA level in SNc neurons [106], and then stored in synaptic vesicles through either a currently unknown VMAT2-dependent or VMAT2-independent mechanism. Against the argument of presynaptic uptake of extracellular GABA supporting GABA release from SNc DAergic neurons are the findings that GAT1 does not show a robust co-localization on DAergic axons of SNc neurons, and that GAT1 inhibition non-selectively depresses vesicular release simply by dampening the excitability of SNc axons [110–113]. However, recent evidence based on the use of conditional ChR2 expression indicates that (*i*) GABA release from SNc neurons is not altered in Aldh1a1$^{-/-}$:DAT$^{IRES\text{-}Cre/+}$ mice; (*ii*) GABA release is not altered by rescuing Aldh1a1 expression in SNc neurons of Aldh1a1$^{-/-}$:DAT$^{IRES\text{-}Cre/+}$ mice; (*iii*) there is no GABA released from SNc neurons in DAT$^{IRES\text{-}Cre/+}$:GAT1$^{fl/+}$ mice, which lack GAT1 in all DAergic neurons [59]. In this case, one would conclude that, despite its limited expression, the membrane GABA transporter GAT1 supports GABA uptake in SNc axons, and that GABA is then loaded into synaptic vesicles by a promiscuous vesicular VMAT2 transporter [59].

How do we then reconcile data, which is collected from mouse models with those obtained from pharmacology, showing that blockade of Aldh1a1 reduces the amplitude of oIPSCs in striatal projection neurons [107, 114]? On one hand, one might be inclined to say that many drugs are not as specific as we would like them to be and may exert potential off-target effects. One way to rule out this hypothesis would be to test the effect of Aldh1a1 antagonists like DEAB on oIPSCs recorded from MSNs in Aldh1a1$^{-/-}$ mice. On the other hand, constitutive knock-out mice are not immune to potential compensatory effects, which may be difficult to account for without knowing all the existing routes for GABA biosynthesis. Lastly, one would also need to rule out GABA$_A$ receptor saturation as a potential confounding factor for the lack of effect of Aldh1a1 rescue on the oIPSC amplitude [59].

Recent evidence suggests that despite the existence of DA-GABA co-transmission, the release of these neurotransmitters may be independent of one another. Accordingly, (*i*) the calcium sensitivity and release probability are different for DA and GABAergic transmission from the same DAergic neuron; (*ii*) GABA release is not controlled by the active zone protein RIM; (*iii*) DA and GABA release are modulated differently by GABA$_B$ and D2 DA receptors [107, 114]. Interesting, in the context of DA-GABA release, is the fact that these neurotransmitters can be released from DAergic axons in an action potential-independent manner [93, 115]. This is thought to occur via synchronous activation of CINs that, by directly activating calcium influx via nicotinic receptors or by indirectly recruiting voltage-gated calcium channels or internal calcium stores, can promote axonal DA-GABA release [115].

The implications of DA-GABA co-release in motivated reward-seeking behaviors have been tested in a mouse model of Angelman syndrome lacking the gene encoding the E3-ubiquitin ligase *Ube3a*, essential for GABA release from DAergic neurons and positively reinforced behaviors [116]. Here, loss of UBE3A was found to enhance optical self-stimulation of VTA-NAc afferents to the ventral striatum, whereas restoring GABA release from these neurons ameliorates reward-seeking behaviors in mice devoid of UBE3A [116]. These findings suggest that DA-GABA co-release might be involved in regulating action selection and initiation through the striatal circuitry. More recently, GABA co-release from DAergic axons has been shown to inhibit DA release by activating pre-synaptic α_3-subunit containing $GABA_A$ receptors in the dorsal striatum and the NAc [47]. Future studies will likely shed light on how GABA co-release might refine motor control not only by altering, for example, action repetition and vigor [117–121].

6 Conclusions

While there is overwhelming evidence of DA-glutamate and DA-GABA from many DAergic neurons in the brain, there is still much unknown about the molecular mechanisms that allow these different transmitters to be co-released. Data showing that the release of these molecules may have different properties and forms of modulation adds further complexity to the dynamics of synaptic transmission. The development of tools that enable selective perturbation of distinct sets of synapses and/or different cell types is likely to be a major contributor to our quest for answers about the physiological relevance and potential pathological implications of these phenomena.

Acknowledgments This work was supported by NSF IOS1655365 (A.S.). The author would like to thank Maurice A. Petroccione and other members of the Scimemi lab for their invaluable discussions and feedbacks on the text of this chapter.

References

1. Strata P, Harvey R (1999) Dale's principle. Brain Res Bull 50:349–350
2. Dale HH, Gaddum JH (1930) Reactions of denervated voluntary muscle, and their bearing on the mode of action of parasympathetic and related nerves. J Physiol 70:109–144
3. Tritsch NX, Granger AJ, Sabatini BL (2016) Mechanisms and functions of GABA co-release. Nat Rev Neurosci 17:139–145
4. Forgac M (2007) Vacuolar ATPases: rotary proton pumps in physiology and pathophysiology. Nat Rev Mol Cell Biol 8:917–929
5. Hnasko TS, Edwards RH (2012) Neurotransmitter corelease: mechanism and physiological role. Annu Rev Physiol 74:225–243
6. Masson J, Sagne C, Hamon M, El Mestikawy S (1999) Neurotransmitter transporters in the central nervous system. Pharmacol Rev 51:439–464
7. Erickson JD, Schafer MK, Bonner TI, Eiden LE, Weihe E (1996) Distinct pharmacological properties and distribution in neurons and endocrine cells of two isoforms of the human vesicular monoamine transporter. Proc Natl Acad Sci USA 93:5166–5171

8. Peter D, Liu Y, Sternini C, de Giorgio R, Brecha N, Edwards RH (1995) Differential expression of two vesicular monoamine transporters. J Neurosci 15:6179–6188
9. Alvarez-Martos I, Ferapontova EE (2017) A DNA sequence obtained by replacement of the dopamine RNA aptamer bases is not an aptamer. Biochem Biophys Res Commun 489:381–385
10. Johnson RG, Carty SE, Scarpa A (1981) Proton: substrate stoichiometries during active transport of biogenic amines in chromaffin ghosts. J Biol Chem 256:5773–5780
11. Knoth J, Zallakian M, Njus D (1981) Stoichiometry of H+-linked dopamine transport in chromaffin granule ghosts. Biochemistry 20:6625–6629
12. Nguyen ML, Cox GD, Parsons SM (1998) Kinetic parameters for the vesicular acetylcholine transporter: two protons are exchanged for one acetylcholine. Biochemistry 37:13400–13410
13. Schenck S, Wojcik SM, Brose N, Takamori S (2009) A chloride conductance in VGLUT1 underlies maximal glutamate loading into synaptic vesicles. Nat Neurosci 12:156–162
14. Eriksen J, Li F, Edwards RH (2020) The mechanism and regulation of vesicular glutamate transport: coordination with the synaptic vesicle cycle. Biochim Biophys Acta Biomembr 1862:183259
15. Fremeau RT Jr, Burman J, Qureshi T, Tran CH, Proctor J, Johnson J, Zhang H, Sulzer D, Copenhagen DR, Storm-Mathisen J, Reimer RJ, Chaudhry FA, Edwards RH (2002) The identification of vesicular glutamate transporter 3 suggests novel modes of signaling by glutamate. Proc Natl Acad Sci USA 99:14488–14493
16. El Mestikawy S, Wallen-Mackenzie A, Fortin GM, Descarries L, Trudeau LE (2011) From glutamate co-release to vesicular synergy: vesicular glutamate transporters. Nat Rev Neurosci 12:204–216
17. Li H, Santos MS, Park CK, Dobry Y, Voglmaier SM (2017) VGLUT2 trafficking is differentially regulated by adaptor proteins AP-1 and AP-3. Front Cell Neurosci 11:324
18. Mackenzie B, Illing AC, Morris ME, Varoqui H, Erickson JD (2008) Analysis of a vesicular glutamate transporter (VGLUT2) supports a cell-leakage mode in addition to vesicular packaging. Neurochem Res 33:238–247
19. Weston MC, Nehring RB, Wojcik SM, Rosenmund C (2011) Interplay between VGLUT isoforms and endophilin A1 regulates neurotransmitter release and short-term plasticity. Neuron 69:1147–1159
20. Bellocchio EE, Reimer RJ, Fremeau RT Jr, Edwards RH (2000) Uptake of glutamate into synaptic vesicles by an inorganic phosphate transporter. Science 289:957–960
21. Moriyama Y, Yamamoto A (1995) Vesicular L-glutamate transporter in microvesicles from bovine pineal glands. Driving force, mechanism of chloride anion activation, and substrate specificity. J Biol Chem 270:22314–22320
22. Naito S, Ueda T (1985) Characterization of glutamate uptake into synaptic vesicles. J Neurochem 44:99–109
23. Takamori S, Rhee JS, Rosenmund C, Jahn R (2001) Identification of differentiation-associated brain-specific phosphate transporter as a second vesicular glutamate transporter (VGLUT2). J Neurosci 21:RC182
24. Arriza JL, Fairman WA, Wadiche JI, Murdoch GH, Kavanaugh MP, Amara SG (1994) Functional comparisons of three glutamate transporter subtypes cloned from human motor cortex. J Neurosci 14:5559–5569
25. Bai L, Xu H, Collins JF, Ghishan FK (2001) Molecular and functional analysis of a novel neuronal vesicular glutamate transporter. J Biol Chem 276:36764–36769
26. Bridges RJ, Esslinger CS (2005) The excitatory amino acid transporters: pharmacological insights on substrate and inhibitor specificity of the EAAT subtypes. Pharmacol Ther 107:271–285
27. Hori T, Takahashi T (2012) Kinetics of synaptic vesicle refilling with neurotransmitter glutamate. Neuron 76:511–517
28. Ishikawa T, Sahara Y, Takahashi T (2002) A single packet of transmitter does not saturate postsynaptic glutamate receptors. Neuron 34:613–621

29. Maycox PR, Deckwerth T, Hell JW, Jahn R (1988) Glutamate uptake by brain synaptic vesicles. Energy dependence of transport and functional reconstitution in proteoliposomes. J Biol Chem 263:15423–15428
30. Wadiche JI, Amara SG, Kavanaugh MP (1995) Ion fluxes associated with excitatory amino acid transport. Neuron 15:721–728
31. Wadiche JI, Arriza JL, Amara SG, Kavanaugh MP (1995) Kinetics of a human glutamate transporter. Neuron 14:1019–1027
32. Wolosker H, de Souza DO, de Meis L (1996) Regulation of glutamate transport into synaptic vesicles by chloride and proton gradient. J Biol Chem 271:11726–11731
33. Chang R, Eriksen J, Edwards RH (2018) The dual role of chloride in synaptic vesicle glutamate transport. elife 7:e34896
34. Disbrow JK, Gershten MJ, Ruth JA (1982) Uptake of L-[3H] glutamic acid by crude and purified synaptic vesicles from rat brain. Biochem Biophys Res Commun 108:1221–1227
35. Chaudhry FA, Reimer RJ, Bellocchio EE, Danbolt NC, Osen KK, Edwards RH, Storm-Mathisen J (1998) The vesicular GABA transporter, VGAT, localizes to synaptic vesicles in sets of glycinergic as well as GABAergic neurons. J Neurosci 18:9733–9750
36. Hell JW, Maycox PR, Jahn R (1990) Energy dependence and functional reconstitution of the gamma-aminobutyric acid carrier from synaptic vesicles. J Biol Chem 265:2111–2117
37. Kish PE, Fischer-Bovenkerk C, Ueda T (1989) Active transport of gamma-aminobutyric acid and glycine into synaptic vesicles. Proc Natl Acad Sci USA 86:3877–3881
38. Juge N, Muroyama A, Hiasa M, Omote H, Moriyama Y (2009) Vesicular inhibitory amino acid transporter is a Cl-/gamma-aminobutyrate co-transporter. J Biol Chem 284:35073–35078
39. Dicken MS, Tooker RE, Hentges ST (2012) Regulation of GABA and glutamate release from proopiomelanocortin neuron terminals in intact hypothalamic networks. J Neurosci 32:4042–4048
40. Gutierrez R, Heinemann U (2001) Kindling induces transient fast inhibition in the dentate gyrus–CA3 projection. Eur J Neurosci 13:1371–1379
41. Kasyanov AM, Safiulina VF, Voronin LL, Cherubini E (2004) GABA-mediated giant depolarizing potentials as coincidence detectors for enhancing synaptic efficacy in the developing hippocampus. Proc Natl Acad Sci USA 101:3967–3972
42. Kim S, Wallace ML, El-Rifai M, Knudsen AR, Sabatini BL (2022) Co-packaging of opposing neurotransmitters in individual synaptic vesicles in the central nervous system. Neuron 110:1371–1384. e1377
43. Walker MC, Ruiz A, Kullmann DM (2001) Monosynaptic GABAergic signaling from dentate to CA3 with a pharmacological and physiological profile typical of mossy fiber synapses. Neuron 29:703–715
44. Chuhma N, Mingote S, Moore H, Rayport S (2014) Dopamine neurons control striatal cholinergic neurons via regionally heterogeneous dopamine and glutamate signaling. Neuron 81:901–912
45. Tecuapetla F, Patel JC, Xenias H, English D, Tadros I, Shah F, Berlin J, Deisseroth K, Rice ME, Tepper JM, Koos T (2010) Glutamatergic signaling by mesolimbic dopamine neurons in the nucleus accumbens. J Neurosci 30:7105–7110
46. Tritsch NX, Ding JB, Sabatini BL (2012) Dopaminergic neurons inhibit striatal output through non-canonical release of GABA. Nature 490:262–266
47. Patel JC, Sherpa AD, Melani R, Witkovsky P, Wiseman MR, O'Neill B, Aoki C, Tritsch NX, Rice ME (2024) GABA co-released from striatal dopamine axons dampens phasic dopamine release through autoregulatory GABA(A) receptors. Cell Rep 43:113834
48. Wojcik SM, Katsurabayashi S, Guillemin I, Friauf E, Rosenmund C, Brose N, Rhee JS (2006) A shared vesicular carrier allows synaptic corelease of GABA and glycine. Neuron 50:575–587
49. Awatramani GB, Turecek R, Trussell LO (2005) Staggered development of GABAergic and glycinergic transmission in the MNTB. J Neurophysiol 93:819–828

50. Jonas P, Bischofberger J, Sandkuhler J (1998) Corelease of two fast neurotransmitters at a central synapse. Science 281:419–424
51. Nabekura J, Katsurabayashi S, Kakazu Y, Shibata S, Matsubara A, Jinno S, Mizoguchi Y, Sasaki A, Ishibashi H (2004) Developmental switch from GABA to glycine release in single central synaptic terminals. Nat Neurosci 7:17–23
52. Munster-Wandowski A, Zander JF, Richter K, Ahnert-Hilger G (2016) Co-existence of functionally different vesicular neurotransmitter transporters. Front Synaptic Neurosci 8:4
53. Kao YH, Lassova L, Bar-Yehuda T, Edwards RH, Sterling P, Vardi N (2004) Evidence that certain retinal bipolar cells use both glutamate and GABA. J Comp Neurol 478:207–218
54. Ottem EN, Godwin JG, Krishnan S, Petersen SL (2004) Dual-phenotype GABA/glutamate neurons in adult preoptic area: sexual dimorphism and function. J Neurosci 24:8097–8105
55. Zander JF, Munster-Wandowski A, Brunk I, Pahner I, Gomez-Lira G, Heinemann U, Gutierrez R, Laube G, Ahnert-Hilger G (2010) Synaptic and vesicular coexistence of VGLUT and VGAT in selected excitatory and inhibitory synapses. J Neurosci 30:7634–7645
56. Aguilar JI et al (2017) Neuronal depolarization drives increased dopamine synaptic vesicle loading via VGLUT. Neuron 95:1074–1088. e1077
57. Hnasko TS, Chuhma N, Zhang H, Goh GY, Sulzer D, Palmiter RD, Rayport S, Edwards RH (2010) Vesicular glutamate transport promotes dopamine storage and glutamate corelease in vivo. Neuron 65:643–656
58. Buck SA, Torregrossa MM, Logan RW, Freyberg Z (2021) Roles of dopamine and glutamate co-release in the nucleus accumbens in mediating the actions of drugs of abuse. FEBS J 288:1462–1474
59. Melani R, Tritsch NX (2022) Inhibitory co-transmission from midbrain dopamine neurons relies on presynaptic GABA uptake. Cell Rep 39:110716
60. Yelin R, Schuldiner S (1995) The pharmacological profile of the vesicular monoamine transporter resembles that of multidrug transporters. FEBS Lett 377:201–207
61. Gonzalez-Reyes LE, Verbitsky M, Blesa J, Jackson-Lewis V, Paredes D, Tillack K, Phani S, Kramer ER, Przedborski S, Kottmann AH (2012) Sonic hedgehog maintains cellular and neurochemical homeostasis in the adult nigrostriatal circuit. Neuron 75:306–319
62. Hokfelt T, Johansson O, Ljungdahl A, Lundberg JM, Schultzberg M (1980) Peptidergic neurones. Nature 284:515–521
63. Kupfermann I (1991) Functional studies of cotransmission. Physiol Rev 71:683–732
64. Chuhma N, Choi WY, Mingote S, Rayport S (2009) Dopamine neuron glutamate cotransmission: frequency-dependent modulation in the mesoventromedial projection. Neuroscience 164:1068–1083
65. Chuhma N, Zhang H, Masson J, Zhuang X, Sulzer D, Hen R, Rayport S (2004) Dopamine neurons mediate a fast excitatory signal via their glutamatergic synapses. J Neurosci 24:972–981
66. Stuber GD, Hnasko TS, Britt JP, Edwards RH, Bonci A (2010) Dopaminergic terminals in the nucleus accumbens but not the dorsal striatum corelease glutamate. J Neurosci 30:8229–8233
67. Sulzer D, Joyce MP, Lin L, Geldwert D, Haber SN, Hattori T, Rayport S (1998) Dopamine neurons make glutamatergic synapses in vitro. J Neurosci 18:4588–4602
68. Dal Bo G, St-Gelais F, Danik M, Williams S, Cotton M, Trudeau LE (2004) Dopamine neurons in culture express VGLUT2 explaining their capacity to release glutamate at synapses in addition to dopamine. J Neurochem 88:1398–1405
69. Gras C, Herzog E, Bellenchi GC, Bernard V, Ravassard P, Pohl M, Gasnier B, Giros B, El Mestikawy S (2002) A third vesicular glutamate transporter expressed by cholinergic and serotoninergic neurons. J Neurosci 22:5442–5451
70. Mendez JA, Bourque MJ, Dal Bo G, Bourdeau ML, Danik M, Williams S, Lacaille JC, Trudeau LE (2008) Developmental and target-dependent regulation of vesicular glutamate transporter expression by dopamine neurons. J Neurosci 28:6309–6318
71. Schafer MK, Varoqui H, Defamie N, Weihe E, Erickson JD (2002) Molecular cloning and functional identification of mouse vesicular glutamate transporter 3 and its expression in subsets of novel excitatory neurons. J Biol Chem 277:50734–50748

72. Yamaguchi T, Sheen W, Morales M (2007) Glutamatergic neurons are present in the rat ventral tegmental area. Eur J Neurosci 25:106–118
73. Kawano M, Kawasaki A, Sakata-Haga H, Fukui Y, Kawano H, Nogami H, Hisano S (2006) Particular subpopulations of midbrain and hypothalamic dopamine neurons express vesicular glutamate transporter 2 in the rat brain. J Comp Neurol 498:581–592
74. Li X, Qi J, Yamaguchi T, Wang HL, Morales M (2013) Heterogeneous composition of dopamine neurons of the rat A10 region: molecular evidence for diverse signaling properties. Brain Struct Funct 218:1159–1176
75. Mingote S, Amsellem A, Kempf A, Rayport S, Chuhma N (2019) Dopamine-glutamate neuron projections to the nucleus accumbens medial shell and behavioral switching. Neurochem Int 129:104482
76. Mongia S, Yamaguchi T, Liu B, Zhang S, Wang H, Morales M (2019) The ventral tegmental area has calbindin neurons with the capability to co-release glutamate and dopamine into the nucleus accumbens. Eur J Neurosci 50:3968–3984
77. Poulin JF, Caronia G, Hofer C, Cui Q, Helm B, Ramakrishnan C, Chan CS, Dombeck DA, Deisseroth K, Awatramani R (2018) Mapping projections of molecularly defined dopamine neuron subtypes using intersectional genetic approaches. Nat Neurosci 21:1260–1271
78. Root DH, Wang HL, Liu B, Barker DJ, Mod L, Szocsics P, Silva AC, Magloczky Z, Morales M (2016) Glutamate neurons are intermixed with midbrain dopamine neurons in nonhuman primates and humans. Sci Rep 6:30615
79. Steinkellner T, Zell V, Farino ZJ, Sonders MS, Villeneuve M, Freyberg RJ, Przedborski S, Lu W, Freyberg Z, Hnasko TS (2018) Role for VGLUT2 in selective vulnerability of midbrain dopamine neurons. J Clin Invest 128:774–788
80. Yamaguchi T, Wang HL, Morales M (2013) Glutamate neurons in the substantia nigra compacta and retrorubral field. Eur J Neurosci 38:3602–3610
81. Hattori T, Takada M, Moriizumi T, Van der Kooy D (1991) Single dopaminergic nigrostriatal neurons form two chemically distinct synaptic types: possible transmitter segregation within neurons. J Comp Neurol 309:391–401
82. Onoa B, Li H, Gagnon-Bartsch JA, Elias LA, Edwards RH (2010) Vesicular monoamine and glutamate transporters select distinct synaptic vesicle recycling pathways. J Neurosci 30:7917–7927
83. Silm K, Yang J, Marcott PF, Asensio CS, Eriksen J, Guthrie DA, Newman AH, Ford CP, Edwards RH (2019) Synaptic vesicle recycling pathway determines neurotransmitter content and release properties. Neuron 102:786–800. e785
84. Cai Y, Ford CP (2018) Dopamine cells differentially regulate striatal cholinergic transmission across regions through corelease of dopamine and glutamate. Cell Rep 25:3148–3157. e3143
85. Chuhma N, Mingote S, Yetnikoff L, Kalmbach A, Ma T, Ztaou S, Sienna AC, Tepler S, Poulin JF, Ansorge M, Awatramani R, Kang UJ, Rayport S (2018) Dopamine neuron glutamate cotransmission evokes a delayed excitation in lateral dorsal striatal cholinergic interneurons. elife 7:e39786
86. Eskenazi D, Malave L, Mingote S, Yetnikoff L, Ztaou S, Velicu V, Rayport S, Chuhma N (2021) Dopamine neurons that cotransmit glutamate, from synapses to circuits to behavior. Front Neural Circuits 15:665386
87. Kabanova A, Pabst M, Lorkowski M, Braganza O, Boehlen A, Nikbakht N, Pothmann L, Vaswani AR, Musgrove R, Di Monte DA, Sauvage M, Beck H, Blaess S (2015) Function and developmental origin of a mesocortical inhibitory circuit. Nat Neurosci 18:872–882
88. Mingote S, Chuhma N, Kusnoor SV, Field B, Deutch AY, Rayport S (2015) Functional connectome analysis of dopamine neuron glutamatergic connections in forebrain regions. J Neurosci 35:16259–16271
89. Perez-Lopez JL, Contreras-Lopez R, Ramirez-Jarquin JO, Tecuapetla F (2018) Direct glutamatergic signaling from midbrain dopaminergic neurons onto pyramidal prefrontal cortex neurons. Front Neural Circuits 12:70

90. Sulzer D, Cragg SJ, Rice ME (2016) Striatal dopamine neurotransmission: regulation of release and uptake. Basal Ganglia 6:123–148
91. Cachope R, Mateo Y, Mathur BN, Irving J, Wang HL, Morales M, Lovinger DM, Cheer JF (2012) Selective activation of cholinergic interneurons enhances accumbal phasic dopamine release: setting the tone for reward processing. Cell Rep 2:33–41
92. Mingote S, Chuhma N, Kalmbach A, Thomsen GM, Wang Y, Mihali A, Sferrazza C, Zucker-Scharff I, Siena AC, Welch MG, Lizardi-Ortiz J, Sulzer D, Moore H, Gaisler-Salomon I, Rayport S (2017) Dopamine neuron dependent behaviors mediated by glutamate cotransmission. elife 6:e27566
93. Threlfell S, Lalic T, Platt NJ, Jennings KA, Deisseroth K, Cragg SJ (2012) Striatal dopamine release is triggered by synchronized activity in cholinergic interneurons. Neuron 75:58–64
94. Warlow SM, Singhal SM, Hollon NG, Faget L, Dowlat DS, Zell V, Hunker AC, Zweifel LS, Hnasko TS (2024) Mesoaccumbal glutamate neurons drive reward via glutamate release but aversion via dopamine co-release. Neuron 112:488–499. e485
95. Nordenankar K, Smith-Anttila CJ, Schweizer N, Viereckel T, Birgner C, Mejia-Toiber J, Morales M, Leao RN, Wallen-Mackenzie A (2015) Increased hippocampal excitability and impaired spatial memory function in mice lacking VGLUT2 selectively in neurons defined by tyrosine hydroxylase promoter activity. Brain Struct Funct 220:2171–2190
96. Papathanou M, Creed M, Dorst MC, Bimpisidis Z, Dumas S, Pettersson H, Bellone C, Silberberg G, Luscher C, Wallen-Mackenzie A (2018) Targeting VGLUT2 in mature dopamine neurons decreases mesoaccumbal glutamatergic transmission and identifies a role for glutamate co-release in synaptic plasticity by increasing baseline AMPA/NMDA ratio. Front Neural Circuits 12:64
97. Dal Bo G, Berube-Carriere N, Mendez JA, Leo D, Riad M, Descarries L, Levesque D, Trudeau LE (2008) Enhanced glutamatergic phenotype of mesencephalic dopamine neurons after neonatal 6-hydroxydopamine lesion. Neuroscience 156:59–70
98. Buck SA et al (2021) Vesicular glutamate transporter modulates sex differences in dopamine neuron vulnerability to age-related neurodegeneration. Aging Cell 20:e13365
99. Kouwenhoven WM, Fortin G, Penttinen AM, Florence C, Delignat-Lavaud B, Bourque MJ, Trimbuch T, Luppi MP, Salvail-Lacoste A, Legault P, Poulin JF, Rosenmund C, Awatramani R, Trudeau LE (2020) VGluT2 expression in dopamine neurons contributes to postlesional striatal reinnervation. J Neurosci 40:8262–8275
100. Shen H, Marino RAM, McDevitt RA, Bi GH, Chen K, Madeo G, Lee PT, Liang Y, De Biase LM, Su TP, Xi ZX, Bonci A (2018) Genetic deletion of vesicular glutamate transporter in dopamine neurons increases vulnerability to MPTP-induced neurotoxicity in mice. Proc Natl Acad Sci USA 115:E11532–E11541
101. Steinkellner T, Conrad WS, Kovacs I, Rissman RA, Lee EB, Trojanowski JQ, Freyberg Z, Roy S, Luk KC, Lee VM, Hnasko TS (2022) Dopamine neurons exhibit emergent glutamatergic identity in Parkinson's disease. Brain 145:879–886
102. Bimpisidis Z, Wallen-Mackenzie A (2019) Neurocircuitry of reward and addiction: potential impact of dopamine-glutamate co-release as future target in substance use disorder. J Clin Med 8:1887
103. Wang DV, Viereckel T, Zell V, Konradsson-Geuken A, Broker CJ, Talishinsky A, Yoo JH, Galinato MH, Arvidsson E, Kesner AJ, Hnasko TS, Wallen-Mackenzie A, Ikemoto S (2017) Disrupting glutamate co-transmission does not affect acquisition of conditioned behavior reinforced by dopamine neuron activation. Cell Rep 18:2584–2591
104. Poulin JF, Zou J, Drouin-Ouellet J, Kim KY, Cicchetti F, Awatramani RB (2014) Defining midbrain dopaminergic neuron diversity by single-cell gene expression profiling. Cell Rep 9:930–943
105. Saunders A, Macosko EZ, Wysoker A, Goldman M, Krienen FM, de Rivera H, Bien E, Baum M, Bortolin L, Wang S, Goeva A, Nemesh J, Kamitaki N, Brumbaugh S, Kulp D, McCarroll SA (2018) Molecular diversity and specializations among the cells of the adult mouse brain. Cell 174:1015–1030. e1016

106. Tritsch NX, Oh WJ, Gu C, Sabatini BL (2014) Midbrain dopamine neurons sustain inhibitory transmission using plasma membrane uptake of GABA, not synthesis. elife 3:e01936
107. Kim JI, Ganesan S, Luo SX, Wu YW, Park E, Huang EJ, Chen L, Ding JB (2015) Aldehyde dehydrogenase 1a1 mediates a GABA synthesis pathway in midbrain dopaminergic neurons. Science 350:102–106
108. Liu G, Yu J, Ding J, Xie C, Sun L, Rudenko I, Zheng W, Sastry N, Luo J, Rudow G, Troncoso JC, Cai H (2014) Aldehyde dehydrogenase 1 defines and protects a nigrostriatal dopaminergic neuron subpopulation. J Clin Invest 124:3032–3046
109. McCaffery P, Drager UC (1994) High levels of a retinoic acid-generating dehydrogenase in the meso-telencephalic dopamine system. Proc Natl Acad Sci USA 91:7772–7776
110. Brodnik ZD, Batra A, Oleson EB, Espana RA (2019) Local GABA(A) receptor-mediated suppression of dopamine release within the nucleus accumbens. ACS Chem Neurosci 10:1978–1985
111. Kramer PF, Twedell EL, Shin JH, Zhang R, Khaliq ZM (2020) Axonal mechanisms mediating gamma-aminobutyric acid receptor type A (GABA-A) inhibition of striatal dopamine release. elife 9:e55729
112. Lopes EF, Roberts BM, Siddorn RE, Clements MA, Cragg SJ (2019) Inhibition of nigrostriatal dopamine release by striatal GABA(A) and GABA(B) receptors. J Neurosci 39:1058–1065
113. Roberts BM, Doig NM, Brimblecombe KR, Lopes EF, Siddorn RE, Threlfell S, Connor-Robson N, Bengoa-Vergniory N, Pasternack N, Wade-Martins R, Magill PJ, Cragg SJ (2020) GABA uptake transporters support dopamine release in dorsal striatum with maladaptive downregulation in a parkinsonism model. Nat Commun 11:4958
114. Zych SM, Ford CP (2022) Divergent properties and independent regulation of striatal dopamine and GABA co-transmission. Cell Rep 39:110823
115. Nelson AB, Hammack N, Yang CF, Shah NM, Seal RP, Kreitzer AC (2014) Striatal cholinergic interneurons drive GABA release from dopamine terminals. Neuron 82:63–70
116. Berrios J, Stamatakis AM, Kantak PA, McElligott ZA, Judson MC, Aita M, Rougie M, Stuber GD, Philpot BD (2016) Loss of UBE3A from TH-expressing neurons suppresses GABA co-release and enhances VTA-NAc optical self-stimulation. Nat Commun 7:10702
117. Dudman JT, Krakauer JW (2016) The basal ganglia: from motor commands to the control of vigor. Curr Opin Neurobiol 37:158–166
118. Hamid AA, Pettibone JR, Mabrouk OS, Hetrick VL, Schmidt R, Vander Weele CM, Kennedy RT, Aragona BJ, Berke JD (2016) Mesolimbic dopamine signals the value of work. Nat Neurosci 19:117–126
119. Hikosaka O (1998) Neural systems for control of voluntary action–a hypothesis. Adv Biophys 35:81–102
120. Nelson AB, Kreitzer AC (2014) Reassessing models of basal ganglia function and dysfunction. Annu Rev Neurosci 37:117–135
121. Schultz W (2016) Dopamine reward prediction-error signalling: a two-component response. Nat Rev Neurosci 17:183–195

Drosophila as an Animal Model To Determine the Functional and Behavioral Significance of Dopamine Transporter Genetic Variations Associated with Brain Disorders

Samuel J. Mabry, Angela M. Carter, and Aurelio Galli

Contents

Abstract *Drosophila melanogaster,* commonly referred to as the "fruit fly," has been a long-utilized animal model in multiple areas of biological research. It is estimated that 75% of human genes, which are associated with disease, have homologues in *Drosophila*. The conservation of biological systems, the genetic tractability, short generation time, and a broad array of available behavioral assessments make *Drosophila* an especially robust model organism for neuroscience investigations. The dopamine (DA) system, in particular, is highly conserved between mammals and *Drosophila*. Mutations of the DA transporter (DAT), a negative regulator of DA neurotransmission, have been associated with multiple different neuropsychiatric and neurodegenerative disorders, including autism spectrum disorders (ASDs), attention deficit hyperactivity disorder (ADHD), and Parkinson's disease (PD). Utilization of *Drosophila* models demonstrates specific structural and functional

Angela M. Carter and Aurelio Galli contributed equally with all other contributors.

S. J. Mabry
UAB Department of Surgery, Birmingham, AL, USA

A. M. Carter · A. Galli (✉)
UAB Department of Surgery, Birmingham, AL, USA

UAB CINEMA, Birmingham, AL, USA
e-mail: agalli@uabmc.edu

O. V. Mortensen (ed.), *Monoamine Transporters*, Advances in Neurobiology 46,
https://doi.org/10.1007/978-3-031-96364-3_9

alterations in mutated DAT that manifest as unique behavioral phenotypes. Ultimately, combining techniques ranging from biochemistry, electrochemistry, and complex behavioral analyses facilitated a deeper understanding of how transporter function and dysfunction can translate to neurological and neuropsychiatric disorders.

Keywords *Drosophila* · Complex behavior · Dopamine transporter (DAT) · Anomalous dopamine efflux (ADE) · Autism spectrum disorder (ASD) · Dopamine transporter deficiency syndrome (DTDS) · Parkinson's disease (PD)

1 Introduction to *Drosophila melanogaster*

Drosophila is a multifaceted animal model for biological research. From Thomas Hunt Morgan's use of the organism in studies that arguably led to the establishment of genetics as an independent scientific discipline, to Hall, Rosbash, and Young's investigations that uncovered the molecular mechanisms that govern circadian rhythms, Six Nobel laureates in physiology or medicine are recognized for their work conducted in this remarkable little invertebrate [1]. It is estimated that 65% of the *Drosophila* genome is homologous to humans; around 75% of the genes responsible for human disease have homologs in *Drosophila* [2]. This conservation of core biological systems, combined with simplicity, inexpensive cost, short generation time, and genetic tractability, makes *Drosophila* a robust model organism. The genetic repertoire that offers cell-type specific expression of transgenes (e.g., Gal4/UAS (upstream activating sequence) system, see below) coupled with the vast number of available behavioral assays make *Drosophila* an attractive model organism for neuropsychiatric and neurological studies, including basic science investigations into substance-use disorders, neurodevelopmental diseases and neurodegenerative illnesses.

Drosophila is an especially useful model for studying diseases associated with DA dysfunction. Genes involved in the synthesis, transport, metabolism, and signal transduction of DA are largely conserved from flies to humans, with only a few minor differences in DA catabolism [3] (Table 1). Furthermore, DA signaling and many of the behaviors that it modulates are conserved across species [4]. For example, DA plays a critical role in locomotion in both mammals [5] and *Drosophila* [6, 7]. Furthermore, it has been implicated in multiple complex behaviors in *Drosophila*, including courtship [8], decision making [9], reward associated memory [10], and sleep [11], providing numerous behavioral options for phenotyping molecular mechanisms of DA dysfunction.

The *Drosophila* DA transporter (dDAT), the protein responsible for the clearance of synaptically released DA, is 52% identical and has 69% sequence similarity to the human DAT (hDAT) [12]. Both transporters have 5 intracellular loops (ILs), 6 extracellular loops (ELs) [13], and utilize a Na^+ gradient to transport DA inside neurons against its concentration gradient [14, 15]. Although DAT is conserved

Table 1 Major genes involved in dopamine pathway regulation in mammals and *Drosophila*

Mammalian gene	*Drosophila* gene	Function
Tyrosine hydroxylase (*TH*)	*Pale*	Rate limiting enzyme in the production of DA
GTP Cyclohydrolase (*GTPCH*)	*Punch*	Synthesizes BH4, a key cofactor for TH function
Dopamine Decarboxylase (*Ddc*)	*Ddc*	Converts L-Dopa to dopamine
Vesicular Monoamine Transporter 2 (*VMAT2, SLC18A2*)	*VMAT*	Loads dopamine into vesicles
Dopamine D1 Receptors	*Dop1R1* and *Dop1R2*	GPCR that activates the cAMP pathway
Dopamine D2 Receptors	*Dop2R*	GPCR that inhibits the cAMP pathway
Dopamine Transporter (*DAT, SLC6A3*)	*DAT* (*Fumin*)	Reuptakes released DA back into the cell
Monoamine Oxidase (*MAO*)	N/A (no ortholog)	DA catabolism
Catechol-O-Methyltransferase (*COMT*)	N/A (no ortholog)	DA methylation
N/A (no ortholog)	Arylalkylamine N Acetyltransferase (*aaNAT*)	DA acetylation

between *Drosophila* and humans, there are important differences in the amino acid sequence of key regulatory regions, such as the N-terminus. In order to increase the translational relevance of *Drosophila* models for investigations involving divergent regulatory protein moieties, lines expressing transgenes derived from humans are created in *Drosophila*, harboring knockouts of endogenous genes of interest. For example, "fumin" flies possess a heterologous insertion in the dDAT gene that causes inhibition of proper RNA splicing, thereby resulting in truncation of the endogenous dDAT protein and a functional dDAT knockout [11]. These *fumin* flies are unable to properly clear DA from the synapse, leading to multiple behavioral phenotypes, including increased locomotion and decreased sleep [11]. Using the binary Gal4/UAS system, *fumin* flies can be generated that selectively express the hDAT gene, under the control of the UAS (UAS-hDAT; transgene 2), specifically in dopaminergic neurons by expression of Gal4 under the control of the tyrosine hydroxylase (TH) promoter (TH-Gal4; transgene 1) [7, 16] (Fig. 1). With this system, it is possible to express any gene of interest in selective subsets of cells by placing Gal4 at the 3′ end of cell-type-specific promoters, and combining with a second transgene that expresses the gene of interest downstream of the UAS [17]. In this chapter, all *Drosophila* lines discussed are derived from crosses between a TH-GAL4 and UAS-hDAT line, both generated in the *fumin* background, resulting in "humanized" flies that express hDAT specifically in DA neurons.

Multiple techniques can probe transporter function, both in vitro and ex vivo. Quantification of DA reuptake in cultured cells or isolated brains of *Drosophila* can be performed using radioactive uptake assays with tritiated DA ([^{3}H]DA) [18] or fluorescently labeled substrates [19, 20]. Additionally, electrochemical techniques,

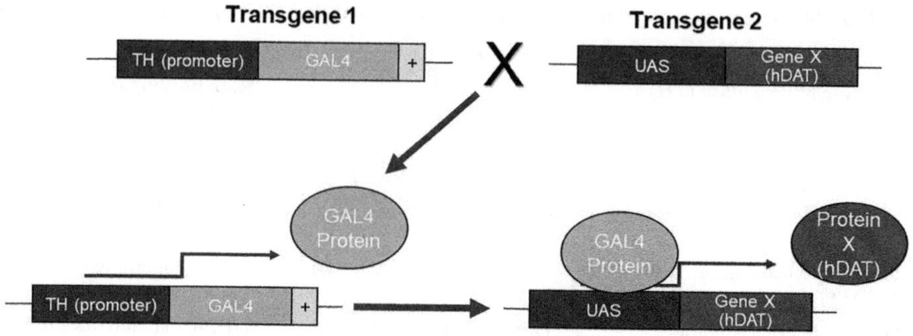

Fig. 1 Schematic of the *Drosophila* GAL4/UAS bitransgenic system for cell-type-specific protein expression. Transgene 1 encodes expression of the Gal4 transcription factor, downstream of the TH promoter and upstream of dDAT-specific enhancer/silencer sequences (+), resulting in expression of Gal4 specifically in DA neurons [64]. Transgene 2 encodes expression of any gene of interest (X), for example, hDAT, under the control of UAS (upstream activating sequence). As Gal4 binds UAS, crossing of the lines results in expression of the protein of interest in dopamine neurons. For studies involving expression of hDAT, this system was utilized within the *fumin* background (lacks functional dDAT expression) to create "humanized" *Drosophila* by expressing hDAT specifically in dopamine neurons that lack dDAT

such as amperometry [21], are commonly used to probe DA dynamics. Briefly, these techniques utilize carbon fiber electrodes held at voltages above the redox potential of DA. DA is oxidized by the electrodes, generating a current that is directly proportional to the amount of DA molecules released by cells and tissue. Surface localization of DAT can be evaluated biochemically using classic biotinylation assays, visualized roughly using standard fluorescence microscopy, or interrogated finely, including subdomain localization, utilizing advanced microscopy techniques, such as TIRF and STORM [22, 23]. DA dynamics can also be assessed in vivo using flies that express the DA-GRAB sensor (fluorescence changes upon binding to DA) [24]. Changes in DAT function, driven either by changes in transport cycle dynamics or surface localization/expression, can be phenotypically defined by assessment of specific *Drosophila* behaviors. For example, DA plays a critical role in locomotion in both mammals and *Drosophila* [5, 6]. Locomotion in *Drosophila* can be monitored in a simple, quantitative, and high-throughput manner, utilizing commercially available multichamber systems that individually house flies and quantitate locomotion as breaks in a beam of light that is directed across the chamber (Fig. 2). Locomotion can also be quantitated using a range of video-tracking technologies. High-speed cameras make interrogation of fine motor coordination, such as patterns in wing movements and leg coordination during a flight take-off, possible on a millisecond time scale. DA is also critical for food-seeking, sexual motivation, and fear response. Assays that investigate these complex behaviors will be discussed in further detail later in this chapter. Taken together, these techniques permit probing of transporter physiology and dysfunction, and allow direct translation to behavioral outcomes using *Drosophila* as an animal model.

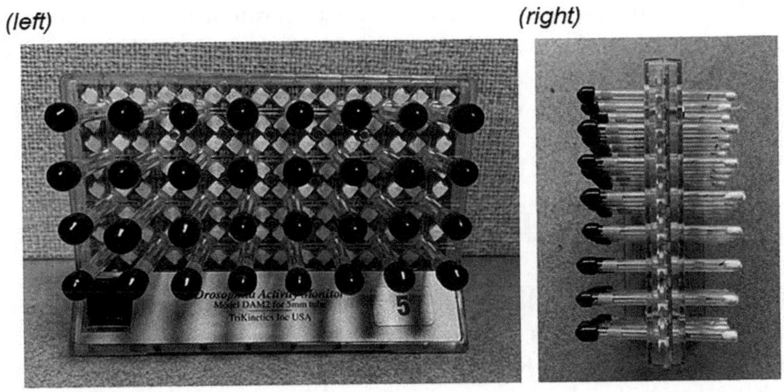

Fig. 2 Photographs of a trikinetics *Drosophila* activity-monitoring (DAM) system. This system is utilized to quantitate *Drosophila* locomotor behavior, in real time, by recording breaks in an infrared light beam as flies move about a clear glass tube

2 DAT Genetic Variations Associated with Brain Disorders

DA plays a critical role in regulating several neurocircuits and behaviors. Therefore, it is not surprising that genetic variations and/or mutations in the DAT gene (SLC6A3) are associated with brain dysfunction. Rare genetic variances in SLC6A3 have been implicated in multiple neurological and neuropsychiatric diseases, including major depressive disorder (MDD), substance-abuse disorder (SUD), attention deficit hyperactivity disorder (ADHD), Parkinson's disease (PD), early dystonia, and autism spectrum disorder (ASD) [25]. In this chapter, we focus on single-nucleotide polymorphisms (SNPs), in-frame deletions, or de novo mutations in hDAT. Some of these rare genetic variations, T356M and A559V, along with an in frame deletion, ΔN336, are associated with ASD [26–29], while another, R445C, has been found in patients with Parkinsonism [18, 30].

Despite being associated with disorders related to DA dysfunction, these genetic variants impair DAT function by vastly different molecular mechanisms. As clearance of extracellular DA is the primary function of DAT [13], some of these SNPs result in impaired DAT-mediated DA uptake [18, 26, 28]. However, the molecular mechanisms by which DA uptake is impaired differ across genetic variations. The R445C variant impairs DA uptake by causing a significant reduction in DAT trafficking to the cell surface [18]. In contrast, the ΔN336 protein has normal surface expression, but decreased DA uptake driven by impaired transport function [28]. Other alterations, such as T356M and A559V, demonstrate a DAT modality called anomalous DA efflux (ADE) [26, 29, 31, 32]. In the ADE mode, the DAT, rather than clearing extracellular DA, runs in reverse and effluxes DA from the cytoplasm to the extracellular space. A similar ADE mode is observed in wild-type DAT upon phosphorylation of syntaxin, a DAT-associated protein, at Ser14 [21]. This phenomenon is defined as constitutive DA efflux (CDE) and/or nonvesicular DA release

(NVDR). It is distinct from ADE as it is driven by posttranslational modifications of the DAT and associated proteins, and not by genetic variants of the transporter [21]. ADE has multiple downstream effects, including altering D2 receptor sensitivity and animal behavior, which will be discussed in more depth later in this chapter.

3 Autism Spectrum Disorder (ASD)

ASD is a neuropsychiatric disorder characterized by impairments in social interaction and communication, as well as patterns of restricted and repetitive behaviors, interests, or activities [33]. Recently, several genetic variations and de novo mutations that are considered risk factors for multiple neuropsychiatric diseases, including ASD, have been identified [25, 34]. Among those are genetic variations in the SLC6A3 gene that lead to DAT dysfunction. Previously, it has been shown that features of ASD may be due to, or exacerbated by, abnormal DA signaling [35]. Thus, the study of dopaminergic dysfunction, as it relates to ASD, is pivotal for understanding the etiology of this disorder. Consistent with the hypothesis that DA dysfunction is either a causative factor or a complication in ASD, changes in the development of the striatum, a brain region enriched with DA projections and DAT expression, are associated with repetitive behaviors in autism. Furthermore, disruptions in the striatal processes associated with social reward have also been found in individuals with ASD [32, 36]. Taken together, this suggests that DAT dysfunction, leading to altered dopaminergic signaling, could support many aspects of the altered behaviors associated with neuropsychiatric disorders, such as ASD. Below we discuss the molecular, functional, and behavioral phenotypes of different DAT genetic variations that were identified in probands with ASD.

3.1 DAT T356M

Studies conducted using whole-exome sequencing (WES) on ASD families have indicated that discrete de novo mutations (single-nucleotide variation [SNV] or small indels) contribute to the overall genetic risk of ASD [34, 37–39]. Among these variations is the first ASD-associated, de novo mutation found in the human SLC6A3 gene [37]. This mutation results in a Thr-to-Met substitution at position 356 (T356M). The T356M variant, identified as a heterozygote in the proband and absent in both parents, was experimentally validated and confirmed to be a de novo mutation that does not appear in the unaffected sibling [26].

The T356 residue is located on TM7, a highly conserved region that plays a role in ion binding [40]. This threonine residue is conserved across species, ranging from *Drosophila* to humans. From initial investigations in a heterologous expression system, hDAT T356M demonstrates a significantly reduced maximum capacity for uptake of $[^3H]$DA [26]. However, while the V_{max} of $[^3H]$DA uptake was much lower in hDAT T356M-expressing cells compared to hDAT-expressing cells, the DA affinity (K_m) was unchanged, demonstrating that the transporter still binds to DA with high affinity. Unlike mutations that will be discussed later in this chapter, the impaired DA uptake of hDAT T356M does not stem from attenuated surface localization, as biotinylation experiments revealed no difference in hDAT T356M levels at the cell surface. Together, these data strongly suggested that the T356M mutation causes impaired transport function. To molecularly define the impact of this T356 mutation, researchers took advantage of the bacterial homolog leucine transporter (LeuT) by substituting A289 (the homologous site to T356) with a Met to perform spin-labeling electron paramagnetic resonance (EPR) spectroscopy [26]. This demonstrated that the Met substitution promotes an outward-facing conformation upon substrate-binding. In the substrate-bound state, an outward-facing transporter conformation is a requirement for substrate efflux, not uptake. In their 2013 paper, Hamilton and colleagues demonstrated, using amperometry on CHO cells expressing hDAT or hDAT T356M, that hDAT T356M supports a cocaine (COC)-sensitive ADE, a phenomenon in which the transporter runs in reverse (detailed earlier in this chapter), supporting an efflux of DA. COC is an established DAT antagonist. When COC was applied to hDAT T356M cells, it generated a significant downward deflection of the amperometric trace, revealing a basal presence of DA efflux (i.e., ADE).

Using the methods previously described, the authors expressed hDAT T356M specifically in DA neurons of *Drosophila* lacking endogenous dDAT expression (DAT KO), in order to determine the behavioral expressions of the T356M mutation [26]. *Drosophila* harboring the hDAT T356M protein are hyperlocomotive compared to their hDAT WT counterparts (Fig. 3). As DA directly modulates locomotion in animals [6], a mutant hDAT that exhibits ADE would be expected to enhance DA neurotransmission, and thereby enhance locomotor activity, consistent with the authors' findings. Interestingly, hDAT T356M *Drosophila* does not demonstrate hyperlocomotion in response to treatment with the psychostimulant amphetamine

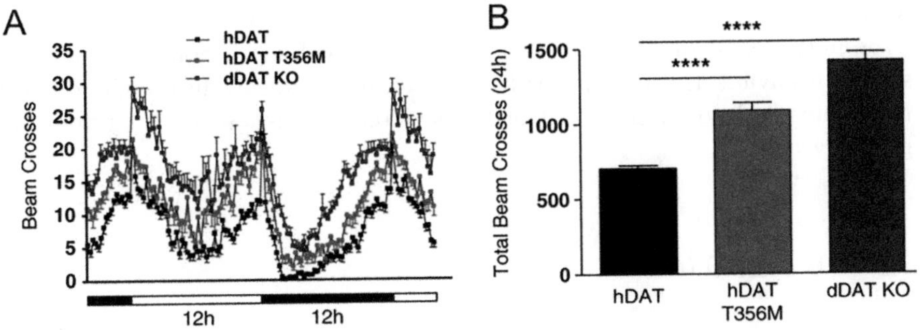

Fig. 3 (**a**) Locomotor activity was assayed over 32 h during the light (horizontal white bars) or dark (horizontal black bars) cycle. Flies expressing hDAT T356M (red squares), as well as dDAT KO flies (blue squares), were hyperactive throughout the 32 h period with respect to flies expressing wild-type hDAT (black squares. (**b**) hDAT T356M and dDAT KO flies both demonstrated significantly increased locomotion over 24 h. (Adapted from Hamilton et al. [26])

(AMPH), unlike *Drosophila* expressing hDAT (Fig. 3). This may be due to the presence of ADE in DAT T356M, as the transporter is in a conformation that is already amenable to efflux DA, and therefore, AMPH may only be able to marginally increase this reverse transport. However, to date, only locomotion, a very basic behavior, has been analyzed in *Drosophila* harboring hDAT T356M. Follow-up studies should look at complex behaviors such as grooming, fear response, and courtship in order to thoroughly understand how this mutation alters animal behavior.

Recently, the DAT T356M mutation was "knocked in" to the endogenous DAT locus in mice [27, 32]. Parallel to results obtained in *Drosophila*, mice homozygous for DAT T356M demonstrate impaired striatal DA clearance and increased extracellular DA levels, possibly due to ADE. Furthermore, DAT T356M mice were significantly hyperlocomotive compared to their WT counterparts, likely due to elevated DA signaling supported by the DAT T356M-induced ADE. To test this hypothesis, the authors exposed mice to GBR12909, a DAT antagonist [41], to inhibit DAT function and, as a consequence, ADE. They found that both GBR and ACT-01, the novel DAT blocker, reduced locomotion in mice homozygous for DAT T356M, supporting the hypothesis that ADE in T356M DAT mice is driving enhanced DA-associated behaviors [32]. Importantly, DAT T356M mice also display social deficits, as demonstrated by loss of social preference in the three-chamber task and loss of social dominance in the tube test. These data support the hypothesis that the DAT T356M-induced ADE sustains aberrant ASD-associated behaviors, which may be due, at least in part, to altered social motivation driven by impaired dopaminergic signaling [32]. Furthermore, the works of DiCarlo and collaborators also demonstrate deficits in marble burying in homozygous DAT T356M mice [27]. Marble burying represents an innate behavior motivated by investigation [42]. The deficit demonstrated by Di Carlo and colleagues can therefore be interpreted as a change in motivated behavior. The behavioral and neurophysiological characteristics discovered in the homozygous DAT T356M mouse provide supporting evidence for a role of DAT dysfunction (e.g., ADE) in the behavioral deficits typically associated with ASD. These data highlight the relevance of previous work, performed in *Drosophila,* that demonstrate that ADE associated with the T356M mutation alters specific behaviors. It also further underscores the translational efficacy of *Drosophila* for probing molecular, functional, and behavioral phenotypes stemming from DAT genetic variants.

3.2 DAT ΔN336

A rare, inherited in-frame deletion of the residue N336 in hDAT was identified in a patient with ASD [28]. The health information of this autism-affected male, in a family of European ancestry, was ascertained as part of the Simons Simplex Collection [28]. The parents and sibling were not diagnosed with ASD; however, the father of the subject transmitted the variant and tested positive for a broad autism phenotype. The broad autism phenotype describes a set of behavioral and language features/deficits, which occur in individuals resulting from a milder expression of the underlying genetic liability for autism. The N336 residue is located in IL3 and is highly conserved across species, from *Drosophila* to humans, indicating that it is likely critical for DAT function.

When expressed in a heterologous expression system, hDAT ΔN336 is localized to the plasma membrane at normal levels but does not demonstrate any measurable DA uptake [28]. Interestingly, this mutant is still capable of AMPH-induced DA efflux, as measured by amperometry, if cells are loaded with DA via a whole-cell patch clamp pipette. However, this DA efflux is significantly lower than the DA efflux measured from hDAT-expressing cells. Interestingly, the characteristic of hDAT ΔN336 cells exhibiting no DA uptake, but retaining the capability to efflux DA, supports a model where the N336 residue can differentially regulate inward versus outward transport of DA. These unique functional perturbations of ΔN336 spurred interest in obtaining structural information to aid additional molecular details on the role of various DAT structural domains in controlling DA transport.

To gain insight into the structural consequences of the ΔN336 mutation, LeuT with the corresponding deletion (ΔV269) was used for EPR studies [28]. Spin-label pairs 309/480 and 7/86 were used to allow monitoring the conformational dynamics of the extracellular (EC) and intracellular (IC) gates, respectively. In LeuT ΔV269, in the absence of substrate and Na^+, the distance between the extracellular pair is significantly shorter than the distance in LeuT WT. This data suggests the collapse of the EC gate in LeuT ΔV269. In conjunction, the distance of the IC pair is longer in LeuT ΔV269, indicating an inward-open conformation. However, once Na^+ and substrate are introduced, there is a shortening of distance between the IC spin-label pairs, while there are no changes in the extracellular side, leading to a unique conformation the authors defined as "half-open and inward facing [28]." This novel conformation has a low probability in the LeuT WT [28]. Thus, the native valine at position 269 in LeuT (hDAT N336) regulates the IC gate dynamics.

LeuT ΔV269 was crystalized in the presence of Na^+ and substrate. The crystal structure revealed additional consequences of the ΔV269 mutation. The largest differences in LeuT WT and ΔV269 structures were observed in the intracellular regions that interact with the loop, which harbors V269. Deletion of V269 causes the translocation of subsequent residues (270–274) in the structure. This causes a change in the side-chain position of Q273, which moves closer to the first turn of TM1 and, most importantly, exists within hydrogen bonding distance with S278. Notably, the corresponding amino acid of S278 in hDAT is D345, a critical residue for the kinetics of the IC gate [21, 43]. This novel hydrogen bond between these residues could potentially disrupt the normal dynamics of the IC gate, contributing to a half-open inward-facing conformation in hDAT ΔN336.

Finally, molecular dynamics (MD) simulation of hDAT ΔN336 revealed that ΔN336 positions K337 in an orientation that enables an interaction between IL3 and TM7, specifically through D345. This interaction is absent in hDAT WT, as K337 faces the intracellular side. Due to the novel interaction between K337 and D345, the K66-D345 salt bridge, which is critical for the dynamics of the IC gate

[43], is lengthened and weakened. The authors hypothesized that the weakening of this salt bridge is a key component of the half-open inward-facing conformation adopted by hDAT ΔN336 [28].

The half-open inward-facing conformation likely contributes to the unique phenomenon of disrupted DA uptake function, but retention of DA efflux capacity. The lack of uptake is possibly a consequence of the conformation of the EC gate, which is locked in a closed position. However, the IC gate, which remains partially open in this mutant, allows for binding of DA from the cytoplasm and, ultimately, DA efflux. This study demonstrates that the intra- and EC gates of DAT are capable of differentially regulating uptake versus efflux, depending on the equilibrium dynamics of their conformational states, and suggests that the two gates of neurotransmitter transporters can act independently during the transport process and be uncoupled by rare variants [28]. Using *Drosophila* as an animal model, the authors defined how disruption of specific gate conformations of hDAT translates into abnormal brain DA function and atypical complex animal behaviors.

Using *Drosophila* genetics as described above, hDAT ΔN336 was expressed specifically in DA neurons of DAT KO flies [28]. Comparable to the data obtained in heterologous expression systems, [H^3]DA uptake in isolated fly brains demonstrated that hDAT ΔN336 has significantly reduced DA uptake capacity with respect to hDAT. Unsurprisingly, hDAT ΔN336 *Drosophila* demonstrated a hyperlocomotive phenotype, likely due to an increase in extracellular DA.

In order to probe how this hDAT variant could potentially contribute to the ASD phenotype seen in the human probands, *Drosophila* expressing hDAT ΔN336 underwent testing to probe complex social-related behaviors such as grooming, fear response, and social interactions. These flies showed increased grooming compared to hDAT WT flies, indicative of enhanced DA neurotransmission [44]. In Campbel et al., the authors then developed novel assays for probing whether ΔN336 alters fear responses, as well as social interactions [28].

Fear A complex number of factors are involved in the detection and evaluation of threat stimuli by animals. Assessment of threat stimuli is fundamental for establishing an appropriate response to threatening situations, such as the presence of a predator. Risk assessment can evoke fear, an evolutionarily conserved behavior, which sparks protective action. Dysfunction of the conserved circuits that regulate fear has been associated with anxiety-related disorders [45, 46], including ASD. In order to determine whether impairments in DA, caused by the variant ΔN336, alter responses to a threat-stimulus (i.e., freezing), Campbell and collaborators exposed either hDAT or hDAT ΔN336 flies to the sound of a predatory wasp. Using a high-speed camera that monitors fly movements in custom printed recording chambers, these investigators determined that in hDAT flies, the sound of the predatory wasp causes a short period of freezing that is followed by a fleeing period, defined by rapid increase in average velocity (Fig. 4). In contrast, hDAT ΔN336 flies display prolonged freezing and delayed fleeing upon the auditory fear stimuli. These results associate DAT impairments promoted by the ΔN336 variant to increased fear and dysfunctional defensive behaviors.

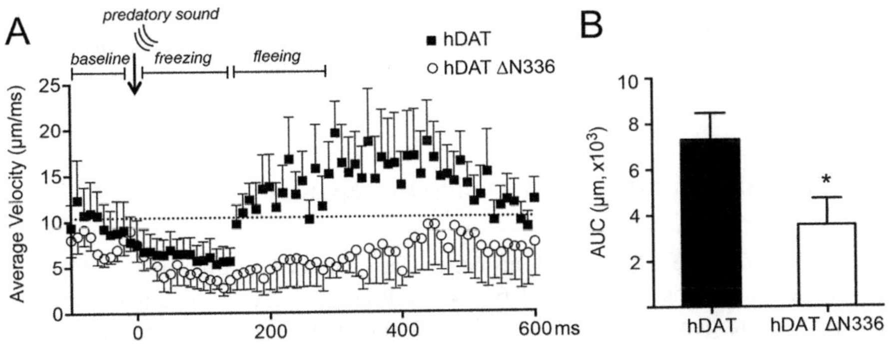

Fig. 4 (**a**) Average velocity was assayed over 600 ms following an auditory stimulus for flies expressing hDAT (■) or hDAT ΔN336 (○). The hashed line indicates baseline velocity. (**b**) hDAT ΔN336 demonstrated decreased velocity in response to the auditory stimulus. (Adapted from Campbell et al. [28])

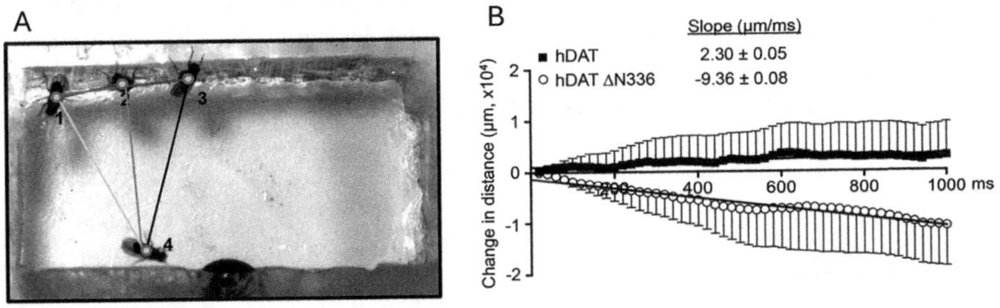

Fig. 5 (**a**) Representative image of quantitation of distances between individual flies (different colored lines) to determine cumulative distances (i.e., flock size). (**b**) hDAT ΔN336 flies show social impairments as measured by proximity to their neighbors during escape responses upon exposure to fear-inducing sounds (predatory wasp). Over time, hDAT flies (■) display increased flock size. In contrast, hDAT ΔN336 flies (○) reveal decreased flock size over time. (Adapted from Campbell et al. [28])

Social Interactions Grouping is a behavior that is adopted by several animal populations as an advantageous survival response. Grouping is coordinated aggregation of animals [47] and can support complex behaviors such as escaping predator attack. During escape, social interactions/distances among the individuals in the flock dictate the flock size [47]. Typically, shortened social distances are assumed by the flock when a predator is detected, and then lengthened social distances (flock size) are adopted when the predator approach crosses a threshold distance from the flock. Since ASD is also characterized by impairments in social interactions and communications, Campbell and collaborators determined changes in flock size upon the sound of the predatory wasp in hDAT and hDAT ΔN336 flies. hDAT WT flies respond to an audible fear sound by increasing their flock size (Fig. 5). However, hDAT ΔN336 flies respond to the sound of a predatory wasp by decreasing their flock size. This indicates an altered social response to fear. This new assay is a powerful tool for probing potential social deficits resulting from genetic variations in the SLC6A3 gene. Future studies looking at other ASD-related genetic variations in multiple genes utilize this novel assay for probing effects on social behaviors.

4 Dopamine Transporter Deficiency Syndrome

Parkinson's disease (PD) is a neurodegenerative disorder, affecting 2–3% of the global population over the age of 65 [48]. Although PD cases occur mainly idiopathically, rare genetic variants have been associated with early-onset PD [18, 25, 49, 50]. Research focused on these inherited forms of PD has shed light on specific cellular pathways and molecular mechanisms that underlie neurodegeneration processes and motor deficits [51].

Mutations in the SLC6A3 gene have been found in patients with a distinct type of infantile parkinsonism-dystonia, referred to as DA transporter deficiency syndrome (DTDS) [52, 53]. Only a few subjects with DTDS survive to adulthood, with a majority of patients dying in childhood or adolescence [52, 53]. Common to DTDS-associated SLC6A3 genetic variants is impaired DAT function that stems either from impaired transporter activity and/or decreased membrane localization [18, 52–55].

4.1 DAT R445C

Recently, missense mutations in the SLC6A3 gene have been identified in a patient who presented with DTDS: a mutation in one allele that resulted in Arg-to-Cys substitution at position 445 (R445C), and a mutation in the second allele that resulted in a Arg-to-Lys substitution at position 85 (R85L) [53]. Each mutation by itself has devastating effects on hDAT activity and expression [53]. In an effort to understand specific hDAT impairments and behavioral phenotypes associated to a specific mutation, Aguilar and collaborators focused on understanding, both mechanistically and structurally, how R445C disrupts transport activity, as well as the phenotypic consequence of this disruption [18].

R445 is located on IL4 near the beginning of TM9, facing the IC vestibule. Critically, this residue forms a salt bridge with D428 that constitutes an integral part of the IC gate of hDAT [30, 43]. Furthermore, it resides in a basic stretch of amino acids, H442-R443-H444-R445, that has been demonstrated to bind phosphatidylinositol 4,5-bisphosphate and to regulate reverse transport of DA [16]. Using the methodologies described above, Aguilar and collaborators expressed hDAT R445C in DA neurons of a DAT null fly and then examined the biochemical, electrophysiological, and behavioral impacts of this mutation [18]. The authors demonstrate that in isolated *Drosophila* brains, the R445C mutation drives decreased transport function. hDAT R445C flies also display reduced DA levels, as well as selective neurodegeneration of protocerebral posterior lateral 1 (PPL1) DA neurons, as measured by reduction in tyrosine hydroxylase (TH)-labeled DA neurons in this brain region. These observations are consistent with various *Drosophila* models of PD [18, 56–58]. A possible mechanism suggested by these studies is that reduced DAT function results in an increase in extracellular DA and depleted presynaptic stores. Elevated synaptic DA leads to overstimulation of presynaptic D2 autoreceptors, which impairs TH activity, thereby decreasing DA synthesis [59]. Interestingly, Aguilar and collaborators noted that despite significant death of DA neurons and decreased DA levels, there were no significant effects on either anxiety-related behaviors (i.e., center time in an open field test) or grooming, behaviors that are, at least in part, regulated by DA [18]. However, hDAT R445C flies have a significant reduction in spontaneous locomotor activity compared to hDAT flies. This reduction was supported by a decrease in movement vigor (fast movement). The expression of this impaired behavioral phenotype is parallel to bradykinesia noticed in both patients with DTDS and PD [48, 52, 53].

Human patients with PD often show impairments in gross motor coordination [60]. To address motor coordination in *Drosophila* hDAT R445C, the authors examined flight initiation patterns. Of note is that, in insects, several monoamines, including DA, modulate flight [61]. In flies, flight initiation is achieved by superb sensory-motor integration. At first, a fly raises its wings to a ready position, and then, extends its mesothoracic legs and depresses its wings simultaneously to coordinate a jump with the initial downstroke [62]. Using a high-speed camera, Aguilar and collaborators studied flight initiation in hDAT R445C flies [18]. In these flies, flight initiation was significantly delayed compared to hDAT flies (Fig. 6). In order to measure flight initiation, the authors of this study quantified the time between initial wing elevation and take-off. In *Drosophila* expressing hDAT R445C, the time to take-off after initial wing movement was almost doubled, demonstrating a robust impairment of motor coordination [18]. This impairment parallels lack of motor coordination observed in early-onset, as well as sporadic PD subjects, and exemplifies the potential strength of the *Drosophila* model for identifying phenotypes that arise from DAT mutations associated with brain disorders.

As discussed earlier, cells expressing hDAT R445C demonstrate a reduced capacity for DA uptake, measured by [^{3}H]DA uptake. However, in contrast to other hDAT mutants that display normal surface expression but adopt a conformational state that is not uptake willing, such as hDAT ΔN336, hDAT R445C exhibits an impairment in trafficking that results in a significantly decreased surface localization, as well as total protein expression [18]. This impaired both trafficking and protein expression, which in turn leads to significantly reduced uptake capacity. Importantly, a charge-neutralizing substitution at R445 (R445A) also leads to significantly impaired surface localization, total protein expression, and DA uptake. These data demonstrate the importance of this Arg residue for normal DAT function and that the effect observed in the R445C mutant is likely due to the absence of a positive charge at position 445 rather than a gain of function of the cysteine mutation.

The authors of this study also show partial rescue of hDAT R445C surface localization by treatment with chloroquine (CQ). Treatment of hDAT R445C with CQ

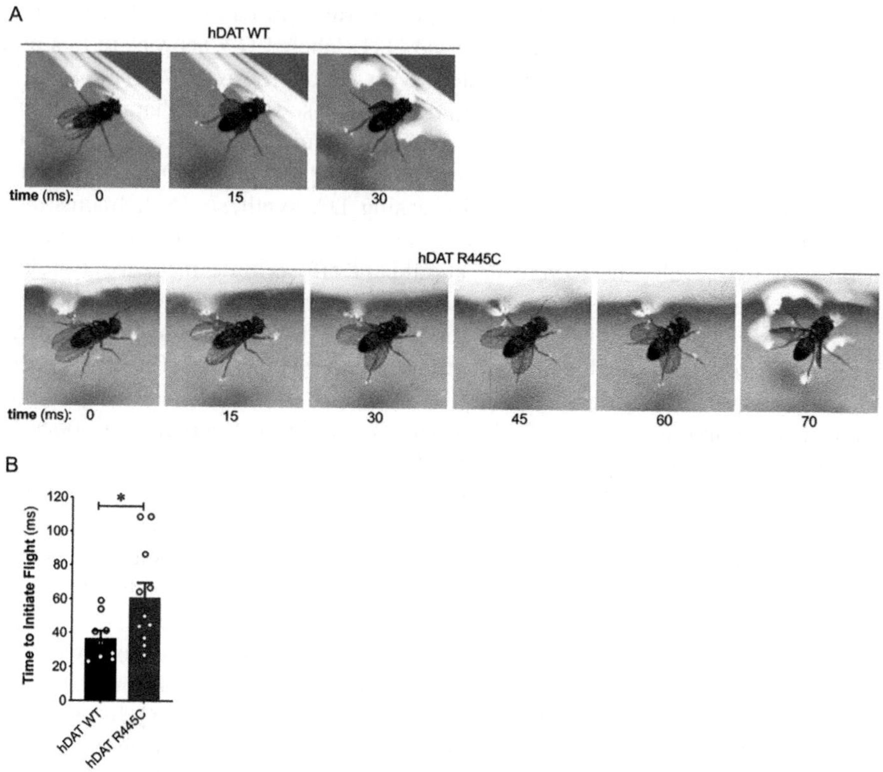

Fig. 6 (**a**) Representative single frames of *Drosophila* hDAT WT (*top*) and hDAT R445C (*bottom*) during various phases of coordinated take-off. (**b**) Flight initiation (take-off) was quantified from the initial phase of wing elevation ($t = 0$) to the second phase of simultaneous wing depression and leg extension. Flight initiation was significantly delayed in hDAT R445C flies relative to hDAT WT. (Adapted from Aguilar et al. [18])

significantly also improved flight initiation [18]. It is important to note that DAT is degraded through lysosomal pathways, and CQ inhibits DAT lysosomal degradation [63]. Together, these data indicate that increasing the level of DAT on the plasma membrane can partially rescue at least one of the parkinsonism-related phenotypes observed in hDAT R445C *Drosophila*, pointing to a potential target mechanism for new treatments for DTDS in humans.

Parallel to what is observed for the DAT mutant A559V, there is evidence to suggest that DAT R445C operates, at least in part, in a channel-like mode [18, 30]. Using both EPR and X-ray crystallography of LeuT, substitution of amino acids corresponding to hDAT R445 (LeuT R375) to neutral amino acids promotes the dissociation of the E6-R375 salt bridge that participates in the IC gate and prompts an inward-facing conformation of LeuT. This conformation alters transport function, which requires LeuT to isomerize toward an outward-facing conformation.

As initial structural work was performed using the DAT bacterial homolog, LeuT, Aguilar and collaborators, using MD simulations, further examined the impact of the R445C mutation on the dynamics of the hDAT IC gate [18]. Similar to results obtained with LeuT, R445C disrupts the interaction of the IC gate network. Unexpectedly, the R445C mutation disrupts the R60-D436 salt bridge and induces intermittent formation of a new salt bridge, E428-R60. These new amino-acid interactions within the IC network give rise to a channel-like intermediate filled with water molecules. This channel-like mode of hDAT has been described for other genetic variants of hDAT [31], as well as in mechanisms involving the psychostimulant amphetamine [21], further supporting the importance of this residue in the functioning of the IC gate and in regulating the channel-like mode of DAT.

5 Conclusions

Drosophila is a remarkably powerful model organism due to their genetic tractability, quick generation time, and a broad array of available behavioral assessments. This animal model is valuable for examining how disease-associated genetic variations impact the DA system, as demonstrated by the three variants of hDAT discussed in this chapter. All three variants discussed here demonstrate loss of DA uptake function; yet, DAT T356M and ΔN336 are associated with ASD, while DAT R445C is linked with DTDS. Interestingly, an ASD-associated variant, DAT T356M, demonstrates anomalous DA efflux, a phenomenon where DAT exhibits reverse transport of DA. This leads to increased extracellular DA levels in animals carrying the DAT T356M mutation, leading to their observed behavioral phenotypes (Table 2). Uniquely, DAT R445C demonstrates a constitutive channel-like mode and impaired trafficking, which contribute to DA cell death, likely leading to the DTDS-/PD-related behavioral phenotypes. Overall, the DA system is highly conserved between *Drosophila* and humans. Combining *Drosophila* modeling studies with traditional biochemical and structural analyses facilitates a deeper understanding of how transporter function translates into behavioral outcomes, and how hDAT genetic variants can underlie the etiology of neurological and neuropsychiatric disorders.

Table 2 hDAT mutants and their associated functional impact and behavioral phenotypes in *Drosophila*

DAT mutation	Associated disease	Surface expression	Dopamine uptake	Anomalous DA efflux	Behavioral phenotypes	Biophysical/ structural changes
DAT T356M	ASD	↔	V_{max} = ↓ K_m = ↔	Yes	1. Hyperlocomotion 2. Decreased AMPH-induced locomotion	Outward facing when bound to DA

(continued)

Table 2 (continued)

DAT mutation	Associated disease	Surface expression	Dopamine uptake	Anomalous DA efflux	Behavioral phenotypes	Biophysical/ structural changes
DAT ΔN336	ASD	↔	V_{max} = ↓ K_m = ↓	No	1. Hyperlocomotion 2. Increased Grooming 3. Increased freezing in a fear assay 4. Perturbed social interactions (decreased flock size)	Half-open inward facing
DAT R445C	DTDS	↓	V_{max} = ↓ K_m = ↓	No	1. Decreased spontaneous locomotion 2. Delayed flight initiation (motor impairments)	Channel-like state due to disruption of the IC gate

↔ represents no change compared to WT

Funding Development of this chapter was supported by the National Institute of Drug Abuse grant RO1DA038058 (A.G.), RO1DA035263 (A.G.), R01DA056484 (A.G. and A.M.C.) and R21DA059884 (A.M.C.).

References

1. Bellen HJ, Tong C, Tsuda H (2010) 100 years of Drosophila research and its impact on vertebrate neuroscience: a history lesson for the future. Nat Rev Neurosci 11:514–522. https://doi.org/10.1038/nrn2839
2. Ugur B, Chen K, Bellen HJ (2016) Drosophila tools and assays for the study of human diseases. Dis Model Mech 9:235–244. https://doi.org/10.1242/dmm.023762
3. Yamamoto S, Seto ES (2014) Dopamine dynamics and signaling in Drosophila: an overview of genes, drugs and behavioral paradigms. Exp Anim 63:107–119. https://doi.org/10.1538/expanim.63.107
4. Yamamoto K, Vernier P (2011) The evolution of dopamine systems in chordates. Front Neuroanat 5:21. https://doi.org/10.3389/fnana.2011.00021
5. Giros B, Jaber M, Jones SR, Wightman RM, Caron MG (1996) Hyperlocomotion and indifference to cocaine and amphetamine in mice lacking the dopamine transporter. Nature 379:606–612
6. Pendleton RG, Rasheed A, Sardina T, Tully T, Hillman R (2002) Effects of tyrosine hydroxylase mutants on locomotor activity in Drosophila: a study in functional genomics. Behav Genet 32:89–94
7. Hamilton PJ, Belovich AN, Khelashvili G, Saunders C, Erreger K, Javitch JA, Sitte HH, Weinstein H, Matthies HJ, Galli A (2014) PIP2 regulates psychostimulant behaviors through its interaction with a membrane protein. Nat Chem Biol 10:582–589. https://doi.org/10.1038/nchembio.1545

8. Zhang SX, Rogulja D, Crickmore MA (2016) Dopaminergic circuitry underlying mating drive. Neuron 91:168–181. https://doi.org/10.1016/j.neuron.2016.05.020
9. Zhang K, Guo JZ, Peng Y, Xi W, Guo A (2007) Dopamine-mushroom body circuit regulates saliency-based decision-making in Drosophila. Science (New York, NY) 316:1901–1904. https://doi.org/10.1126/science.1137357
10. Scaplen KM, Kaun KR (2016) Reward from bugs to bipeds: a comparative approach to understanding how reward circuits function. J Neurogenet 30:133–148. https://doi.org/10.1080/01677063.2016.1180385
11. Kume K, Kume S, Park SK, Hirsh J, Jackson FR (2005) Dopamine is a regulator of arousal in the fruit fly. J Neurosci 25:7377–7384. [pii]. https://doi.org/10.1523/JNEUROSCI.2048-05.2005
12. Altschul SF, Gish W, Miller W, Myers EW, Lipman DJ (1990) Basic local alignment search tool. J Mol Biol 215:403–410. https://doi.org/10.1016/s0022-2836(05)80360-2
13. Giros B, el Mestikawy S, Godinot N, Zheng K, Han H, Yang-Feng T, Caron MG (1992) Cloning, pharmacological characterization, and chromosome assignment of the human dopamine transporter. Mol Pharmacol 42:383–390
14. Penmatsa A, Wang KH, Gouaux E (2013) X-ray structure of dopamine transporter elucidates antidepressant mechanism. Nature 503:85–90. https://doi.org/10.1038/nature12533
15. Giros B, Caron MG (1993) Molecular characterization of the dopamine transporter. Trends Pharmacol Sci 14:43–49
16. Belovich AN, Aguilar JI, Mabry SJ, Cheng MH, Zanella D, Hamilton PJ, Stanislowski DJ, Shekar A, Foster JD, Bahar I et al (2019) A network of phosphatidylinositol (4,5)-bisphosphate (PIP2) binding sites on the dopamine transporter regulates amphetamine behavior in Drosophila melanogaster. Mol Psychiatry 26:4417. https://doi.org/10.1038/s41380-019-0620-0
17. Busson D, Pret AM (2007) GAL4/UAS targeted gene expression for studying Drosophila hedgehog signaling. Methods Mol Biol 397:161–201. https://doi.org/10.1007/978-1-59745-516-9_13
18. Aguilar JI, Cheng MH, Font J, Schwartz AC, Ledwitch K, Duran A, Mabry SJ, Belovich AN, Zhu Y, Carter AM et al (2021) Psychomotor impairments and therapeutic implications revealed by a mutation associated with infantile Parkinsonism-dystonia. elife 10:e68039. https://doi.org/10.7554/eLife.68039
19. Klein Herenbrink C, Støier JF, Reith WD, Dagra A, Gregorek MAC, Cola RB, Patriarchi T, Li Y, Tian L, Gether U et al (2022) Multimodal detection of dopamine by sniffer cells expressing genetically encoded fluorescent sensors. Commun Biol 5:578. https://doi.org/10.1038/s42003-022-03488-5
20. Aguilar JI, Dunn M, Mingote S, Karam CS, Farino ZJ, Sonders MS, Choi SJ, Grygoruk A, Zhang Y, Cela C et al (2017) Neuronal depolarization drives increased dopamine synaptic vesicle loading via VGLUT. Neuron 95:1074–1088. e1077. https://doi.org/10.1016/j.neuron.2017.07.038
21. Shekar A, Mabry SJ, Cheng MH, Aguilar JI, Patel S, Zanella D, Saleeby DP, Zhu Y, Romanazzi T, Ulery-Reynolds P et al (2023) Syntaxin 1 Ser14 phosphorylation is required for nonvesicular dopamine release. Sci Adv 9:eadd8417. https://doi.org/10.1126/sciadv.add8417
22. Rahbek-Clemmensen T, Lycas MD, Erlendsson S, Eriksen J, Apuschkin M, Vilhardt F, Jørgensen TN, Hansen FH, Gether U (2017) Super-resolution microscopy reveals functional organization of dopamine transporters into cholesterol and neuronal activity-dependent nanodomains. Nat Commun 8:740. https://doi.org/10.1038/s41467-017-00790-3
23. Lycas MD, Ejdrup AL, Sørensen AT, Haahr NO, Jørgensen SH, Guthrie DA, Støier JF, Werner C, Newman AH, Sauer M et al (2022) Nanoscopic dopamine transporter distribution and conformation are inversely regulated by excitatory drive and D2 autoreceptor activity. Cell Rep 40:111431. https://doi.org/10.1016/j.celrep.2022.111431
24. Sun F, Zeng J, Jing M, Zhou J, Feng J, Owen SF, Luo Y, Li F, Wang H, Yamaguchi T et al (2018) A genetically encoded fluorescent sensor enables rapid and specific detection of dopamine in flies, fish, and mice. Cell 174:481–496. e419. https://doi.org/10.1016/j.cell.2018.06.042

25. Reith MEA, Kortagere S, Wiers CE, Sun H, Kurian MA, Galli A, Volkow ND, Lin Z (2022) The dopamine transporter gene SLC6A3: multidisease risks. Mol Psychiatry 27:1031–1046. https://doi.org/10.1038/s41380-021-01341-5
26. Hamilton PJ, Campbell NG, Sharma S, Erreger K, Herborg Hansen F, Saunders C, Belovich AN, Sahai MA, Cook EH, Gether U et al (2013) De novo mutation in the dopamine transporter gene associates dopamine dysfunction with autism spectrum disorder. Mol Psychiatry 18:1315–1323. https://doi.org/10.1038/mp.2013.102
27. DiCarlo GE, Mabry SJ, Cao X, McMillan C, Woynaroski TG, Harrison FE, Reddy IA, Matthies HJG, Flynn CR, Wallace MT et al (2021) Autism-associated variant in the SLC6A3 gene alters the oral microbiome and metabolism in a murine model. Front Psych 12:655451. https://doi.org/10.3389/fpsyt.2021.655451
28. Campbell NG, Shekar A, Aguilar JI, Peng D, Navratna V, Yang D, Morley AN, Duran AM, Galli G, O'Grady B et al (2019) Structural, functional, and behavioral insights of dopamine dysfunction revealed by a deletion in SLC6A3. Proc Natl Acad Sci USA 116:3853–3862. https://doi.org/10.1073/pnas.1816247116
29. Mazei-Robison MS, Bowton E, Holy M, Schmudermaier M, Freissmuth M, Sitte HH, Galli A, Blakely RD (2008) Anomalous dopamine release associated with a human dopamine transporter coding variant. J Neurosci 28:7040–7046. [pii]. https://doi.org/10.1523/JNEUROSCI.0473-08.2008
30. Reith MEA, Jones KT, Zhen J, Topiol S (2018) Latch and trigger role for R445 in DAT transport explains molecular basis of DTDS. Bioorg Med Chem Lett 28:470–475. https://doi.org/10.1016/j.bmcl.2017.12.016
31. Bowton E, Saunders C, Erreger K, Sakrikar D, Matthies HJ, Sen N, Jessen T, Colbran RJ, Caron MG, Javitch JA et al (2010) Dysregulation of dopamine transporters via dopamine D2 autoreceptors triggers anomalous dopamine efflux associated with attention-deficit hyperactivity disorder. J Neurosci 30:6048–6057. [pii]. https://doi.org/10.1523/JNEUROSCI.5094-09.2010
32. DiCarlo GE, Aguilar JI, Matthies HJ, Harrison FE, Bundschuh KE, West A, Hashemi P, Herborg F, Rickhag M, Chen H et al (2019) Autism-linked dopamine transporter mutation alters striatal dopamine neurotransmission and dopamine-dependent behaviors. J Clin Invest 129:130. https://doi.org/10.1172/JCI127411
33. American Psychiatric Association. 5th edn. Washington; 2013
34. Sanders SJ, Murtha MT, Gupta AR, Murdoch JD, Raubeson MJ, Willsey AJ, Ercan-Sencicek AG, DiLullo NM, Parikshak NN, Stein JL et al (2012) De novo mutations revealed by whole-exome sequencing are strongly associated with autism. Nature 485:237–241. https://doi.org/10.1038/nature10945
35. Anderson BM, Schnetz-Boutaud N, Bartlett J, Wright HH, Abramson RK, Cuccaro ML, Gilbert JR, Pericak-Vance MA, Haines JL (2008) Examination of association to autism of common genetic variationin genes related to dopamine. Autism Res 1:364–369. https://doi.org/10.1002/aur.55
36. Kohls G, Chevallier C, Troiani V, Schultz RT (2012) Social 'wanting' dysfunction in autism: neurobiological underpinnings and treatment implications. J Neurodev Disord 4:10. https://doi.org/10.1186/1866-1955-4-10
37. Neale BM, Kou Y, Liu L, Ma'ayan A, Samocha KE, Sabo A, Lin CF, Stevens C, Wang LS, Makarov V et al (2012) Patterns and rates of exonic de novo mutations in autism spectrum disorders. Nature 485:242–245. https://doi.org/10.1038/nature11011
38. O'Roak BJ, Vives L, Girirajan S, Karakoc E, Krumm N, Coe BP, Levy R, Ko A, Lee C, Smith JD et al (2012) Sporadic autism exomes reveal a highly interconnected protein network of de novo mutations. Nature 485:246–250. https://doi.org/10.1038/nature10989
39. Iossifov I, Ronemus M, Levy D, Wang Z, Hakker I, Rosenbaum J, Yamrom B, Lee YH, Narzisi G, Leotta A et al (2012) De novo gene disruptions in children on the autistic spectrum. Neuron 74:285–299. https://doi.org/10.1016/j.neuron.2012.04.009

40. Yamashita A, Singh SK, Kawate T, Jin Y, Gouaux E (2005) Crystal structure of a bacterial homologue of Na+/Cl--dependent neurotransmitter transporters. Nature 437:215–223. https://doi.org/10.1038/nature03978
41. Prisinzano T, Rice CK, Baumann HM, Rothman BR (2004) Development of neurochemical normalization ("agonist substitution") therapeutics for stimulant abuse: focus on the dopamine uptake inhibitor, GBR12909. Curr Med Chem Cent Nerv Syst Agents 4:47–59. https://doi.org/10.2174/1568015043477630
42. Londei T, Valentini AM, Leone VG (1998) Investigative burying by laboratory mice may involve non-functional, compulsive, behaviour. Behav Brain Res 94:249–254. https://doi.org/10.1016/s0166-4328(97)00162-9
43. Cheng MH, Bahar I (2015) Molecular mechanism of dopamine transport by human dopamine transporter. Structure 23:2171–2181. https://doi.org/10.1016/j.str.2015.09.001
44. Chang HY, Grygoruk A, Brooks ES, Ackerson LC, Maidment NT, Bainton RJ, Krantz DE (2006) Overexpression of the Drosophila vesicular monoamine transporter increases motor activity and courtship but decreases the behavioral response to cocaine. Mol Psychiatry 11:99–113. https://doi.org/10.1038/sj.mp.4001742
45. LeDoux J (2003) The emotional brain, fear, and the amygdala. Cell Mol Neurobiol 23:727–738. https://doi.org/10.1023/a:1025048802629
46. Steimer T (2002) The biology of fear- and anxiety-related behaviors. Dialogues Clin Neurosci 4:231–249. https://doi.org/10.31887/DCNS.2002.4.3/tsteimer
47. Lee SH, Pak HK, Chon TS (2006) Dynamics of prey-flock escaping behavior in response to predator's attack. J Theor Biol 240:250–259. https://doi.org/10.1016/j.jtbi.2005.09.009
48. Chai C, Lim KL (2013) Genetic insights into sporadic Parkinson's disease pathogenesis. Curr Genomics 14:486–501. https://doi.org/10.2174/1389202914666131210195808
49. Lill CM (2016) Genetics of Parkinson's disease. Mol Cell Probes 30:386–396. https://doi.org/10.1016/j.mcp.2016.11.001
50. Herborg F, Jensen KL, Tolstoy S, Arends NV, Posselt LP, Shekar A, Aguilar JI, Lund VK, Erreger K, Rickhag M et al (2021) Identifying dominant-negative actions of a dopamine transporter variant in patients with parkinsonism and neuropsychiatric disease. JCI Insight 6:6. https://doi.org/10.1172/jci.insight.151496
51. Trinh J, Farrer M (2013) Advances in the genetics of Parkinson disease. Nat Rev Neurol 9:445–454. https://doi.org/10.1038/nrneurol.2013.132
52. Kurian MA, Gissen P, Smith M, Heales S Jr, Clayton PT (2011) The monoamine neurotransmitter disorders: an expanding range of neurological syndromes. Lancet Neurol 10:721–733. https://doi.org/10.1016/s1474-4422(11)70141-7
53. Ng J, Zhen J, Meyer E, Erreger K, Li Y, Kakar N, Ahmad J, Thiele H, Kubisch C, Rider NL et al (2014) Dopamine transporter deficiency syndrome: phenotypic spectrum from infancy to adulthood. Brain 137:1107–1119. https://doi.org/10.1093/brain/awu022
54. Beerepoot P, Lam VM, Salahpour A (2016) Pharmacological chaperones of the dopamine transporter rescue dopamine transporter deficiency syndrome mutations in heterologous cells. J Biol Chem 291:22053–22062. https://doi.org/10.1074/jbc.M116.749119
55. Asjad HMM, Kasture A, El-Kasaby A, Sackel M, Hummel T, Freissmuth M, Sucic S (2017) Pharmacochaperoning in a Drosophila model system rescues human dopamine transporter variants associated with infantile/juvenile parkinsonism. J Biol Chem 292:19250–19265. https://doi.org/10.1074/jbc.M117.797092
56. Barone MC, Sykiotis GP, Bohmann D (2011) Genetic activation of Nrf2 signaling is sufficient to ameliorate neurodegenerative phenotypes in a Drosophila model of Parkinson's disease. Dis Model Mech 4:701–707. https://doi.org/10.1242/dmm.007575
57. Cackovic J, Gutierrez-Luke S, Call GB, Juba A, O'Brien S, Jun CH, Buhlman LM (2018) Vulnerable Parkin loss-of-function Drosophila dopaminergic neurons have advanced mitochondrial aging, mitochondrial network loss and transiently reduced autophagosome recruitment. Front Cell Neurosci 12:39. https://doi.org/10.3389/fncel.2018.00039

58. Trinh K, Moore K, Wes PD, Muchowski PJ, Dey J, Andrews L, Pallanck LJ (2008) Induction of the phase II detoxification pathway suppresses neuron loss in drosophila models of Parkinson's disease. J Neurosci 28:465–472. https://doi.org/10.1523/jneurosci.4778-07.2008
59. Ford CP (2014) The role of D2-autoreceptors in regulating dopamine neuron activity and transmission. Neuroscience 282:13–22. https://doi.org/10.1016/j.neuroscience.2014.01.025
60. van den Berg C, Beek PJ, Wagenaar RC, van Wieringen PC (2000) Coordination disorders in patients with Parkinson's disease: a study of paced rhythmic forearm movements. Exp Brain Res 134:174–186. https://doi.org/10.1007/s002210000441
61. Sadaf S, Reddy OV, Sane SP, Hasan G (2015) Neural control of wing coordination in flies. Curr Biol 25:80–86. https://doi.org/10.1016/j.cub.2014.10.069
62. Card G, Dickinson MH (2008) Visually mediated motor planning in the escape response of Drosophila. Curr Biol 18:1300–1307. https://doi.org/10.1016/j.cub.2008.07.094
63. Daniels GM, Amara SG (1999) Regulated trafficking of the human dopamine transporter. Clathrin-mediated internalization and lysosomal degradation in response to phorbol esters. J Biol Chem 274:35794–35801
64. Friggi-Grelin F, Coulom H, Meller M, Gomez D, Hirsh J, Birman S (2003) Targeted gene expression in Drosophila dopaminergic cells using regulatory sequences from tyrosine hydroxylase. J Neurobiol 54:618–627. https://doi.org/10.1002/neu.10185

The Ins and Outs of Dopamine Transporter Gene Manipulation: In Vivo Models of DAT Dysfunction

Adele Stewart and Randy D. Blakely

Contents

Abstract The clinical efficacy of psychostimulant drugs, which target monoamine transporters, in treating attention-deficit/hyperactivity disorders (ADHDs) has stimulated interest on the role of transporter proteins like the dopamine (DA) transporter (DAT) in neurotransmission as well as the potential utility of DAT knockout organisms as models for neuropsychiatric disorders. Indeed, the study of DAT-deficient worms, flies, fish, mice, and rats has revealed a conserved role for DAT in the control of motor behavior as well as repetitive behavior, threat aversion, social behavior, and cognition in mammals. However, the disconnect between phenotypes

A. Stewart (✉)
Department of Neuroscience and Pharmacology, University of Iowa, Iowa City, IA, USA

Iowa Neuroscience Institute, University of Iowa, Iowa City, IA, USA
e-mail: adele-stewart@uiowa.edu

R. D. Blakely (✉)
Department of Biomedical Science, Florida Atlantic University, Jupiter, FL, USA

Stiles-Nicholson Brain Institute, Florida Atlantic University, Jupiter, FL, USA
e-mail: rblakely@health.fau.edu

O. V. Mortensen (ed.), *Monoamine Transporters*, Advances in Neurobiology 46,
https://doi.org/10.1007/978-3-031-96364-3_10

observed in DAT-deficient model organisms and humans, which exhibit an early-onset syndrome characterized by Parkinsonism/dystonia and premature death, challenges the construct validity of DAT knockout models with respect to modeling neurobehavioral disorders. As an alternate approach, several groups have utilized coding variants in the *SLC6A3* gene linked to psychiatric conditions, which display divergent molecular phenotypes. This chapter reviews the development and characterization of models of DAT gene deletion and mutation with a particular emphasis on comparing/contrasting the functional impact of DAT deficiency to DAT dysregulation triggered by neuropsychiatric disorder-linked DAT mutants in vivo. Ultimately, the study of DAT knockout and mutant models has revealed novel functions for DA in the mammalian brain, uncovered a dynamic interplay between the monoaminergic systems, highlighted sex differences in the DA system that determine the behavioral trajectory of DAT deregulation, and allowed for the screening of potential leads for therapeutics to treat disorders linked to aberrant dopaminergic neurotransmission.

Keywords Dopamine transporter · Model organism · Psychostimulants · Knockout · Animal behavior

1 Introduction

Though comprising a comparatively small number of neurons in the human brain (~450,000) [1], dopamine (DA)-releasing neurons provide dense innervation to the striatum as well as other cortical and subcortical structures with some DA neurons forming hundreds of thousands synapses within the basal ganglia [2]. Two predominant dopaminergic nuclei originate in the ventral midbrain, each of which innervate distinct forebrain regions and subserve unique functions. The nigrostriatal circuit comprises DA-synthesizing cells in the substantia nigra pars compacta (SNc) that project to the caudate nucleus and putamen (CPu) in the dorsal striatum and play a critical role in the control of voluntary movement, procedural learning, and the formation of habitual behaviors [3]. Two additional dopaminergic circuits originate in the ventral tegmental area (VTA) and innervate the ventral striatum (nucleus accumbens, NAc) or prefrontal cortex (PFC) forming the mesolimbic and mesocorticolimbic circuits, respectively. Although mesolimbic dopaminergic neurons are critical for reward prediction and response, reinforcement, and social behavior [4], mesocorticolimbic neurons control working memory, attention, action selection, and cognitive flexibility [4]. A growing body of evidence has also alluded to sex-specific patterns of synthesis, release, and reuptake of DA [5–10], which may explain why virtually all DA-linked neuropsychiatric disorders display some degree of sex bias in prevalence, age of onset, and/or clinical symptomology [11–14].

As with most chemical neurotransmitters, the primary extracellular mechanism terminating DA action is via the active transport of the neurotransmitter back into the presynaptic DA neuron where it can be repackaged for future release [15].

Molecular cloning of the gene (*SLC6A3*) encoding the cocaine-sensitive DA transporter (DAT) first in rodents [16–18] and later in humans [19] helped usher in a new era in research into the molecular basis of DA signaling. In humans, the *SLC6A3* gene maps to chromosome 5 [19] and contains 15 exons and 14 introns with the start codon located in exon 2 and the stop codon located 23 bp into the 2 kb long exon 15 (GenBank Accession # NM_001044). As a result, the *SLC6A3* gene possesses long 5′ and 3′ untranslated regions (UTRs), both of which contain variable number tandem repeat (VNTR) segments whose length differs across the human population. In addition, several coding variants, single nucleotide polymorphisms (SNPs) in the intronic and promoter regions, and intronic VNTRs have been identified in the *SLC6A3* gene and linked to the etiology of neuropsychiatric and neurological disorders including substance-use disorders, attention-deficit/hyperactivity disorder (ADHD), autism spectrum disorder (ASD), bipolar disorder (BPD), schizophrenia, Parkinson's Disease (PD), and posttraumatic stress disorder (PTSD) (for review, see [20]). DAT is also the direct target of multiple clinically efficacious drugs used in the treatment of ADHD and narcolepsy (e.g., D-amphetamine and methylphenidate) as well as drugs of abuse (e.g., cocaine and methamphetamine). Notably, biallelic loss of function in *SLC6A3* is associated with a rare, early-onset Parkinsonism/dystonia syndrome referred to as DA Transporter Deficiency Syndrome (DTDS). Symptoms of DTDS include delayed motor development, hyperkinetic movements, bradykinesia, rigidity, tremor, dystonia, and breathing difficulties, and many children with infantile onset DTDS do not survive past early adolescence [21–24]. Though fewer than 30 cases have been reported [21–24], DTDS disease severity has been shown to negatively correlate with residual DAT activity [21, 22] emphasizing the critical importance of DAT in the control of DA homeostasis. Though *SLC6A3* gene deletion in mouse [25] or rat [26] fails to faithfully recapitulate behavioral phenotypes observed in humans with DTDS, DA persists up to 100 times longer in the extracellular space in homozygous knockouts, emphasizing the role of DAT as the primary mechanism of DA clearance in mammals in vivo.

The DAT protein belongs to a family of Na^+/Cl^- cotransporters, also referred to as neurotransmitter Na^+ solute symporters (NSSs or SLC6), that couple chemiosmotic ion gradients to neurotransmitter transport and thereby shuttle their respective substrates across the plasma membrane against a cytoplasmic concentration gradient [27]. Generation of a crystal structure for the *Drosophila melanogaster* (*D. melanogaster*) DAT (dDAT) in 2014 [28] confirmed key structural elements of the DAT protein that had previously been inferred from study of its prokaryotic relative LeuT [29–31] including 12 transmembrane-spanning domains linked by 6 extracellular and 5 intracellular loops as well as intracellular N- and C-terminal tails. Engineered or DTDS-associated mutations located in or in close proximity to regions of DAT required for substrate binding or ion transport [28] frequently result in loss of transport function, a consequence attributed to blockade of transport capacity, loss of protein maturation, or impaired surface trafficking [21, 22, 24, 32]. In contrast, many neuropsychiatric-disorder-associated DAT-coding variants have also been identified (Fig. 1) with altered microdomain localization (R615C [32, 33]), loss of conformational flexibility (ΔN336 [34]), disrupted protein–protein interactions

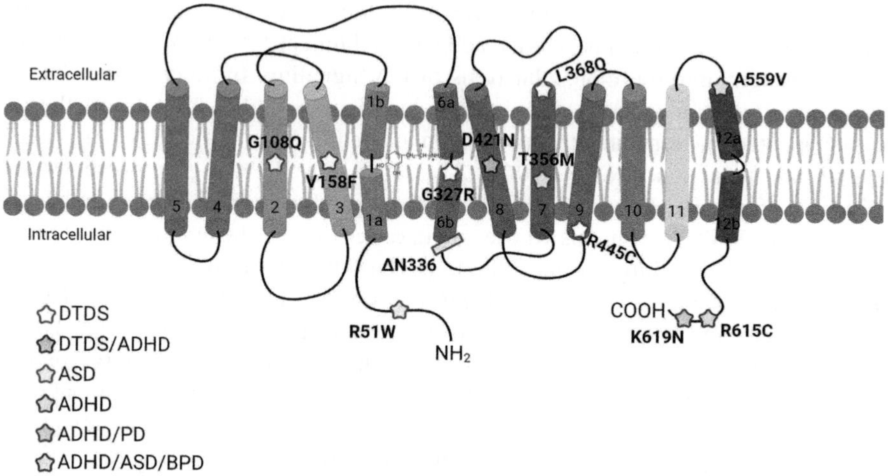

Fig. 1 Location of DAT mutations cited in this chapter superimposed on a diagram of the DAT protein. Mutations are color-coded based on associated disorders. ADHD, attention-deficit/hyperactivity disorder; ASD, autism spectrum disorder; BPD, bipolar disorder; DTDS, dopamine transporter deficiency syndrome; PD, Parkinson's disease

(K619N [35], R51W [36]), or a unique gain of function resulting in aberrant reverse transport termed anomalous or constitutive DA efflux (A559V [32, 37], D421N [32], T356M [32, 38, 39]). Significant insights into structure–function relationships across several domains of the human DAT (hDAT) protein have been gleaned from study of these variants in vitro. Importantly, from a translational perspective, identification of psychiatric-disease-linked DAT mutants has afforded an opportunity to generate construct valid models to investigate mechanisms driving risk for neuropsychiatric disorders since neurobehavioral conditions including ASD [34, 36, 38–40], ADHD [21, 33, 37, 41], and BPD [32, 42] feature these variants rather than a complete loss of DAT expression and function. These efforts have led to the hypothesis that behavioral changes in these disorders, as well as the clinical efficacy of DAT blockers, may derive from aberrant DAT regulation or function in addition to, or rather than, a simple loss of DA clearance per se. To test this hypothesis, several groups have generated genetically modified organisms harboring disease-linked DAT-coding variants. Here, we review and compare changes in neural signaling and behavior arising from the study of both DAT gene deletion and mutation across species from invertebrate nematodes to mammalian models, with an eye on how findings compare across specific behavioral domains. With a few notable exceptions, studies in model organisms wherein DAT activity is eliminated or altered have primarily utilized male subjects. Where key sex differences in the impact of DAT gene deletion or mutation have been found, they are highlighted below.

2 DAT Protein-Encoding Gene Across Species

Though evolutionary divergence has resulted in a lack of a DAT-encoding gene in birds, crocodiles, and lizards [43], the genomes of other ancestral vertebrates (e.g., zebrafish, *Danio rerio*) and invertebrates (e.g., *C. elegans*, *D. dorotocephala*, *D. melanogaster*) contain a DAT gene that subserves DA clearance. In this regard, genes encoding proteins required for DA biosynthesis, vesicular release, reuptake, and signaling have been identified in the nematode *C. elegans* [44] and define a unique population of 8 DA-releasing neurons in the hermaphrodite [45]. Using conserved amino-acid sequences from mammalian DAT proteins, Jayanthi and colleagues [46] cloned a *C. elegans* cDNA that encoded a DA transporter (*dat-1*) [46] and shares 46% identity with human DAT [46]. Heterologous expression of the cDNA induced cocaine-sensitive transport of both DA and norepinephrine (NE), though efficiency of transport for DA is higher than that observed for NE. As worms do not synthesize NE, and noncoding genomic sequences upstream of the transcription unit drive reporter expression selectively in worm DA neurons [47], the gene was designated *dat-1*. Notably, most DTDS-causing *SLC6A3* mutations occur in highly conserved amino-acid residues shared between hDAT and *C. elegans* DAT [22, 24], supporting the in vivo activity of *dat-1* and a rationale for use of the nematode to model human DAT regulation and structure–function relationships. *C. elegans dat-1* knockouts have also been used to define a fundamental role for DAT in feedback control of vesicular release of synaptic DA [48], and rescue experiments in *dat-1* null mutants allowed for the identification of conserved C-terminal motifs required for axonal export and synaptic localization of DAT in vivo [49]. The first crystal structure of a DA transporter was generated utilizing the *Drosophila* DAT (dDAT) [28], which exhibits ~50% identity with the human transporter. As with its mammalian homolog, dDAT substrate specificity favors DA tenfold over NE and dDAT has low affinity for serotonin (5-HT) [50]. However, as with *C. elegans* DAT-1 [46], the pharmacological profile of dDAT more closely parallels that of the human NE transporter (NET) [50].

The machinery responsible for dopaminergic neurotransmission is highly conserved between zebrafish and mammals, and this model has proven particularly useful in monitoring the development of catecholaminergic systems due to embryo transparency and development occurring external to the mother where direct observation is possible. As in mammals, DA neurons in zebrafish can be distinguished by the presence of tyrosine hydroxylase (TH), expression of a zebrafish DAT homolog (zDAT) and the absence of DA β hydroxylase, the enzyme responsible for conversion of DA to NE [51, 52]. The susceptibility of zebrafish to the neurotoxic actions of both 1-methyl-4-phenyl-1,2,3,6-tetrahydropyridine (MPTP) [53, 54] and 6-hydroxydopamine (6-OHDA) [55, 56], which require DAT for intracellular transport, confirmed the presence of functional zDAT in the zebrafish brain.

The mouse DAT knockout (mDAT$^{-/-}$) represented the first mammalian model of DAT disruption [25] and has been extensively studied over the past two decades. Cloning of mDAT yielded cDNAs encoding a polypeptide that is 93.5% identical to hDAT [57, 58], and allowed for the design of targeting constructs for homologous recombination-dependent gene deletion and the generation of the first mammalian DAT knockout model [25]. Though generation of genetically modified rat models has historically presented a challenge due to the unavailability of embryonic rat stem cell lines suitable for gene targeting, recent advances have allowed for modification of the genome of outbred Wistar-Han rats with targeted deletion of a portion of exon2 of the *SLC6A3* gene using zinc-finger nuclease (ZFN) technology, resulting in effective gene knockout [26]. A second line of DAT knockout Wistar rats was generated using CRISPR/Cas9-mediated gene editing, creating a frameshift deletion mutation in exon 4 and resulting in a prematurely truncated protein product [59].

Regardless of the model organism studied, loss of DAT function results in markedly delayed synaptic DA clearance. Across species, an evolutionarily conserved role for DAT in control of locomotion has been identified through the study of DAT knockout model organisms with a more diverse array of phenotypes evident with increasing species complexity (Fig. 2). As we will note below, altered DAT function or expression triggers changes in social behavior in invertebrate *D. melanogaster*, which rely on social dynamics to coordinate escape from predation, as well as in mammalian species where social behaviors are also key to species' survival and propagation. Studies in rodents have also unmasked the critical role of DAT in learning, habit formation, motivation, and cognitive flexibility. Intriguingly, neuropsychiatric-disorder-associated DAT variants recapitulate some, but not all, of the phenotypes observed in DAT knockouts, emphasizing the inherent intricacy of DAT action across brain circuits and behavioral states.

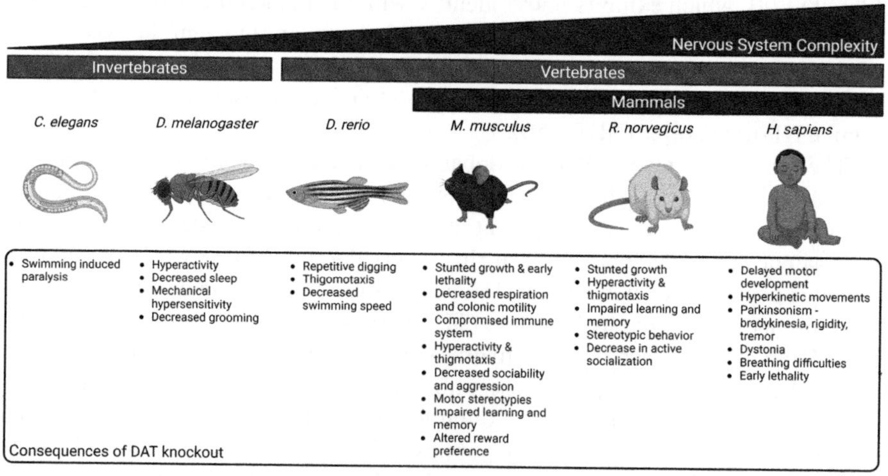

Fig. 2 Summary of key phenotypes resulting from DAT gene deletion across species. To date, DAT null model organisms have been identified or generated in nematodes (*C. elegans*), fruit flies (*D. melanogaster*), zebrafish (*D. rerio*), mice (*M. musculus*), and rats (*R. norvegicus*). A rare early-onset Parkinsonism/dystonia syndrome (DTDS) results from DAT deficiency in humans

3 Functional Consequences of DAT Gene Deletion or Mutation on DA Dynamics

3.1 DAT Knockout

Though it is difficult to measure DA clearance in humans where investigators are limited to noninvasive approaches, single-photon emission computerized tomography (SPECT) with ^{123}I (Ioflupane; DaTScan) imaging did reveal absent or significantly impaired tracer uptake in the basal nuclei of children with DTDS consistent with loss of DAT function [24]. DA accumulation in isolated striatal synaptosomes is almost completely abolished in the mDAT knockout [25]. Utilizing fast-scan cyclic voltammetry (FSCV) in slices from the striatum of wildtype (WT) or mDAT knockout mice, Giros and colleagues demonstrated that both the peak and time course of electrically evoked DA release were altered in the absence of mDAT, reflecting both a depletion of releasable DA pools and impaired DA clearance [25]. Comparable results were obtained from the dorsal striatum and NAc in vivo, and, notably, DA overflow evoked following electrical stimulation of the medial forebrain bundle rapidly decayed in the striatum of DAT$^{-/-}$ mice following TH inhibition, indicating increased reliance on ongoing DA synthesis [60]. However, TH levels were found to be decreased in the CPu and NAc in DAT$^{-/-}$ mice, with a corresponding decrease in TH regulatory phosphorylation at Ser31 but an increase in Ser40 phosphorylation, indicative of ongoing dysregulation in synthetic capacity, particularly at projection sites [61]. Overall, these data demonstrated conclusively that DAT is the primary mechanism of DA clearance and recycling in the mouse brain and, thus, a critical determinant of DA signaling capacity and dynamics.

Loss of mDAT protein also leads to multiple compensatory changes in the DA system including downregulation of postsynaptic striatal DA D1 (D1R) and D2 (D2R) receptors [25, 62, 63]. However, functional readouts of inhibitory D2R and stimulatory D1R signaling remain elevated as evidenced by decreased expression of preproenkephalin A and increased prodynorphin, respectively, in the basal ganglia of DAT$^{-/-}$ mice [25, 62, 63]. D2R binding is also decreased in the ventral midbrain of DAT knockout mice indicative of a loss of autoreceptor expression in dopaminergic cell bodies. Consistent with a loss of autoreceptor expression, the ability of D2 agonists to block DA synthesis, suppress DA neuron firing and DA release, and decrease extracellular DA is absent in mice lacking DAT expression [64]. Intriguingly, while D1Rs and D2Rs are depleted in DAT knockouts, expression of D3 receptors (D3Rs) is increased across striatal subregions [65]. In addition, whereas, overall, the density of medium spiny neurons (MSNs) is unaltered in mice lacking DAT, there is a selective increase in the proportion of neurons expressing D3R [65]. Expression of brain-derived neurotrophic factor (BDNF), a neurotrophin important for neuronal development and synaptic plasticity, is also decreased in the cortex of DAT knockout mice [66, 67] and may contribute to the diminished functional connectivity in cortico-striatalthalamic circuitry observed via manganese-enhanced magnetic resonance imaging (meMRI) in mice lacking DAT [68].

FSCV revealed a large increase in the half-life of electrically evoked DA in striatal brain slices from DAT knockout rats [26, 59] that corresponded to an eightfold increase in extracellular DA levels in striatum [26]. In vivo, the half-life of synaptic DA is also increased, and DA clearance is insensitive to cocaine in the CPu or SNc of DAT knockout rats [59]. Striatal DA and 5-HT tissue contents are decreased in DAT knockout rats, an effect accompanied by a decrease in TH expression and downregulation of D2R and D1R mRNA [26]. Loss of D2 autoreceptor (D2AR) function resulting from rat DAT (rDAT) gene deletion is evident in the SNc where the impact of D2R agonists of DA neuron firing frequency is significantly attenuated and in the CPu where DA clearance is largely insensitive to D2R stimulation [59]. Similar to trends observed in mice [66], BDNF levels are decreased in the cortex and VTA of rats lacking DAT, though DAT knockout rats also show an increase in BDNF in the dorsolateral striatum with no change in the NAc [26, 69].

Deregulation in DA dynamics is also evident in invertebrate models lacking DAT. For example, a *Drosophila* mutant named *fumin* was identified that expresses two nonfunctional copies of the DAT gene and displays markedly delayed DA clearance [70]. dDAT deficiency also triggers a reduction in D1R expression in the mushroom body of flies, an effect attributed to compensatory receptor downregulation in the presence of excess extracellular DA [71]. Zebrafish lacking zDAT expression also display a decrease in whole brain DA content and the number of midbrain TH+ neurons, though it remains to be determined if this difference reflects neurodegeneration in the adult or results from impaired development of the dopaminergic system [72].

3.2 Disease-Associated DAT Variants

Perturbations in DA signaling have been implicated in several polygenic psychiatric disorders [20] wherein complex gene/gene or gene/environment interactions determine a trajectory of behavioral manifestations and pairwise comorbidity is common. Of the mammalian models expressing neuropsychiatric-disorder-associated DAT variants, mice harboring the DAT Ala559Val (A559V) substitution are the best characterized. This mutation, identified in individuals with ADHD [41], ASD [40], and BPD [42], does not impact DAT expression, trafficking, or uptake in vitro, but instead supports nonvesicular, transporter-mediated DA release [37, 41]. In mice, expression of the DAT A559V variant also fails to change DAT expression, TH levels, tissue DA content, or synaptosomal DA uptake [73]. However, DAT Val559 mice do display increased extracellular DA levels in the dorsal striatum consistent with tonic DA leak [73].

Heterologous expression studies in vitro indicate that the anomalous DA efflux (ADE) of hDAT A559V is sustained by D2R-dependent, Ca^{2+}/calmodulin-dependent kinase II (CaMKII) activation that drives tonic hyperphosphorylation of the transporter [74], with DA efflux inhibited by DAT blockers, amphetamine or methylphenidate [37]. Ex vivo, there is also evidence of constitutive D2AR stimulation resulting from expression of the DAT A559V variant. First, in slices containing the SNc from DAT Val559 mice, the duration of spontaneous and electrically evoked, D2-dependent inhibitory postsynaptic currents (IPSCs) is enhanced relative to controls [73]. Second, depolarization-driven DA release is blunted in striatal slices from DAT Val559 mice as compared to slices from WT mice, an effect that can be normalized by D2AR blockade [73]. Finally, DAT phosphorylation at Thr53 (p-Thr53) and transporter surface expression is selectively increased in slices from the dorsal striatum, but not ventral striatum, of male DAT Val559 mice due to circuit-selective signaling between D2ARs and DAT [75]. Here again, a D2R antagonist can restore DAT surface trafficking and phosphorylation to WT levels [75]. Intriguingly, the circuit selectivity of D2AR-DAT coupling is reversed in female mice, with D2AR stimulation promoting DAT phosphorylation and trafficking in the ventral striatum, but not the dorsal striatum of females [76]. As a result, an increase in DAT surface expression and DAT p-Thr53 is observed in slices from the ventral striatum of DAT Val559 females [76]. In vivo chronoamperometry revealed compromised DA clearance in DAT Val559 mutants that parallels sex-biased impacts on DAT regulation occurring in the CPu in males and NAc in females [75, 76].

In 2013, Hamilton et al described the identification of the first de novo hDAT coding variant associated with ASD, the Thr356Met point mutant. Mutation of this site, conserved in humans, rodents, zebrafish, fruit flies, and nematodes, results in a reduction in transport capacity (Vmax) and ADE [32, 39]. Whereas WT hDAT expression suppresses the hyperlocomotion of dDAT knockout flies, animals expressing hDAT T356M remain hyperactive, indicating that this mutation compromises DA clearance in vivo [39]. Introduction of the DAT T356M mutation did not change DAT expression in murine striatum but did slow DA clearance in striatal slices [38]. In addition, striatal DA tissue content and phosphorylation of TH were decreased in DAT Met356 mice indicative of compromised DA synthesis [38], possibly a compensatory consequence of excess presynaptic D2AR stimulation driven by ADE. A second ASD-associated hDAT variant was described in 2015, and this substitution (Arg51Trp; R51W) was found to decrease the interaction between DAT and the presynaptic membrane protein syntaxin 1a (SYN1A), though it did not impact baseline uptake [36]. A third ASD-linked hDAT variant, the in-frame deletion at N336 (ΔN336), has also been characterized in *Drosophila.* The hDAT ΔN336 mutation results in a truncated transporter protein that, in vitro, displays impaired DA transport [34]. Molecular dynamic stimulations suggest impaired DA uptake by the hDAT ΔN336 may result from formation of an aberrant conformation in the intracellular gate of the transporter [34].

The hDAT coding variant K619N was identified in two unrelated patients, one with early-onset parkinsonism and comorbid ADHD and another with ASD [35]. DAT binding in the striatum, as measured by [^{123}I]FP-CIT single-photon emission computerized tomography (SPECT), was found to be decreased in both subjects [35]. hDAT K619N appears to function in a dominant-negative manner by impairing DA uptake when coexpressed with WT hDAT [35]. When expressed in *fumin* flies, hDAT K619N displays decreased rescue of DA uptake and amphetamine-stimulated DA efflux when compared to WT hDAT [35]. To test the functional impact of the DAT K619N mutation in a mammalian system, Herborg et al. injected a virus containing WT hDAT or hDAT K619N unilaterally into the SNc of mice [35]. Intriguingly, while overexpression of hDAT did not change uptake capacity, expression of hDAT K619N decreased striatal DA uptake to 50% compared to vector controls, confirming that hDAT K619N functions in a dominant-negative manner to impair functionality of endogenous mDAT [35]. The K619 residue is located on the distal C-terminus in a PSD-95/Discs-large/ZO-1 homology (PDZ) binding domain, a region previously linked to surface trafficking and axonal targeting of DAT [77]. Disruption of this region in mice (DAT-AAA, 3 terminal a.a. residues: LLV → AAA) impairs the interaction of DAT with the PDZ domain-containing protein PICK1 and results in constitutive transporter internalization [77, 78]. In fact, mutation of the DAT PDZ domain (DAT-AAA) in mice results in modest reductions in DAT expression in midbrain DA cell bodies, though a 70% reduction in DAT protein is observed in striatum, with a corresponding loss of DA uptake and a hyperactive behavioral phenotype [77]. In vitro, hDAT K619N phenocopies DAT-AAA with accelerated cellular turnover, decreased surface expression, and reduced uptake, though the net impact of the K619N mutation is not a robust as elimination of the entire PDZ motif [35].

4 Physiological Consequences of DAT Gene Deletion or Mutation

In addition to the parkinsonism-like motor symptoms, human patients with DTDS have significantly compromised longevity often not surviving past adolescence [21–24]. The severity of DAT deficiency on overall health and survival is less profound in model organisms, but rodent and invertebrate DAT knockout models do exhibit signs of body-wide physiological dysfunction. For example, in addition to their overt hyperactivity, *fumin* flies are hyperresponsive to mechanical stimuli [79], prefer lower temperatures and exhibit an increased metabolic rate [80], groom their heads less frequently [81], and exhibit lower fecundity [82]. Despite drastic changes to their behavioral patterns, dDAT knockout Drosophila display a normal lifespan irrespective of sex or genetic background [83]. However, when reared on a high calorie diet, the survival of *fumin* flies is decreased relative to WT controls [83].

Homozygous DAT knockout mice display stunted growth and an increased propensity for premature death [25]. In addition to altered central nervous system function, deletion of mDAT also impacts DA homeostasis in peripheral organ systems. Indeed, $DAT^{-/-}$ mice exhibit altered colonic motility [84], decreased respiratory frequency during normoxic and hypoxic conditions [85], lower body temperature [85], and aberrant acquired and innate immune responses [86, 87]. Unlike DAT knockout mice [25], DAT knockout rats do not exhibit early lethality, but do gain weight at a slower rate than their WT littermates [26, 59] and display behavioral changes consistent with impaired wellbeing such as decreased burrowing in the home cage [88] and inflexible grooming habits [89]. rDAT knockouts also display evidence of impaired peripheral antioxidant defense detectable as early as 1 week postnatally [90] and altered sensitivity to anesthetics [91].

In contrast to effects seen in DAT deficient rodents, the introduction of DAT point mutations does not seem to have overt impacts on mouse health. Male homozygous DAT Val559 mice grow and develop normally and exhibit no obvious alterations in general health, motor capability, or reflexes [92]. Female DAT Val559 homozygotes are similarly healthy in appearance, retain fertility, and display normal maternal care [92]. Though the impact of the DAT T356M mutation on DA clearance is not as drastic as in DAT knockouts, male, but not female, DAT Met356 mice do exhibit stunted growth relative to controls though this occurs without deficits in strength, coordination, motor learning, or anxiety [38]. Reduced weight gain in DAT Met356 mice might rather be a consequence of their increased caloric expenditure, improved glucose handling, and reduced body fat compared to WT mice, which correlate with altered oral microbiota [93].

5 DAT-Dependent Behaviors

5.1 Locomotor Activity and Sleep

The characteristic delayed motor development and progressive parkinsonism-like motor disturbances in individuals with DTDS speak to a key role for DAT in the control of locomotor behaviors in humans [94]. Although a deletion mutation of *C. elegans dat-1* (*ok157*) did not result in the overt hyperactivity seen in DAT knockout rodents [25, 26], extrasynaptic spillover of DA onto cholinergic motor neuron D2-type DA receptors results from insufficient DA clearance, triggering a locomotor phenotype when these worms are placed in water termed swimming-induced paralysis (Swip) [95]. The Swip phenotype has proven extremely useful as an easily visible and quantifiable readout of altered dopaminergic tone, leading to the use of Swip for forward genetic screens seeking to establish conserved features of dopaminergic neurotransmission [96]. Utilizing this approach, several novel non-functional *dat-1* alleles have been identified [96] in addition to a transcription factor controlling expression of *dat-1* mRNA [97] and a kinase responsible for regulation of DAT surface trafficking [98]. More recently, Illiano et al. generated genetically humanized *C. elegans* lines by expressing wild type hDAT or two DTDS-linked hDAT variants in *dat-1* null animals [99]. While hDAT expression rescued Swip in *dat-1* knockout animals, these capacities were impaired when the DTDS DAT mutants were expressed [99]. Similarly, elimination of the *SLC6A3* gene from zebrafish utilizing CRISPR/Cas9 technology resulted in only modest impacts on locomotor behavior, manifesting as a slight reduction in swimming speed [72].

Further insight into the role of dDAT in fly neurophysiology was gleaned from the study of a hyperactive *Drosophila* mutant named *fumin*, which expresses two nonfunctional copies of the DAT gene and displays markedly delayed DA clearance [70]. The increased locomotion in *fumin* flies is so profound that they display little to no rest behavior, a phenotype that inspired the line name, as fumin in Japanese translates to "sleepless" [79, 100]. DAT variants have also been knocked into the *Drosophila* model resulting in perturbations in locomotion. Introduction of either the ASD-linked hDAT ΔN336 [34] or the ADHD/PD-associated hDAT K619N [35] variant led to profound hyperactivity in flies, effects attributed to compromised DA clearance. hDAT K619N knock-in flies also exhibit age-dependent changes in climbing behavior that may result from compromised DA neuron survival over the *Drosophila* lifespan [35].

As in other invertebrate models of DAT gene deletion, mice lacking DAT are profoundly hyperactive [25], even under circumstances designed to induce freezing [101, 102], as in the forced swim test or tail suspension test. Hyperactivity resulting from DAT gene deletion has been reproduced across groups and mouse strains [103] and is detectable in DAT knockout mice as early as P22 [104]. $DAT^{-/-}$ mice also display non-focal perseverative patterns of locomotion, decreased rearing, alterations in gait, and impaired motor coordination [62, 105]. DAT deletion also disrupts diurnal changes in motor activity, reflecting important DA signaling contributions to the sleep/wake cycle [106]. As in mice [25, 107], DAT null rats are hyperactive though they display increased thigmotaxis and a decreased preference for exploring new enviornments [26, 59, 108].

Compromised DA clearance in DAT Met356 mice does lead to persistent hyperactivity and increased rearing comparable to that observed in mDAT knockouts. However, unlike DAT knockouts, neither male [73] nor female [76] DAT Val559 mice are hyperactive, though they both display a hyper-reactive response to imminent handling termed "darting." These results are consistent with compromised or intact DA uptake observed in cells expressing hDAT T356M or hDAT A559V, respectively [32]. Notably, when measured in vivo utilizing chronoamperometry, disrupted DA clearance was observed in a region-restricted and sex-biased manner in DAT Val559 mice [75, 76], which may explain the more subtle alterations in locomotor behavior observed in the DAT Val559 model. In contrast to results obtained in *Drosophila*, mice expressing hDAT K619N are not hyperactive at baseline, but, whereas expression of WT hDAT induces a rotational motor behavior when expressed unilaterally, a similar impact is not seen following overexpression of hDAT K619N alluding to altered DAT function upon introduction of the K619N mutation [35].

5.2 Repetitive Behavior and Threat Aversion

Elimination or mutation of DAT also triggers abnormal behaviors that are characterized by repetition, rigidity, inappropriateness, and lack of adaptability. For example, *Drosophila* expressing the ASD-associated hDATΔN336 mutation display prolonged freezing and impaired fleeing upon presentation of a predatory sound and increased, repetitive grooming behavior [34]. The behavior of zDAT knockout zebrafish is easily distinguishable from WT fish as they display a unique propensity to hover near the bottom of the tank and exhibit repetitive digging-like behavior [72]. Thigmotaxis is present in zDAT knockout zebrafish regardless of day/night period, indicating aberrant adaptation to environmental stimuli [109].

Whereas mice with one intact DAT allele spend more time exploring novel environments and objects, homozygous DAT knockouts exhibit enhanced thigmotaxis and neophobia that habituates overtime [101, 110]. It has been proposed that the contextual dependency of locomotor and exploratory behavior in DAT knockout mice may reflect asynchronous coordination in corticostriatal circuits rather than a simple tonic increase in circuit activity [111, 112]. In addition, whereas DAT knockout mice typically ignore marbles in the marble buying test [113], DAT knockout rats spend more time frantically burying marbles compared to WTs [114]. This phenotype likely does not reflect increased anxiety in DAT knockout rats, but possibly compulsivity, as the knockout rats display an increase in light occupancy time in the light/dark transition test, indicative of decreased threat aversion [114].

While neither DAT Met356 [38] nor DAT Val559 [73] males display altered anxiety in the elevated plus or zero mazes, male DAT Val559 mutants exhibit aberrant patterns of exploration in the light/dark transition test, displaying higher light occupancy and an overall increase in the latency to move between the light and dark compartments [76]. Female DAT Val559 mice, on the other hand, display light aversion, spending more time in the dark compartment and exhibiting an increase in the latency to return to the light [76]. This sex-biased phenotype in the light/dark transition test observed in the DAT Val559 model may reflect divergent neuroadaptation to DAT dysfunction in each sex as, for example, male, but not female, DAT Val559 mice display increased apomorphine-driven stereotypic motor movements [115].

5.3 *Learning, Memory, and Cognition*

Speech delays are common in humans with DTDS, but these patients generally show appropriate cognitive development including good working memory, communication with eye-gaze devices, and accurate recognition of familiar information [24]. It is important to note, however, that limited clinical assessment of cognition has been performed in DTDS patients due to their confounding symptoms and compromised long-term health. A major strength of rodents as model organisms lies in the broad repertoire of behaviors with human parallels that can be studied to evaluate the functional impact of gene disruption. In this regard, DAT knockout mice display several readily observable and contextually dependent behavioral changes that have provided valuable insight into the role of DA neurotransmission in learning, memory, and higher cognitive functions. Homozygous DAT KO mice display impaired spatial learning and memory in the 8-arm radial maze [107, 116], Morris water maze [117], non-matching-to-place version of a water T-maze task [118],

H-maze [119], and Y maze [67] in addition to impaired motor learning on the rotarod [62] and novel object recognition [120]. DAT heterozygosity also impairs associative learning, consolidation, and memory recall, but only for partially cued responses [121]. $DAT^{+/-}$ mice also exhibit increased impulsivity and impaired attention in the 5-choice serial reaction time task (5-CSRTT) [122] and lack trace fear conditioning, in which cue and shock stimulus presentation are temporally discontinuous [118]. Both homozygous mDAT knockouts and heterozygotes show impaired reversal learning in an attentional set shifting task reflecting behavioral inflexibility [123, 124]. In addition, DAT null mice display an almost complete absence of learning in a conditioned avoidance task [125]. Sensorimotor gating as measured by pre-pulse inhibition of the startle response is also disrupted in mice lacking DAT [105], a phenotype that can be reversed with antipsychotic treatment [126]. DAT knockouts are impulsive, with around 50% of animals lacking a cliff avoidance response, a behavior typical of WT mice [127]. Whereas DAT knockouts can acquire a simple operant-based task with food reinforcement, they are resistant to response extinction and will continue nosepoking for reward even when the reinforcer is no longer delivered [128, 129]. Homozygous DAT knockouts are also unable to modulate temporal response rates to match associative cues in either a fixed interval or peak interval timing procedure and are largely insensitive to the time accelerating actions of methamphetamine [130]. As a caveat to studies utilizing food reward to drive operant responding in DAT knockouts, these mice also display altered taste preference [131], changes in patterns of consummatory behavior [129], and deficits in olfactory discrimination [132].

Phenotypes of DAT knockout mice and rats align in several cognitive tasks where both species display a decrease in spontaneous alternation in the Y Maze [26, 67], disrupted PPI [26, 105], and impaired fear conditioning [114, 118]. Rats are generally superior to mice at maze-learning and can learn more sophisticated strategies and retain stable performance over time. In the Hebb–Williams maze, used to assess spatial learning and memory, DAT knockout rats make more errors, exhibit perseverative returns, and take longer to reach the goal box compared to WT controls [133]. In the T maze, DAT knockout rats also exhibit increased perseverative behavior that impairs learning and task performance during both the acquisition and reversal learning phases [89]. This propensity for perseveration by rDAT knockouts was observed in the T Maze even upon removal of a reinforcer and evident in operant tasks where DAT knockout rats demonstrated poor extinction of operant lever pressing [89].

In relation to reward sensing, sucrose and saccharin preference have been found to be decreased in rats lacking DAT [88, 134], perhaps explaining why, when their choice behavior was tested in an Intolerance-to-Delay Task (IDT), the DAT knockout rats chose the larger reward more frequently than WT rats, especially for longer delays. These results allude to reward hyposensivity in DAT knockout rats that can be counterbalanced by increasing the magnitude of reward acquired, but nevertheless reflects a more rigid choice pattern that is accompanied by compulsive stereotypies [134]. Notably, while it does suppress hyperactivity in rDAT knockouts, amphetamine fails to improve performance of DAT knockout rats in the IDT task

[134]. DAT knockout rats also display cognitive deficits and compulsivity in a conditioned associative learning task [135]. Although DAT null rats learn stimulus/response associations, they generally display lower response rates despite exhibiting increased, possibly perseverative, visits to the food delivery magazine compared to WT rats [135]. In an object recognition task wherein rats were trained to move an object to retrieve food from familiar objects and not to move the non-rewarded novel objects, the DAT knockouts explored novel objects significantly longer resulting in an increased time to decision, but made fewer choice errors as compared to their WT littermates [136]. These data speak to decreased cognitive flexibility in DAT knockouts or dysregulated motivation for reward. Consistent with changes in motivational drive, DAT knockout rats also display higher levels of sexual motivation and activity [69].

The impact of DAT variants on cognitive function is generally more subtle than effects observed in the global knockout models. In the 5-CSRTT, male DAT Val559 mutants acquire the task faster than their WT littermates, and although they adapt well to a variable timing delay between the start of a trial and stimulus presentation, they display more premature responses with a fixed long delay indicative of a waiting-specific impulsive phenotype [137]. Notably, DAT T356M mice also display enhanced learning on the rotarod [38], though their improved performance relative to WTs could also reflect superior motor coordination. In contrast to the impact of homozygous DAT gene deletion [130], interval timing was found to be unaltered in DAT Val559 mice, though the latter mice do display an increase response rate on unrewarded trials after a long delay. Further, compared to WTs, DAT Val559 mice are better able to withhold a response if there is a probability they will be rewarded for a correct rejection as in the Go/No-Go task [137], and DAT Val559 mutants can also successfully acquire and re-learn a visual pairwise discrimination task [115]. Data indicate that the impulsivity of DAT Val559 mutants is likely due to enhanced motivation for reward as they display an increase break point in a progressive ratio task relative to WTs [137]. Additionally, whereas DAT Val559 mutants discontinue operant responding when reward is devalued, they are less impacted by reward devaluation following the implementation of random interval training designed to bias animals toward habitual behavior [115]. However, male DAT Val559 mice do display compulsive reward seeking, returning to the reward delivery magazine to check for reward even when they fail to properly engage in operant responding [115]. Thus, mice expressing the DAT Val559 variant are prone to both impulsive and compulsive behaviors that are associated with striatal dysfunction, a conclusion supported by increased dendritic spine density in striatal MSNs in the dorsomedial striatum of male DAT Val559 mice [115]. Male DAT knockout mice exhibit decreased spontaneous alternation in the Y Maze [67] and a lack of novel object recognition [120], phenotypes selectively displayed by male or female DAT Val559 mice, respectively. Notably, the sex-specific changes in cognitive function displayed by DAT Val559 mice can be rescued with the antipsychotic and D2R antagonist sulpiride, linking aberrant behavior in DAT Val559 mutants to sex and region specific hyperactive DA2Rs [115].

5.4 Social Behavior

Children with DTDS exhibit empathy and social awareness, but, as with measures of cognition, compromised speech, motor disturbances, and premature death have made more sophisticated assessment of social behavior difficult in humans lacking DAT expression [24]. Activation of mesolimbic DA neurons drives social interaction in rodents [138, 139]. However, even in invertebrates, impaired DA clearance alters social behaviors as flies expressing hDAT ΔN336 show social impairments as measured by proximity to their neighbors during the predator escape response [34]. Though fertile, female DAT knockout rodents display impaired maternal behavior and foster mothers are necessary to rear pups from homozygous parents [25, 101]. Male $DAT^{-/-}$ mice are also more aggressive than WT littermates, a phenotype exacerbated following social isolation and characterized by inflexible perseverative and repetitive reactivity to introduction of a social stimulus [140]. Consequently, social transmission of food preference is completely abolished in $DAT^{-/-}$ mice [120]. Whereas DAT knockout mice are largely asocial and reactive aggressively when approached by a foreign mouse [140], DAT knockout rats only shift their social preference from active to passive interactions [88, 114]. As rats tend to be less territorial than mice, this difference may reflect species intrinsic differences in social cognition.

Both DAT Val559 and DAT Met356 mice also display altered sociability. DAT Met356 mice spend less time exploring a novel social stimulus and exhibit fewer wins in the tube test [38], a test of social dominance [141], though the mutant animals encounter novel WTs rather than males from a community where social hierarchy has been established. A decrease in social stimulus exploration is also evident in DAT Val559 males [76]. However, in contrast to behavior displayed by DAT Met356 mice, DAT Val559 males exhibit increased wins in the tube test [76]. Interestingly, social phenotypes are absent in DAT Val559 females, indicating that female sex confers resilience to the detrimental impact of DAT dysfunction on circuits subserving social behaviors [76].

6 Impact of DAT Mutation on Drug Responses

6.1 Role of DAT in Entry of Neurotoxins into DA Neurons

DAT is required for intracellular accumulation of the neurotoxin 6-OHDA utilized to trigger selective degeneration of DA neurons in rodent models of PD [142]. Consistent with these findings, loss of *dat-1* renders nematodes completely resistant to loss of DA neurons following 6-OHDA exposure [47]. Resistance to 6-OHDA-induced neurotoxicity was also observed in *C. elegans* lines expressing DTDS-linked hDAT mutants [99]. Utilizing degeneration of green fluorescent protein (GFP)-tagged DA neurons as an endpoint, a second forward genetic screen in nematodes exposed to 6-OHDA identified additional *dat-1* alleles whose encoded protein displayed altered stability, surface trafficking, and/or maturation in addition to compromised uptake kinetics [143]. As has been observed in nematodes, zebrafish, and fruit flies, DAT is also the primary mechanism responsible for intracellular accumulation of neurotoxic drugs such as methamphetamine and MPTP in mammals, as evidenced by complete mitigation of DA neuron loss in $DAT^{-/-}$ mice treated with these agents [144, 145]. Acute lethality from methamphetamine overdose was also significantly attenuated in mice lacking DAT [146].

6.2 Psychostimulant Responses in DAT Mutants

DAT is a major target for psychostimulants, a class of drugs that includes therapeutic agents used in the treatment of ADHD (e.g., amphetamine, methylphenidate) and drugs of abuse such as cocaine and methamphetamine. A subset of drugs that serve as substrates for or block mammalian DAT also target DAT-1 proteins in nematodes, including the psychostimulant amphetamine, producing Swip [147]. Amphetamine-dependent Swip and DA efflux are completely abolished in *dat-1* knockouts confirming the potential utility of the nematode for the study of molecular determinants of amphetamine action in vivo [148]. DAT antagonists such as amphetamine also increase locomotion in *Drosophila*, an effect lost in mutants lacking dDAT or that express a phosphorylation-deficient dDAT mutant [149, 150]. In fact, while amphetamine disrupts sleep in WT flies, the psychostimulant partially restores sleep during the night cycle in dDAT null animals [150]. Amphetamine is a DAT substrate that, in addition to blocking uptake, also triggers nonvesicular DA efflux. Several disease-associated DAT variants impair the ability of amphetamine to induce DA release in vitro or in vivo in *Drosophila* including the hDAT ΔN336 [34], K619N [35], T356M [39], A559V [40], and R51W [36] mutants. Prior work demonstrated that the interaction between SYN1A and DAT facilitates DAT phosphorylation, surface trafficking, and uptake [151]. The R51W mutation, located in close proximity to the binding site for SYN1A on dDAT at the transporter's N-terminal domain (aa 1–33), appears to selectively impact reverse transport, implicating this domain in in amphetamine-induced DA efflux [152]. As expected, flies expressing hDAT R51W are not hyperactive, but do display blunted amphetamine-driven locomotion [36]. Similarly, whereas expression of WT hDAT in the SNc of mice increases amphetamine-dependent hyperlocomotion, this effect was absent following introduction of the hDAT K619N mutant [35].

Homozygous DAT knockout mice are known to be insensitive to the acute locomotor stimulating effects of cocaine, amphetamine, and methamphetamine [25, 153]. Indeed, amphetamine-dependent DA release is lost in striatal slices from DAT null mice. Similarly, in DAT knockout mice, amphetamine and 3,4-methylenedioxy-N-methylamphetamine (MDMA) fail to elicit increases in DA in dorsal striatum [25, 154, 155]. In addition, amphetamine does not affect clearance of electrically evoked DA in the NAc or suppress firing of DA neurons in the VTA of DAT null mice [156]. Though cocaine does not promote DA efflux, elevations in DA due to afferent vesicular DA release as well as cocaine-dependent induction of immediate early genes (e.g., c-fos) are absent in the striatum of DAT knockouts [25, 157] consistent with a lack of cocaine-dependent inhibition of DA clearance [158]. In fact, brain-wide patterns of neuronal activation following amphetamine, methylphenidate, or cocaine exposure are largely disrupted or absent in DAT$^{-/-}$ mice [159–161].

Intriguingly, whereas inhibition of DA receptors with haloperidol or clozapine suppresses hyperactivity in DAT knockout mice, presumably by counteracting tonically increased dopaminergic tone in the context of a lack of high affinity DA clearance [101], the same outcome can be achieved by treating DAT$^{-/-}$ mice with psychostimulants (e.g., amphetamine, methylphenidate, cocaine) [107]. This observation is seemingly paradoxical as psychostimulants increase locomotor activity in WT mice but suppress hyperactivity in DAT knockouts. Phosphorylation of the signal transducers extracellular signal related kinase (ERK) 1/2, a key driver of psychostimulant-dependent locomotor activation [162, 163], is elevated in mDAT$^{-/-}$ mice. Whereas psychostimulants increase ERK 1/2 activation in WT mice, the opposite effect was observed in mice lacking mDAT paralleling behavioral outcomes [164]. Though cocaine fails to increase locomotion in DAT knockout mice, the reinforcing properties of cocaine remain intact in DAT$^{-/-}$ mice, though DAT knockout mice do acquire conditioned place preference (CPP) at a more restricted range of doses and self-administer cocaine at a reduced rate compared to WTs [157, 165, 166]. In addition, whereas DAT knockouts lack locomotor sensitization following repeated amphetamine administration [167], locomotor sensitization to cocaine remains intact [168, 169]. Intriguingly, whereas bottom dwelling behavior increases with aging in WT zebrafish and could be stimulated by exposing animals to cocaine, fish lacking *SLC6A3* were insensitive to acute cocaine as well as the impact of age [109]. The impact of *SLC6A3* gene deletion or cocaine on swimming behavior could be mitigated via inhibition of DA D1 receptors (D1Rs), indicating that this behavior is driven by excessive DA signaling [109].

The hyperlocomotion in rDAT knockouts can also be suppressed by treatment with the D2R antagonist haloperidol or with amphetamine [59]. As in mDAT$^{-/-}$ mice, amphetamine also fails to elevate extracellular DA levels in the striatum of DAT knockout rats [26]. Unlike results obtained in DAT knockout mice [107], however, cocaine fails to influence locomotion in homozygous DAT knockout rats even following repeated administration [170]. Heterozygous rDAT knockouts, however, do display increased cocaine-driven locomotion relative to WTs with female rDAT heterozygotes exhibiting increased cocaine sensitivity as compared to male littermates [170].

DAT Val559 mice also display key differences in psychostimulant-driven phenotypes, though distinct from phenotypes observed in mDAT knockouts. In vitro, hDAT A559V displays impaired amphetamine transport [40]. The ability of amphetamine to stimulate D2-IPSCs in slices containing the SNc, promote DA release in striatal slices, or increase extracellular DA in the striatum in vivo are all blunted in DAT Val559 mice compared to WT littermates [73]. Similarly, in DAT Val559 males, the locomotor response to amphetamine, including horizontal, vertical, and stereotyped motor behaviors, though intact in DAT Val559 mice, is decreased relative to WTs [73]. Notably, amphetamine-dependent locomotion in female DAT Val559 mice is unperturbed [76] and, while the persistence of the drug/context association following amphetamine CPP is shorter in DAT Val559 males relative to WT males, the opposite trend was observed in females [76]. Further, unlike DAT knockout mice, DAT Val559 mice of both sexes [76] demonstrate intact amphetamine locomotor sensitization. In searching for mechanisms to explain the divergent impact of DAT Val559 genotype on amphetamine-dependent behaviors, it was noted that the ability of amphetamine to promote phosphorylation of ERK 1/2 in striatum is unique to males and lost in DAT Val559 mutants, which may explain sex-specific disruption of D1R-driven locomotion and possibly other male-specific behaviors [76]. Remarkably, whereas amphetamine- and methylphenidate-triggered locomotor activation is blunted in DAT Val559 males, the locomotor response to cocaine is completely absent even at high doses of cocaine [73, 76, 171].

6.3 Drug Reinforcement in DAT Mutants

$DAT^{-/-}$ mice display enhanced morphine CPP and increased morphine-stimulated DA release and neuronal activation in the NAc shell but decreased naloxone-precipitated withdrawal behavior [172]. Further, although DAT genotype did not alter the analgesic properties of morphine, morphine treatment paradoxically suppressed locomotion in DAT mutants [172]. DAT deletion does not universally promote increased consumption of addictive substances, however, as ethanol preference is decreased or increased in DAT knockout females or males, respectively [131, 173]. In contrast, the ability of ethanol to induce acute hyperactivity and locomotor sensitization is increased in mixed sex cohorts of DAT knockout mice [169]. Indeed, though not extensively studied in the DAT knockout model, there is evidence of differential sensitivity of male versus female mice to DAT haploinsufficiency with respect to striatal monoamine content, DA release, psychostimulant sensitivity, and behavior [174, 175].

7 Impact of DAT Mutation on Nondopaminergic Neurotransmitter Systems

7.1 Serotonergic and Noradrenergic Systems

Decades of research have shown that cellular and circuit plasticities are likely to arise following deletion of key genes involved in controlling neurotransmission. Thus, behavioral phenotypes in DAT knockouts may derive from acute potentiation of DA's synaptic occupancy or result from neural adaptations to altered dopaminergic signaling throughout development. In addition to triggering alterations in dopaminergic neurotransmission, a series of studies in mDAT knockouts linked loss of DAT to aberrant, but likely compensatory, changes in other neurotransmitter systems including dysregulation in signaling via the monoamines 5-HT and NE. Homozygous DAT gene deletion results in decreased tissue content of DA, NE, and 5-HT in the CPu and NAc [110]. MDMA, a 5-HT releaser, or drugs that elevate extracellular 5-HT also decrease locomotion in DAT knockouts, a dataset that provided the first evidence for alterations in the 5-HT system resulting from DAT gene deletion [107, 176]. Antagonists for 5-HT 2A receptors (5-HT_{2A}R) [177] or 5-HT 1B receptors (5-HT_{1B}R) [178] suppress hyperactivity in $DAT^{-/-}$ mice, and the ability of nicotine to suppress hyperactivity and restore PPI in DAT knockouts also requires 5-HT_{1A}R stimulation, a phenotype that may involve altered expression of nicotinic acetylcholine receptor subunits on converging serotonergic inputs into the PFC [63, 179]. DAT knockout mice also display increased behavioral stereotypies following 5-HT 1A receptor (5-HT_{1A}R) activation and decreased DOI (5-$HT_{2A/2C}$R agonist)-induced head twitch [113]. Further, DAT knockout mice exhibit multiple phenotypes typically associated with altered 5-HT neurotransmission including decreased marble burying [113] and decreased immobility in the tail suspension and forced swim tests [101, 102]. Though not extensively studied, there are also initial hints at emergent serotonergic plasticities in DAT knockout rats, which exhibit decreases in 5-HT tissue content in the midbrain, PFC, medulla oblongata, cerebellum, and spinal cord [180].

The disconnect between actions of cocaine versus amphetamine in $mDAT^{-/-}$ animals has been attributed to pharmacological differences between the two drugs as cocaine blocks DAT, NET, and the 5-HT transporter (SERT), whereas amphetamine lacks the high-affinity SERT interaction [181]. Indeed, elimination of cocaine CPP requires simultaneous deletion of both DAT and SERT [182]. However, studies in mice expressing a cocaine-insensitive DAT mutant (DAT-CI; L104V, F105C, A109V) confirmed that targeting of DAT is required for both cocaine-dependent locomotor activation and CPP [183] with expression of WT DAT in either the midbrain or NAc restoring cocaine-dependent behaviors in DAT-CI mice [184]. As an alternate interpretation, the ability of SERT blockade to drive CPP may only be engaged in the context of lifelong DAT knockout.

Comparing the impact of cocaine in DAT knockout versus DAT-CI mice emphasizes the dramatic consequences that long-term hyperdopaminergia generates with respect to ongoing monoaminergic neurotransmission. For example, whereas cocaine-induced DA elevations are lost in the NAc of DAT-CI mice [183], cocaine elevates extracellular DA in the NAc and PFC, but not CPu, of DAT null mice [185, 186]. Initially, this observation was attributed to compensatory recruitment of either NET or SERT to modulate DA clearance in the ventral striatum in the absence of DAT. However, though systemic administration of NET or SERT blockers will stimulate DA release in the NAc of DAT knockout mice [185, 187], inhibitors of either transporter fail to alter DA clearance in the NAc core or shell [188]. Consistent with the emergence of an adaptive extrastriatal mechanism that modulates extracellular

DA levels following DAT gene deletion, cocaine infusion into the VTA, but not the NAc, triggers DA release in the absence of DAT [187]. However, once again, neither NET nor SERT inhibitors influence somatodendritic DA clearance in VTA slices from DAT knockout mice [189]. Thus, the prevailing hypothesis posits that 5-HT promotes DA release in the NAc of DAT knockouts by modulating, either directly or indirectly, VTA neuron activity. Indeed, full elimination of cocaine-stimulated DA elevations in the NAc requires deletion of both DAT and SERT [186]. The molecular determinants underlying the aberrant ability of 5-HT to promote NAc DA release in mice lacking DAT remain unelucidated but have clear behavioral consequences. For example, though blockers of SERT and NET do not elicit CPP in WT mice, they are reinforcing in DAT knockouts [190]. DAT null mice also display accelerated amphetamine CPP acquisition and markedly delayed extinction compared to WTs, a phenotype that can be blocked via inhibition of 5-HT_{1A}Rs [156].

Though retention of cocaine-reward in DAT null mice can be attributed to a 5-HT-centric mechanism, the observation that both methylphenidate and amphetamine, which target DAT and NET but not SERT, suppress hyperactivity in $DAT^{-/-}$ mice remains unexplained. More recently, Harris et al. showed that elevations in DA can be stimulated in the PFC of $DAT^{-/-}$ mice with inhibitors of NET, but not SERT and, that, whereas systemic administration of SERT or NET inhibitors suppresses hyperlocomotion in DAT knockouts, when these drugs were infused directly into the PFC, only NET blockers were effective [191]. Further, systemic administration or PFC infusion of a DA β hydroxylase inhibitor, which depletes NE and increases DA, also decreased hyperactivity in DAT knockout mice [191]. Together, these data suggest that hyperlocomotion in mice lacking DAT can be mitigated either by increasing cortical DA or by restoring midbrain dopaminergic neuron firing.

DAT Val559 mice display blunted, but intact amphetamine-dependent hyperactivity but completely lack a locomotor response to cocaine [73, 171]. Given that cocaine, unlike amphetamine and methylphenidate, targets SERT with high affinity, it was hypothesized that a change in SERT-governed 5-HT signaling might have occurred that suppresses locomotor activation by cocaine in DAT Val559 mice. In support of this hypothesis, though tissue content of DA and NE are unaltered in DAT Val559 mice, there is a small, but significant increase in striatal and cortical 5-HT levels [73]. Moreover, a cocaine analog (RTI-113) with a higher selectivity for DAT over SERT regained the ability to induce hyperactivity in DAT Val559 mice [192]. In addition, introduction of the SERT I172M mutation, which eliminates high-affinity interactions between cocaine and SERT [192], or blockade of 5-HT 2C receptors (5-HT_{2C}Rs) restored cocaine-induced locomotion in DAT Val559 mice [171]. Though the influence of 5-HT on locomotion in DAT knockout mice occurs via modulation of DA release, introduction of the SERT I172M mutation in DAT Val559 mice rescued locomotor output without restoring the absent elevations in striatal DA following cocaine administration observed in DAT Val559 mice [171], suggesting a plasticity arises cell nonautonomously, either through changes in 5-HT innervation and/or release, altered sensitivity of postsynaptic 5-HT receptors, or both, in DAT Val559 mice. Interestingly, although DAT Val559 mice demonstrate a loss of cocaine-induced locomotor activation, they retain cocaine CPP, though extinction is delayed in mutant males relative to WT controls [171]. Changes observed in DAT Val559 mice with respect to cocaine action support an altered DA/5-HT balance for regulation of motor and reward circuits arising in the knock-in mice.

Together, data generated on the study of DAT knockout and knock-in mice speak to a critical role for DAT in the control of dopaminergic neurotransmission as well as the development and functional crosstalk between monoaminergic systems. Thus, although homozygous loss of DAT in humans is not mimicked well by behavioral changes seen in DAT KO mice, the work has provided valuable insight into the molecular mechanisms underlying the efficacy of a widely used and abused drug class (e.g., psychostimulants) while simultaneously aiding in the identification of additional targets with potential clinical utility in the treatment of neuropsychiatric disorders. In particular, the changes that arise in 5-HT systems in the context of DAT loss raise possibilities that unrecognized serotonergic plasticities constitute critical targets for restoring the consequences of excessive DA release.

7.2 Glutamatergic System

There is also evidence for altered glutamatergic neurotransmission in DAT knockout rodent models. For example, the pharmacological effect of psychostimulants on hyperactivity in mDAT knockouts can be enhanced by inhibiting *N*-methyl-D-aspartate (NMDA) receptors or blocked in the presence of L-a-amino-3-hydroxy-5-methylisoxazole-4-propionate (AMPA) receptor–positive modulators, indicating that glutamatergic neurotransmission suppresses locomotor activation in DAT knockout mice [193]. Recent evidence also points to broad suppression in glutamatergic neurotransmission in the medial PFC of DAT knockout rats with increased expression of glutamate transporters and loss of synaptic and extrasynaptic expression of AMPA receptor subunits and anchoring proteins [194]. Further, whereas the NMDA receptor blocker MK-801 increases locomotor activity in WT rats, it paradoxically suppresses hyperactivity in rDAT knockouts, providing further evidence that disrupted glutamatergic neurotransmission follow DAT depletion [195]. Functional magnetic resonance imaging (fRMI) analysis in DAT knockouts has largely paralleled molecular findings identifying PFC-midbrain decoupling and striato-cerebellar neuroplastic compensation as the major disruptions resulting from chronic hyperdopaminergia in the rat brain [196]. Electroencephalogram (EEG) recordings support large-scale disruptions in corticostriatal networks in DAT knockout rats with prominent decreases in power in the theta range in PFC, motor cortex, and striatum and lower signal coherence, indicating a loss of functional synchronization between brain areas [133].

8 Predictive Validity of DAT Knockout and Humanized DAT KI Models

Despite the limited behavioral endpoints and the comparative simplicity of the dopaminergic circuitry, *C. elegans* have nevertheless proven particularly useful in the rapid characterization and screening of disease-associated DAT variants and in exploration of structural determinants of DAT function, mechanisms of DAT regulation, and the pharmacological properties of DAT-targeted drugs. A particularly interesting example of the latter opportunity comes from work of Refai and colleagues who utilized the worm model to demonstrate the possibility of a pharmacological reduction of amphetamine action via the allosteric DAT agent KM822 in vivo, raising the possibility that similar agents active at human DAT might be useful in treating methamphetamine overdose [197].

Drosophila expressing human DAT mutations have also been used to model pharmacological approaches to restore DAT function, a goal in the treatment of humans expressing DTDS-causing DAT mutations. For example, the hDAT R445C substitution was identified in a patient with classical DTDS [22], a mutation located in the cytoplasmic face of TM9. Molecular modeling demonstrated that R445 forms a salt bridge with E428 creating part of the intracellular gate [28, 198] and substitution of this residue results in a channel-like transporter state predicted to compromise hDAT function [199]. In vitro, hDAT R445C expression, maturation, and surface trafficking are compromised relative to WT hDAT, resulting in a complete loss of DA uptake [199]. Treatment of flies expressing hDAT R445C with the lysosomal inhibitor chloroquine to block lysosomal degradation of DAT improved motor coordination as measured by flies' take-off time [199]. Of the mutations identified in DTDS patients, many result in retention of DAT in the endoplasmic reticulum due to defective protein folding, an effect that would also impact trafficking of WT proteins due to transporter oligomerization [200]. Operating under the hypothesis that functionality might be restored for these mutants in the presence of pharmacochaperones that stabilize the transition state and allow for more facile protein folding, investigators exposed flies expressing dDAT G108Q to the pharmacochaperone noribogaine and/or HSP70 inhibitor pifithrin-μ, a finding that restored axonal transport of dDAT and sleep time in mutant flies [201]. Sleep time, activity, and grooming behavior were also restored by noribogaine or pifithrin-μ in Drosophila expressing the DTDS-linked point mutants hDAT-V158F or hDAT-G327R [202], which display an almost complete loss of DA uptake in the absence of intervention. In contrast, this approach was not effective in restoring sleep behavior for *fumin* mutants in which the endogenous dDAT is truncated after TM6 and is thus nonfunctional [79]. Notably, whereas hDAT-V158F and hDAT-G327R expressing flies phenocopy *fumin* mutants, expression of the DTDS-associated hDAT variant DAT-L368Q is lethal and mutants do not hatch, an impact that may be due to dopaminergic neurotoxicity during development [202]. As an alternate strategy to restore DAT function, adeno-associated viral (AAV) delivery of DAT into the SNc is sufficient to suppress hyperactivity, decrease anxiety, improve motor coordination, and prolong survival of DAT knockout mice [203], a strategy that could be employed in DTDS patients.

The observation that chronic treatment with clozapine, a DA/5HT receptor antagonist and atypical antipsychotic, suppresses aberrant swimming behavior in zDAT knockouts and restores TH+ neuron number emphasizes the plasticity intrinsic to this system in zebrafish brain [72]. Interestingly, the typical antipsychotic sulpiride rescues select phenotypes in DAT Val559 mice including deficits in spatial working memory in the Y Maze and impaired novel object recognition in male and female mutants, respectively [76]. Given evidence of tonic stimulation of D2ARs in the DAT Val559 mutants, it is unclear if the impact of sulpiride in DAT Val559 mice is due to normalization of autoreceptor-dependent enhanced surface trafficking of leaky transporters or targeting of postsynaptic D2R populations responding to deregulated DA efflux [75, 76].

Psychostimulants decrease hyperactivity in DAT knockout mice and rats [59, 107], a somewhat puzzling observation given DAT is believed to be a key therapeutic target for these medications. One possible explanation is that DAT gene deletion triggers compensatory changes in the other monoamine systems (5-HT, NE). In support of this hypothesis, inhibitors of SERT also suppress hyperactivity in rodents lacking DAT expression [107]. Consistent with impaired task performance, neuronal activation during learning tasks is decreased in DAT null mice relative to WT controls [123]. Remarkably, long-term potentiation (LTP), a cellular correlate of learning and memory, is completely absent in slices from the PFC of $DAT^{-/-}$ mice, and LTP was restored following systemic amphetamine treatment [204] consistent with the rescue of cognitive performance in an H-maze observed in amphetamine-treated DAT knockout mice [119]. In support of a critical role for cortical dysfunction in the cognitive phenotypes of $DAT^{-/-}$ mice, inhibition of NET in the PFC, but not the NAc, restores PPI responses in DAT knockout mice [205]. Interestingly, whereas atomoxetine, a NET blocker used in ADHD treatment, improved PPI deficits in DAT null rats, it failed to improve their performance in the Hebb-Williams maze or suppress hyperactivity [133]. Also, though blockade of α2 adrenergic receptors (α2AR) exacerbated hyperactivity and further compromised spatial memory in DAT knockout rats, the α2AR agonist guanfacine did decrease perseverative return runs without improving maze completion, indicating that the molecular substrates underlying the cognitive and compulsive behaviors in this task may be distinct [206]. Guanfacine also suppresses hyperactivity and improved PPI in DAT knockout rats, with a chronic treatment proving more efficacious [207]. Together, these data speak to a critical role for dysregulated 5-HT/SERT in locomotor hyperactivity in DAT knockout rodents, while altered NE signaling and/or NET activity contributes to cognitive phenotypes.

9 Conclusions

As the complexity of the nervous system, and consequently the elaboration of the dopaminergic circuitry, expands, so does the list of neurochemical, neurophysiological, and behavioral consequences of DAT deficiency detected in animal models (Fig. 2). Regardless of the species studied, however, certain themes have emerged from the study of DAT null organisms, most notably an evolutionarily conserved role for DA in motor control. In mammals, nevertheless, there is a clear disconnect between phenotypes in DAT knockout rodents as compared to humans that lack two functional copies of the *SLC6A3* gene. For example, whereas rodents lacking DAT exhibit hyperactivity, they fail to show the progressive neurodegeneration and Parkinsonism/dystonia developed by humans with DTDS. Conversely, while intellect and cognitive development appear intact in humans with DTDS, mice and rats show clear deficiencies in learning and memory, though, as few children with DTDS survive past adolescence, this does not preclude an impact of DAT insufficiency on later cortical development in humans. Though often touted as a model of ADHD [208], this lack of construct validity for DAT knockout rodents has led to the generation of several new murine models of DAT dysfunction, DAT Val559 and DAT Met356 mice, which express neuropsychiatric-disorder-associated DAT variants. Both models exhibit face validity, expressing altered disease-relevant traits such as impaired sociability, impulsivity, and cognitive deficits, but generally lack the drastic changes in neurodevelopment, growth, and longevity associated with DAT knockouts (Fig. 3). Notably, even in the absence of overt hyperactivity, knock-in models such as the DAT Val559 mouse recapitulate circuit level plasticities such as altered serotonergic neurotransmission that drive behavioral alterations and can be therapeutically targeted. Overt differences in the phenotypic manifestation of DAT

	DAT Knockout Rat	DAT Knockout Mouse	DAT Val559 Mouse	DAT Met356 Mouse
General Health	Reduced weight gain	Stunted growth, Premature death	None	Reduced weight gain
Locomotion	Hyperactive	Hyperactive	Darting	Hyperactive
Repetitive Behavior & Anxiety	↑ Light exploration ↑ Marble burying	Thigmotaxis, neophobia, ↓ Marble burying	Males: ↑ Light exploration Females: ↓ Light exploration	N/D
Cognitive Function	Impaired spatial working memory Cognitive inflexibility Impulsivity & inattention Disrupted sensorimotor gating Increased motivation	Impaired spatial working memory Cognitive inflexibility Impulsivity & inattention Disrupted sensorimotor gating Increased motivation	Impaired spatial working memory Perseveration Impulsivity Increased motivation	N/D
Social Behavior	↓Active social interaction	Aggression, ↓ Sociability	↑ Dominance, ↓ Interaction (Males Only)	↓ Dominance, ↓ Interaction

Fig. 3 Comparison of behavior phenotypes in existing rodent models of DAT gene deletion or mutation. N/D = not determined

deletion or mutation between the sexes also highlight the prevalence of sex differences in the DA system and emphasize the critical importance of including sex as a biological variable in studies seeking to define the impact of DA dysfunction on animal behavior. Having already been utilized to define novel neural mechanisms impacted by DAT deletion or deregulation, these models represent a new frontier in research on how dysregulation of DA clearance contributes to the etiology of ASD, ADHD, BPD, schizophrenia, and other diseases linked to aberrant dopaminergic neurotransmission and will serve as tools to screen for more efficacious interventions for these common neuropsychiatric disorders.

Acknowledgments Support for this contribution derives from a Brain and Behavioral Research Foundation Young Investigator Award to A.S. and a grant from the National Institute of Mental Health (MH105994). Figures were generated with BioRender.com.

References

1. German DC, Schlusselberg DS, Woodward DJ (1983) Three-dimensional computer reconstruction of midbrain dopaminergic neuronal populations: from mouse to man. J Neural Transm 57:243–254
2. Arbuthnott GW, Wickens J (2007) Space, time and dopamine. Trends Neurosci 30:62–69
3. Cox J, Witten IB (2019) Striatal circuits for reward learning and decision-making. Nat Rev Neurosci 20:482–494
4. Morales M, Margolis EB (2017) Ventral tegmental area: cellular heterogeneity, connectivity and behaviour. Nat Rev Neurosci 18:73–85
5. Walker QD, Rooney MB, Wightman RM, Kuhn CM (2000) Dopamine release and uptake are greater in female than male rat striatum as measured by fast cyclic voltammetry. Neuroscience 95:1061–1070
6. Calipari ES et al (2017) Dopaminergic dynamics underlying sex-specific cocaine reward. Nat Commun 8:13877
7. Brundage JN et al (2021) Regional and sex differences in spontaneous striatal dopamine transmission. J Neurochem 160:598
8. Dewing P et al (2006) Direct regulation of adult brain function by the male-specific factor SRY. Curr Biol 16:415–420
9. Dluzen DE, McDermott JL (2008) Sex differences in dopamine- and vesicular monoamine-transporter functions. Ann N Y Acad Sci 1139:140–150
10. Walker QD, Ray R, Kuhn CM (2006) Sex differences in neurochemical effects of dopaminergic drugs in rat striatum. Neuropsychopharmacology 31:1193–1202
11. Gogos A, Ney LJ, Seymour N, Van Rheenen TE, Felmingham KL (2019) Sex differences in schizophrenia, bipolar disorder, and post-traumatic stress disorder: are gonadal hormones the link? Br J Pharmacol 176:4119–4135
12. Mowlem FD et al (2019) Sex differences in predicting ADHD clinical diagnosis and pharmacological treatment. Eur Child Adolesc Psychiatry 28:481–489
13. Becker JB, McClellan ML, Reed BG (2017) Sex differences, gender and addiction. J Neurosci Res 95:136–147
14. Gottgens I et al (2020) The impact of sex and gender on the multidisciplinary management of care for persons with Parkinson's disease. Front Neurol 11:576121
15. Iversen LL (1973) Catecholamine uptake processes. Br Med Bull 29:130–135

16. Shimada S et al (1991) Cloning and expression of a cocaine-sensitive dopamine transporter complementary DNA. Science 254:576–578
17. Kilty JE, Lorang D, Amara SG (1991) Cloning and expression of a cocaine-sensitive rat dopamine transporter. Science 254:578–579
18. Giros B, el Mestikawy S, Bertrand L, Caron MG (1991) Cloning and functional characterization of a cocaine-sensitive dopamine transporter. FEBS Lett 295:149–154
19. Giros B et al (1992) Cloning, pharmacological characterization, and chromosome assignment of the human dopamine transporter. Mol Pharmacol 42:383–390
20. Reith MEA et al (2022) The dopamine transporter gene SLC6A3: multidisease risks. Mol Psychiatry 27:1031–1046
21. Hansen FH et al (2014) Missense dopamine transporter mutations associate with adult parkinsonism and ADHD. J Clin Invest 124:3107–3120
22. Ng J et al (2014) Dopamine transporter deficiency syndrome: phenotypic spectrum from infancy to adulthood. Brain 137:1107–1119
23. Kurian MA et al (2009) Homozygous loss-of-function mutations in the gene encoding the dopamine transporter are associated with infantile parkinsonism-dystonia. J Clin Invest 119:1595–1603
24. Kurian MA et al (2011) Clinical and molecular characterisation of hereditary dopamine transporter deficiency syndrome: an observational cohort and experimental study. Lancet Neurol 10:54–62
25. Giros B, Jaber M, Jones SR, Wightman RM, Caron MG (1996) Hyperlocomotion and indifference to cocaine and amphetamine in mice lacking the dopamine transporter. Nature 379:606–612
26. Leo D et al (2018) Pronounced hyperactivity, cognitive dysfunctions, and BDNF dysregulation in dopamine transporter knock-out rats. J Neurosci Off J Soc Neurosci 38:1959–1972
27. Rudnick G (1998) Ion-coupled neurotransmitter transport: thermodynamic vs. kinetic determinations of stoichiometry. Methods Enzymol 296:233–247
28. Penmatsa A, Wang KH, Gouaux E (2013) X-ray structure of dopamine transporter elucidates antidepressant mechanism. Nature 503:85–90
29. Yamashita A, Singh SK, Kawate T, Jin Y, Gouaux E (2005) Crystal structure of a bacterial homologue of Na+/Cl−-dependent neurotransmitter transporters. Nature 437:215–223
30. Singh SK, Piscitelli CL, Yamashita A, Gouaux E (2008) A competitive inhibitor traps LeuT in an open-to-out conformation. Science 322:1655–1661
31. Krishnamurthy H, Gouaux E (2012) X-ray structures of LeuT in substrate-free outward-open and apo inward-open states. Nature 481:469–474
32. Herborg F, Andreassen TF, Berlin F, Loland CJ, Gether U (2018) Neuropsychiatric disease-associated genetic variants of the dopamine transporter display heterogeneous molecular phenotypes. J Biol Chem 293:7250–7262
33. Sakrikar D et al (2012) Attention deficit/hyperactivity disorder-derived coding variation in the dopamine transporter disrupts microdomain targeting and trafficking regulation. J Neurosci Off J Soc Neurosci 32:5385–5397
34. Campbell NG et al (2019) Structural, functional, and behavioral insights of dopamine dysfunction revealed by a deletion in SLC6A3. Proc Natl Acad Sci USA 116:3853–3862
35. Herborg F et al (2021) Identifying dominant-negative actions of a dopamine transporter variant in patients with parkinsonism and neuropsychiatric disease. JCI Insight 6:e151496
36. Cartier E et al (2015) Rare autism-associated variants implicate syntaxin 1 (STX1 R26Q) phosphorylation and the dopamine transporter (hDAT R51W) in dopamine neurotransmission and behaviors. EBioMedicine 2:135–146
37. Mazei-Robison MS et al (2008) Anomalous dopamine release associated with a human dopamine transporter coding variant. J Neurosci Off J Soc Neurosci 28:7040–7046
38. DiCarlo GE et al (2019) Autism-linked dopamine transporter mutation alters striatal dopamine neurotransmission and dopamine-dependent behaviors. J Clin Invest 129:3407–3419

39. Hamilton PJ et al (2013) De novo mutation in the dopamine transporter gene associates dopamine dysfunction with autism spectrum disorder. Mol Psychiatry 18:1315–1323
40. Bowton E et al (2014) SLC6A3 coding variant Ala559Val found in two autism probands alters dopamine transporter function and trafficking. Transl Psychiatry 4:e464
41. Mazei-Robison MS, Couch RS, Shelton RC, Stein MA, Blakely RD (2005) Sequence variation in the human dopamine transporter gene in children with attention deficit hyperactivity disorder. Neuropharmacology 49:724–736
42. Grunhage F et al (2000) Systematic screening for DNA sequence variation in the coding region of the human dopamine transporter gene (DAT1). Mol Psychiatry 5:275–282
43. Lovell PV, Kasimi B, Carleton J, Velho TA, Mello CV (2015) Living without DAT: loss and compensation of the dopamine transporter gene in sauropsids (birds and reptiles). Sci Rep 5:14093
44. Chase DL, Pepper JS, Koelle MR (2004) Mechanism of extrasynaptic dopamine signaling in Caenorhabditis elegans. Nat Neurosci 7:1096–1103
45. Sulston J, Dew M, Brenner S (1975) Dopaminergic neurons in the nematode Caenorhabditis elegans. J Comp Neurol 163:215–226
46. Jayanthi LD et al (1998) The Caenorhabditis elegans gene T23G5.5 encodes an antidepressant- and cocaine-sensitive dopamine transporter. Mol Pharmacol 54:601–609
47. Nass R, Hall DH, Miller DM 3rd, Blakely RD (2002) Neurotoxin-induced degeneration of dopamine neurons in Caenorhabditis elegans. Proc Natl Acad Sci USA 99:3264–3269
48. Formisano R, Rosikon KD, Singh A, Dhillon HS (2022) The dopamine membrane transporter plays an active modulatory role in synaptic dopamine homeostasis. J Neurosci Res 100:1551–1559
49. Robinson SB et al (2017) Sequence determinants of the Caenhorhabditis elegans dopamine transporter dictating in vivo axonal export and synaptic localization. Mol Cell Neurosci 78:41–51
50. Porzgen P, Park SK, Hirsh J, Sonders MS, Amara SG (2001) The antidepressant-sensitive dopamine transporter in Drosophila melanogaster: a primordial carrier for catecholamines. Mol Pharmacol 59:83–95
51. Guo S et al (1999) Mutations in the zebrafish unmask shared regulatory pathways controlling the development of catecholaminergic neurons. Dev Biol 208:473–487
52. Holzschuh J, Ryu S, Aberger F, Driever W (2001) Dopamine transporter expression distinguishes dopaminergic neurons from other catecholaminergic neurons in the developing zebrafish embryo. Mech Dev 101:237–243
53. Lam CS, Korzh V, Strahle U (2005) Zebrafish embryos are susceptible to the dopaminergic neurotoxin MPTP. Eur J Neurosci 21:1758–1762
54. Sallinen V et al (2009) MPTP and MPP+ target specific aminergic cell populations in larval zebrafish. J Neurochem 108:719–731
55. Feng CW et al (2014) Effects of 6-hydroxydopamine exposure on motor activity and biochemical expression in zebrafish (Danio rerio) larvae. Zebrafish 11:227–239
56. Vijayanathan Y et al (2017) 6-OHDA-lesioned adult zebrafish as a useful Parkinson's disease model for dopaminergic neuroregeneration. Neurotox Res 32:496–508
57. Wu X, Gu HH (1999) Molecular cloning of the mouse dopamine transporter and pharmacological comparison with the human homologue. Gene 233:163–170
58. Bruss M, Wieland A, Bonisch H (1999) Molecular cloning and functional expression of the mouse dopamine transporter. J Neural Transm (Vienna) 106:657–662
59. Lloyd JT et al (2022) Dopamine dysregulation and altered responses to drugs affecting dopaminergic transmission in a new dopamine transporter knockout (DAT-KO) rat model. Neuroscience 491:43–64
60. Benoit-Marand M, Jaber M, Gonon F (2000) Release and elimination of dopamine in vivo in mice lacking the dopamine transporter: functional consequences. Eur J Neurosci 12:2985–2992

61. Salvatore MF, Calipari ES, Jones SR (2016) Regulation of tyrosine hydroxylase expression and phosphorylation in dopamine transporter-deficient mice. ACS Chem Neurosci 7:941–951
62. Fernagut PO et al (2003) Motor behaviour deficits and their histopathological and functional correlates in the nigrostriatal system of dopamine transporter knockout mice. Neuroscience 116:1123–1130
63. Weiss S et al (2007) Functional alterations of nicotinic neurotransmission in dopamine transporter knock-out mice. Neuropharmacology 52:1496–1508
64. Jones SR et al (1999) Loss of autoreceptor functions in mice lacking the dopamine transporter. Nat Neurosci 2:649–655
65. Fauchey V, Jaber M, Caron MG, Bloch B, Le Moine C (2000) Differential regulation of the dopamine D1, D2 and D3 receptor gene expression and changes in the phenotype of the striatal neurons in mice lacking the dopamine transporter. Eur J Neurosci 12:19–26
66. Fumagalli F, Racagni G, Colombo E, Riva MA (2003) BDNF gene expression is reduced in the frontal cortex of dopamine transporter knockout mice. Mol Psychiatry 8:898–899
67. Li B, Arime Y, Hall FS, Uhl GR, Sora I (2010) Impaired spatial working memory and decreased frontal cortex BDNF protein level in dopamine transporter knockout mice. Eur J Pharmacol 628:104–107
68. Zhang X et al (2010) Altered neurocircuitry in the dopamine transporter knockout mouse brain. PLoS One 5:e11506
69. Sanna F et al (2020) Altered sexual behavior in dopamine transporter (DAT) knockout male rats: a behavioral, neurochemical and intracerebral microdialysis study. Front Behav Neurosci 14:58
70. Makos MA, Kim YC, Han KA, Heien ML, Ewing AG (2009) In vivo electrochemical measurements of exogenously applied dopamine in Drosophila melanogaster. Anal Chem 81:1848–1854
71. Ueno T, Kume K (2014) Functional characterization of dopamine transporter in vivo using Drosophila melanogaster behavioral assays. Front Behav Neurosci 8:303
72. Wang G et al (2019) Abnormal behavior of zebrafish mutant in dopamine transporter is rescued by clozapine. iScience 17:325–333
73. Mergy MA et al (2014) The rare DAT coding variant Val559 perturbs DA neuron function, changes behavior, and alters in vivo responses to psychostimulants. Proc Natl Acad Sci USA 111:E4779–E4788
74. Bowton E et al (2010) Dysregulation of dopamine transporters via dopamine D2 autoreceptors triggers anomalous dopamine efflux associated with attention-deficit hyperactivity disorder. J Neurosci Off J Soc Neurosci 30:6048–6057
75. Gowrishankar R et al (2018) Region-specific regulation of presynaptic dopamine homeostasis by D2 autoreceptors shapes the in vivo impact of the neuropsychiatric disease-associated DAT variant Val559. J Neurosci Off J Soc Neurosci 38:5302–5312
76. Stewart A et al (2022) Behaviorally penetrant, anomalous dopamine efflux exposes sex and circuit dependent regulation of dopamine transporters. Mol Psychiatry 27:4869–4880
77. Rickhag M et al (2013) A C-terminal PDZ domain-binding sequence is required for striatal distribution of the dopamine transporter. Nat Commun 4:1580
78. Torres GE et al (2001) Functional interaction between monoamine plasma membrane transporters and the synaptic PDZ domain-containing protein PICK1. Neuron 30:121–134
79. Kume K, Kume S, Park SK, Hirsh J, Jackson FR (2005) Dopamine is a regulator of arousal in the fruit fly. J Neurosci Off J Soc Neurosci 25:7377–7384
80. Ueno T, Tomita J, Kume S, Kume K (2012) Dopamine modulates metabolic rate and temperature sensitivity in Drosophila melanogaster. PLoS One 7:e31513
81. King LB et al (2016) Neurofibromin loss of function drives excessive grooming in Drosophila. G3 (Bethesda) 6:1083–1093
82. Potdar S, Daniel DK, Thomas FA, Lall S, Sheeba V (2018) Sleep deprivation negatively impacts reproductive output in Drosophila melanogaster. J Exp Biol 221:jeb174771

83. Yamazaki M et al (2012) High calorie diet augments age-associated sleep impairment in Drosophila. Biochem Biophys Res Commun 417:812–816
84. Walker JK, Gainetdinov RR, Mangel AW, Caron MG, Shetzline MA (2000) Mice lacking the dopamine transporter display altered regulation of distal colonic motility. Am J Physiol Gastrointest Liver Physiol 279:G311–G318
85. Vincent SG, Waddell AE, Caron MG, Walker JK, Fisher JT (2007) A murine model of hyper-dopaminergic state displays altered respiratory control. FASEB J 21:1463–1471
86. Kavelaars A, Cobelens PM, Teunis MA, Heijnen CJ (2005) Changes in innate and acquired immune responses in mice with targeted deletion of the dopamine transporter gene. J Neuroimmunol 161:162–168
87. Gopinath A et al (2023) Who knew? Dopamine transporter activity is critical in innate and adaptive immune responses. Cells 12:269
88. Mallien AS et al (2022) Dopamine transporter knockout rats show impaired wellbeing in a multimodal severity assessment approach. Front Behav Neurosci 16:924603
89. Belskaya A et al (2024) Rats lacking the dopamine transporter display inflexibility in innate and learned behavior. Biomedicines 12:1270
90. Apryatin SA et al (2023) Antioxidant system alterations and physiological characteristics of neonatal and juvenile DAT-KO rats. J Neurosci Res 101:1651–1661
91. Belov D, Fesenko Z, Efimov A, Lakstygal A, Efimova E (2022) Different sensitivity to anesthesia according to ECoG data in dopamine transporter knockout and heterozygous rats. Neurosci Lett 788:136839
92. Mergy MA et al (2014) Genetic targeting of the amphetamine and methylphenidate-sensitive dopamine transporter: on the path to an animal model of attention-deficit hyperactivity disorder. Neurochem Int 73:56–70
93. DiCarlo GE et al (2021) Autism-associated variant in the SLC6A3 gene alters the oral microbiome and metabolism in a murine model. Front Psych 12:655451
94. Ng J, Barral S, Waddington SN, Kurian MA (2023) Dopamine transporter deficiency syndrome (DTDS): expanding the clinical phenotype and precision medicine approaches. Cells 12:1737
95. McDonald PW et al (2007) Vigorous motor activity in Caenorhabditis elegans requires efficient clearance of dopamine mediated by synaptic localization of the dopamine transporter DAT-1. J Neurosci Off J Soc Neurosci 27:14216–14227
96. Hardaway JA et al (2012) Forward genetic analysis to identify determinants of dopamine signaling in Caenorhabditis elegans using swimming-induced paralysis. G3 (Bethesda) 2:961–975
97. Robinson SB et al (2019) Dopamine-dependent, swimming-induced paralysis arises as a consequence of loss of function mutations in the RUNX transcription factor RNT-1. PLoS One 14:e0216417
98. Bermingham DP et al (2017) The atypical MAP kinase SWIP-13/ERK8 regulates dopamine transporters through a rho-dependent mechanism. J Neurosci Off J Soc Neurosci 37:9288–9304
99. Illiano P et al (2017) A Caenorhabditis elegans model to study dopamine transporter deficiency syndrome. Eur J Neurosci 45:207–214
100. Ueno T, Masuda N, Kume S, Kume K (2012) Dopamine modulates the rest period length without perturbation of its power law distribution in Drosophila melanogaster. PLoS One 7:e32007
101. Spielewoy C et al (2000) Behavioural disturbances associated with hyperdopaminergia in dopamine-transporter knockout mice. Behav Pharmacol 11:279–290
102. Perona MT et al (2008) Animal models of depression in dopamine, serotonin, and norepinephrine transporter knockout mice: prominent effects of dopamine transporter deletions. Behav Pharmacol 19:566–574

103. Morice E, Denis C, Giros B, Nosten-Bertrand M (2004) Phenotypic expression of the targeted null-mutation in the dopamine transporter gene varies as a function of the genetic background. Eur J Neurosci 20:120–126
104. Carpenter AC, Saborido TP, Stanwood GD (2012) Development of hyperactivity and anxiety responses in dopamine transporter-deficient mice. Dev Neurosci 34:250–257
105. Ralph RJ, Paulus MP, Fumagalli F, Caron MG, Geyer MA (2001) Prepulse inhibition deficits and perseverative motor patterns in dopamine transporter knock-out mice: differential effects of D1 and D2 receptor antagonists. J Neurosci Off J Soc Neurosci 21:305–313
106. Ferris MJ et al (2014) Dopamine transporters govern diurnal variation in extracellular dopamine tone. Proc Natl Acad Sci USA 111:E2751–E2759
107. Gainetdinov RR et al (1999) Role of serotonin in the paradoxical calming effect of psychostimulants on hyperactivity. Science 283:397–401
108. Adinolfi A et al (2018) Novelty-related behavior of young and adult dopamine transporter knockout rats: implication for cognitive and emotional phenotypic patterns. Genes Brain Behav 17:e12463
109. Kacprzak V et al (2017) Dopaminergic control of anxiety in young and aged zebrafish. Pharmacol Biochem Behav 157:1–8
110. Pogorelov VM, Rodriguiz RM, Insco ML, Caron MG, Wetsel WC (2005) Novelty seeking and stereotypic activation of behavior in mice with disruption of the Dat1 gene. Neuropsychopharmacology 30:1818–1831
111. Costa RM et al (2006) Rapid alterations in corticostriatal ensemble coordination during acute dopamine-dependent motor dysfunction. Neuron 52:359–369
112. Berlanga ML et al (2011) Multiscale imaging characterization of dopamine transporter knockout mice reveals regional alterations in spine density of medium spiny neurons. Brain Res 1390:41–49
113. Fox MA, Panessiti MG, Hall FS, Uhl GR, Murphy DL (2013) An evaluation of the serotonin system and perseverative, compulsive, stereotypical, and hyperactive behaviors in dopamine transporter (DAT) knockout mice. Psychopharmacology 227:685–695
114. Adinolfi A et al (2019) Behavioral characterization of DAT-KO rats and evidence of asocial-like phenotypes in DAT-HET rats: the potential involvement of norepinephrine system. Behav Brain Res 359:516–527
115. Stewart A et al (2022) Male DAT Val559 mice exhibit compulsive behavior under devalued reward conditions accompanied by cellular and pharmacological changes. Cells 11:4059
116. Kurzina NP, Aristova IY, Volnova AB, Gainetdinov RR (2020) Deficit in working memory and abnormal behavioral tactics in dopamine transporter knockout rats during training in the 8-arm maze. Behav Brain Res 390:112642
117. Weiss S, Nosten-Bertrand M, McIntosh JM, Giros B, Martres MP (2007) Nicotine improves cognitive deficits of dopamine transporter knockout mice without long-term tolerance. Neuropsychopharmacology 32:2465–2478
118. Deng S et al (2015) A behavioral defect of temporal association memory in mice that partly lack dopamine reuptake transporter. Sci Rep 5:17461
119. Del'Guidice T et al (2014) Dissociations between cognitive and motor effects of psychostimulants and atomoxetine in hyperactive DAT-KO mice. Psychopharmacology 231:109–122
120. Wong P, Sze Y, Chang CC, Lee J, Zhang X (2015) Pregnenolone sulfate normalizes schizophrenia-like behaviors in dopamine transporter knockout mice through the AKT/GSK3beta pathway. Transl Psychiatry 5:e528
121. Li F, Wang LP, Shen X, Tsien JZ (2010) Balanced dopamine is critical for pattern completion during associative memory recall. PLoS One 5:e15401
122. Mereu M et al (2017) Dopamine transporter (DAT) genetic hypofunction in mice produces alterations consistent with ADHD but not schizophrenia or bipolar disorder. Neuropharmacology 121:179–194

123. Cybulska-Klosowicz A, Laczkowska M, Zakrzewska R, Kaliszewska A (2017) Attentional deficits and altered neuronal activation in medial prefrontal and posterior parietal cortices in mice with reduced dopamine transporter levels. Mol Cell Neurosci 85:82–92
124. Cybulska-Klosowicz A, Dabrowska J, Niedzielec S, Zakrzewska R, Rozycka A (2017) Potential role of dopamine transporter in behavioral flexibility. Acta Neurobiol Exp (Wars) 77:176–189
125. Takamatsu Y et al (2015) Improvement of learning and increase in dopamine level in the frontal cortex by methylphenidate in mice lacking dopamine transporter. Curr Mol Med 15:245–252
126. Powell SB, Young JW, Ong JC, Caron MG, Geyer MA (2008) Atypical antipsychotics clozapine and quetiapine attenuate prepulse inhibition deficits in dopamine transporter knockout mice. Behav Pharmacol 19:562–565
127. Yamashita M et al (2013) Impaired cliff avoidance reaction in dopamine transporter knockout mice. Psychopharmacology 227:741–749
128. Hironaka N, Ikeda K, Sora I, Uhl GR, Niki H (2004) Food-reinforced operant behavior in dopamine transporter knockout mice: enhanced resistance to extinction. Ann N Y Acad Sci 1025:140–145
129. Rossi MA, Yin HH (2015) Elevated dopamine alters consummatory pattern generation and increases behavioral variability during learning. Front Integr Neurosci 9:37
130. Meck WH et al (2012) Gene-dose dependent effects of methamphetamine on interval timing in dopamine-transporter knockout mice. Neuropharmacology 62:1221–1229
131. Savelieva KV, Caudle WM, Findlay GS, Caron MG, Miller GW (2002) Decreased ethanol preference and consumption in dopamine transporter female knock-out mice. Alcohol Clin Exp Res 26:758–764
132. Tillerson JL et al (2006) Olfactory discrimination deficits in mice lacking the dopamine transporter or the D2 dopamine receptor. Behav Brain Res 172:97–105
133. Ptukha M et al (2022) Effects of atomoxetine on motor and cognitive behaviors and brain electrophysiological activity of dopamine transporter knockout rats. Biomolecules 12:1484
134. Cinque S et al (2018) Behavioral phenotyping of dopamine transporter knockout rats: compulsive traits, motor stereotypies, and anhedonia. Front Psych 9:43
135. Savchenko A et al (2022) The lack of dopamine transporter is associated with conditional associative learning impairments and striatal proteomic changes. Front Psych 13:799433
136. Kurzina NP, Volnova AB, Aristova IY, Gainetdinov RR (2021) A new paradigm for training hyperactive dopamine transporter knockout rats: influence of novel stimuli on object recognition. Front Behav Neurosci 15:654469
137. Davis GL et al (2018) Functional coding variation in the presynaptic dopamine transporter associated with neuropsychiatric disorders drives enhanced motivation and context-dependent impulsivity in mice. Behav Brain Res 337:61–69
138. Gunaydin LA et al (2014) Natural neural projection dynamics underlying social behavior. Cell 157:1535–1551
139. Solie C, Girard B, Righetti B, Tapparel M, Bellone C (2022) VTA dopamine neuron activity encodes social interaction and promotes reinforcement learning through social prediction error. Nat Neurosci 25:86–97
140. Rodriguiz RM, Chu R, Caron MG, Wetsel WC (2004) Aberrant responses in social interaction of dopamine transporter knockout mice. Behav Brain Res 148:185–198
141. Fan Z et al (2019) Using the tube test to measure social hierarchy in mice. Nat Protoc 14:819–831
142. Javitch JA, D'Amato RJ, Strittmatter SM, Snyder SH (1985) Parkinsonism-inducing neurotoxin, N-methyl-4-phenyl-1,2,3,6-tetrahydropyridine: uptake of the metabolite N-methyl-4-phenylpyridine by dopamine neurons explains selective toxicity. Proc Natl Acad Sci USA 82:2173–2177

143. Nass R et al (2005) A genetic screen in Caenorhabditis elegans for dopamine neuron insensitivity to 6-hydroxydopamine identifies dopamine transporter mutants impacting transporter biosynthesis and trafficking. J Neurochem 94:774–785
144. Fumagalli F, Gainetdinov RR, Valenzano KJ, Caron MG (1998) Role of dopamine transporter in methamphetamine-induced neurotoxicity: evidence from mice lacking the transporter. J Neurosci Off J Soc Neurosci 18:4861–4869
145. Bezard E et al (1999) Absence of MPTP-induced neuronal death in mice lacking the dopamine transporter. Exp Neurol 155:268–273
146. Numachi Y et al (2007) Methamphetamine-induced hyperthermia and lethal toxicity: role of the dopamine and serotonin transporters. Eur J Pharmacol 572:120–128
147. Carvelli L, Matthies DS, Galli A (2010) Molecular mechanisms of amphetamine actions in Caenorhabditis elegans. Mol Pharmacol 78:151–156
148. Torres Valladares D, Kudumala S, Hossain M, Carvelli L (2020) Caenorhabditis elegans as an in vivo model to assess amphetamine tolerance. Brain Behav Evol 95:247–255
149. Pizzo AB et al (2013) The membrane raft protein Flotillin-1 is essential in dopamine neurons for amphetamine-induced behavior in Drosophila. Mol Psychiatry 18:824–833
150. Karam CS, Williams BL, Jones SK, Javitch JA (2022) The role of the dopamine transporter in the effects of amphetamine on sleep and sleep architecture in Drosophila. Neurochem Res 47:177–189
151. Cervinski MA, Foster JD, Vaughan RA (2010) Syntaxin 1A regulates dopamine transporter activity, phosphorylation and surface expression. Neuroscience 170:408–416
152. Binda F et al (2008) Syntaxin 1A interaction with the dopamine transporter promotes amphetamine-induced dopamine efflux. Mol Pharmacol 74:1101–1108
153. Fukushima S et al (2007) Methamphetamine-induced locomotor activity and sensitization in dopamine transporter and vesicular monoamine transporter 2 double mutant mice. Psychopharmacology 193:55–62
154. Jones SR, Gainetdinov RR, Wightman RM, Caron MG (1998) Mechanisms of amphetamine action revealed in mice lacking the dopamine transporter. J Neurosci Off J Soc Neurosci 18:1979–1986
155. Hagino Y et al (2011) Effects of MDMA on extracellular dopamine and serotonin levels in mice lacking dopamine and/or serotonin transporters. Curr Neuropharmacol 9:91–95
156. Budygin EA et al (2004) Dissociation of rewarding and dopamine transporter-mediated properties of amphetamine. Proc Natl Acad Sci USA 101:7781–7786
157. Rocha BA et al (1998) Cocaine self-administration in dopamine-transporter knockout mice. Nat Neurosci 1:132–137
158. Budygin EA, John CE, Mateo Y, Jones SR (2002) Lack of cocaine effect on dopamine clearance in the core and shell of the nucleus accumbens of dopamine transporter knock-out mice. J Neurosci 22:RC222
159. Trinh JV, Nehrenberg DL, Jacobsen JP, Caron MG, Wetsel WC (2003) Differential psychostimulant-induced activation of neural circuits in dopamine transporter knockout and wild type mice. Neuroscience 118:297–310
160. Thanos PK, Michaelides M, Benveniste H, Wang GJ, Volkow ND (2008) The effects of cocaine on regional brain glucose metabolism is attenuated in dopamine transporter knockout mice. Synapse 62:319–324
161. Perles-Barbacaru TA et al (2011) Quantitative pharmacologic MRI: mapping the cerebral blood volume response to cocaine in dopamine transporter knockout mice. NeuroImage 55:622–628
162. Shi X, McGinty JF (2006) Extracellular signal-regulated mitogen-activated protein kinase inhibitors decrease amphetamine-induced behavior and neuropeptide gene expression in the striatum. Neuroscience 138:1289–1298
163. Valjent E, Corvol JC, Trzaskos JM, Girault JA, Herve D (2006) Role of the ERK pathway in psychostimulant-induced locomotor sensitization. BMC Neurosci 7:20

164. Beaulieu JM, Sotnikova TD, Gainetdinov RR, Caron MG (2006) Paradoxical striatal cellular signaling responses to psychostimulants in hyperactive mice. J Biol Chem 281:32072–32080
165. Medvedev IO et al (2005) Characterization of conditioned place preference to cocaine in congenic dopamine transporter knockout female mice. Psychopharmacology 180:408–413
166. Thomsen M, Hall FS, Uhl GR, Caine SB (2009) Dramatically decreased cocaine self-administration in dopamine but not serotonin transporter knock-out mice. J Neurosci Off J Soc Neurosci 29:1087–1092
167. Mead AN, Rocha BA, Donovan DM, Katz JL (2002) Intravenous cocaine induced-activity and behavioural sensitization in norepinephrine-, but not dopamine-transporter knockout mice. Eur J Neurosci 16:514–520
168. Hall FS et al (2009) Cocaine-conditioned locomotion in dopamine transporter, norepinephrine transporter and 5-HT transporter knockout mice. Neuroscience 162:870–880
169. Morice E, Denis C, Giros B, Nosten-Bertrand M (2010) Evidence of long-term expression of behavioral sensitization to both cocaine and ethanol in dopamine transporter knockout mice. Psychopharmacology 208:57–66
170. Pardo M, Martin M, Gainetdinov RR, Mash DC, Izenwasser S (2022) Heterozygote dopamine transporter knockout rats display enhanced cocaine locomotion in adolescent females. Int J Mol Sci 23:15414
171. Stewart A et al (2019) Serotonin transporter inhibition and 5-HT(2C) receptor activation drive loss of cocaine-induced locomotor activation in DAT Val559 mice. Neuropsychopharmacology 44:994–1006
172. Spielewoy C et al (2000) Increased rewarding properties of morphine in dopamine-transporter knockout mice. Eur J Neurosci 12:1827–1837
173. Hall FS, Sora I, Uhl GR (2003) Sex-dependent modulation of ethanol consumption in vesicular monoamine transporter 2 (VMAT2) and dopamine transporter (DAT) knockout mice. Neuropsychopharmacology 28:620–628
174. Costa KM, Schenkel D, Roeper J (2021) Sex-dependent alterations in behavior, drug responses and dopamine transporter expression in heterozygous DAT-Cre mice. Sci Rep 11:3334
175. Ji J, Dluzen DE (2008) Sex differences in striatal dopaminergic function within heterozygous mutant dopamine transporter knock-out mice. J Neural Transm (Vienna) 115:809–817
176. Powell SB et al (2004) MDMA "ecstasy" alters hyperactive and perseverative behaviors in dopamine transporter knockout mice. Psychopharmacology 173:310–317
177. Barr AM et al (2004) The selective serotonin-2A receptor antagonist M100907 reverses behavioral deficits in dopamine transporter knockout mice. Neuropsychopharmacology 29:221–228
178. Hall FS, Sora I, Hen R, Uhl GR (2014) Serotonin/dopamine interactions in a hyperactive mouse: reduced serotonin receptor 1B activity reverses effects of dopamine transporter knockout. PLoS One 9:e115009
179. Uchiumi O et al (2013) Serotonergic involvement in the amelioration of behavioral abnormalities in dopamine transporter knockout mice by nicotine. Neuropharmacology 64:348–356
180. Traktirov DS et al (2023) Alterations in serotonin neurotransmission in hyperdopaminergic rats lacking the dopamine transporter. Biomedicines 11:2881
181. Han DD, Gu HH (2006) Comparison of the monoamine transporters from human and mouse in their sensitivities to psychostimulant drugs. BMC Pharmacol 6:6
182. Sora I et al (2001) Molecular mechanisms of cocaine reward: combined dopamine and serotonin transporter knockouts eliminate cocaine place preference. Proc Natl Acad Sci USA 98:5300–5305
183. Chen R et al (2006) Abolished cocaine reward in mice with a cocaine-insensitive dopamine transporter. Proc Natl Acad Sci USA 103:9333–9338
184. Wu H et al (2014) Restoration of cocaine stimulation and reward by reintroducing wild type dopamine transporter in adult knock-in mice with a cocaine-insensitive dopamine transporter. Neuropharmacology 86:31–37

185. Carboni E et al (2001) Cocaine and amphetamine increase extracellular dopamine in the nucleus accumbens of mice lacking the dopamine transporter gene. J Neurosci Off J Soc Neurosci 21(RC141):141–144
186. Shen HW et al (2004) Regional differences in extracellular dopamine and serotonin assessed by in vivo microdialysis in mice lacking dopamine and/or serotonin transporters. Neuropsychopharmacology 29:1790–1799
187. Mateo Y, Budygin EA, John CE, Jones SR (2004) Role of serotonin in cocaine effects in mice with reduced dopamine transporter function. Proc Natl Acad Sci USA 101:372–377
188. Mateo Y, Budygin EA, John CE, Banks ML, Jones SR (2004) Voltammetric assessment of dopamine clearance in the absence of the dopamine transporter: no contribution of other transporters in core or shell of nucleus accumbens. J Neurosci Methods 140:183–187
189. John CE, Jones SR (2006) Exocytotic release of dopamine in ventral tegmental area slices from C57BL/6 and dopamine transporter knockout mice. Neurochem Int 49:737–745
190. Hall FS et al (2002) Cocaine mechanisms: enhanced cocaine, fluoxetine and nisoxetine place preferences following monoamine transporter deletions. Neuroscience 115:153–161
191. Harris SS, Green SM, Kumar M, Urs NM (2022) A role for cortical dopamine in the paradoxical calming effects of psychostimulants. Sci Rep 12:3129
192. Simmler LD et al (2017) Blockade of the 5-HT transporter contributes to the behavioural, neuronal and molecular effects of cocaine. Br J Pharmacol 174:2716–2738
193. Gainetdinov RR, Mohn AR, Bohn LM, Caron MG (2001) Glutamatergic modulation of hyperactivity in mice lacking the dopamine transporter. Proc Natl Acad Sci USA 98:11047–11054
194. Targa G et al (2023) Dysregulation of AMPA receptor trafficking and intracellular vesicular sorting in the prefrontal cortex of dopamine transporter knock-out rats. Biomolecules 13:516
195. Hlavacova N et al (2023) Altered responsiveness to glutamatergic modulation by MK-801 and to repeated stress of immune challenge in female dopamine transporter knockout rats. Prog Neuro-Psychopharmacol Biol Psychiatry 126:110804
196. Reinwald JR et al (2022) Dopamine transporter silencing in the rat: systems-level alterations in striato-cerebellar and prefrontal-midbrain circuits. Mol Psychiatry 27:2329–2339
197. Refai O et al (2022) Allosteric modulator KM822 attenuates behavioral actions of amphetamine in Caenorhabditis elegans through interactions with the dopamine transporter DAT-1. Mol Pharmacol 101:123–131
198. Reith MEA, Jones KT, Zhen J, Topiol S (2018) Latch and trigger role for R445 in DAT transport explains molecular basis of DTDS. Bioorg Med Chem Lett 28:470–475
199. Aguilar JI et al (2021) Psychomotor impairments and therapeutic implications revealed by a mutation associated with infantile Parkinsonism-Dystonia. elife 10:e68039
200. Beerepoot P, Lam VM, Salahpour A (2016) Pharmacological chaperones of the dopamine transporter rescue dopamine transporter deficiency syndrome mutations in heterologous cells. J Biol Chem 291:22053–22062
201. Sucic S et al (2016) When transporters fail to be transported: how to rescue folding-deficient SLC6 transporters. J Neurol Neuromedicine 1:34–40
202. Asjad HMM et al (2017) Pharmacochaperoning in a Drosophila model system rescues human dopamine transporter variants associated with infantile/juvenile parkinsonism. J Biol Chem 292:19250–19265
203. Illiano P et al (2017) Recombinant adeno-associated virus-mediated rescue of function in a mouse model of dopamine transporter deficiency syndrome. Sci Rep 7:46280
204. Xu TX, Ma Q, Spealman RD, Yao WD (2010) Amphetamine modulation of long-term potentiation in the prefrontal cortex: dose dependency, monoaminergic contributions, and paradoxical rescue in hyperdopaminergic mutant. J Neurochem 115:1643–1654
205. Arime Y, Kasahara Y, Hall FS, Uhl GR, Sora I (2012) Cortico-subcortical neuromodulation involved in the amelioration of prepulse inhibition deficits in dopamine transporter knockout mice. Neuropsychopharmacology 37:2522–2530

206. Kurzina N et al (2022) Modulation of spatial memory deficit and hyperactivity in dopamine transporter knockout rats via alpha2A-adrenoceptors. Front Psych 13:851296
207. Volnova A et al (2023) Noradrenergic modulation of learned and innate behaviors in dopamine transporter knockout rats by guanfacine. Biomedicines 11:222
208. Regan SL, Williams MT, Vorhees CV (2022) Review of rodent models of attention deficit hyperactivity disorder. Neurosci Biobehav Rev 132:621–637

Novel Ways of Targeting the Dopamine Transporter

Yibin Xu, Hanming Zeng, Shaili Aggarwal, and Ole Valente Mortensen

Contents

Abstract Dopamine (DA) is an important modulatory neurotransmitter that is involved in daily activities such as movement, memory, and reward-oriented learning of essential behaviors and needs. DA signaling is initiated by the release of DA into the synaptic cleft that will bind to dopamine receptors to mediate the physiological response. To terminate the DA response, the DA is taken up by the dopamine transporter (DAT), a surface membrane protein. Psychostimulants, like cocaine and amphetamine, both target DAT and interfere with the DA uptake process, resulting in an increased amount of DA in the synaptic cleft. Continuous use of psychostimulants can lead to psychostimulant use disorders (PUDs), which are marked by uncontrollable psychostimulant craving and misuse. Because of the unmet need for treatment options for PUDs, novel strategies for discovering therapies are essential. Over the years, DAT-targeting ligands have been identified with atypical properties such as reduced abuse liability compared to cocaine. These compounds have been proposed to bind to different sites from cocaine and/or prefer and stabilize specific conformations of DAT. In addition, some of these compounds can interfere with psychostimulant-DAT binding and may have therapeutic potential in treating PUDs. This chapter introduces the role of DAT in PUDs, presents the mechanism of action

Y. Xu · H. Zeng · S. Aggarwal · O. V. Mortensen (✉)
Department of Pharmacology and Physiology, Drexel University College of Medicine, Philadelphia, PA, USA
e-mail: ovm23@drexel.edu

O. V. Mortensen (ed.), *Monoamine Transporters*, Advances in Neurobiology 46,
https://doi.org/10.1007/978-3-031-96364-3_11

of novel DAT-binding ligands, and discusses the therapeutic potential of atypical DAT-binding ligands for PUDs.

Keywords Dopamine transporter · Psychostimulants · Cocaine · Amphetamine · Allosteric modulation · Substance use disorders

1 Psychostimulant Use Disorders

According to the DSM-V, psychostimulant use disorders (PUDs) are disorders where individuals continue psychostimulant use ignoring the harmful substance-related problems associated with this use [1]. PUD is marked by uncontrollable drug-consuming and drug-seeking behavior and people who suffer from PUDs often display risky behaviors and social impairment [2, 3]. In 2020, approximately 5.1 million individuals aged 12 or above were found to have engaged in psychostimulant misuse according to the National Survey on Drug Use and Health (NSDUH). In addition to cocaine, the number of people using other psychostimulants with misuse potential, such as methamphetamine (METH) and amphetamine (AMPH), is also skyrocketing [4]. Because psychostimulants impair self-control, adolescents are more likely to misuse drug as this age group is known to be highly vulnerable to external stimuli [5]. Although most people can make voluntary decisions at the beginning of psychostimulant use and some can obtain abstinence after treatment interventions such as cognitive behavioral therapy, alterations in the central nervous system (CNS) after repeated psychostimulant are thought to produce changes in the individual's ability of self-control and decision making. These structural changes in the CNS can be irreversible and will further contribute to relapse [6].

Psychostimulant use also causes comorbid mental health problems. Evidence shows that with excessive or prolonged use, cocaine can cause itching, increased heart rate, hallucinations, and paranoid delusions [7]. A significant proportion of patients may be vulnerable to psychotic episodes associated with drug use, as 29 out of 55 individuals with cocaine dependence reported experiencing transient episodes of cocaine-induced psychosis in Lydiard's study [8]. Moreover, individuals with 5 years of cocaine dependence show specific and polymorphic clinical features with the presence of psychotic symptoms, including auditory hallucinations [9, 10].

After long-term use, individuals who misuse psychostimulants can also develop tolerance to the euphoric effects. This tolerance is developed due to changes in pharmacokinetic and pharmacodynamic mechanisms including attenuation of dopamine responses to psychostimulant exposure. Due to the drug tolerance, the dose needed to generate the same rewarding effect will increase over time, which then can lead to overdose. According to Centers for Disease Control and prevention, the overdose death involving psychostimulants increased by 37% from 2020 to 2021, and the number of deaths have reached 33,000 in the United States [11]. The high relapse rate exacerbates this situation with around an 85% rate of patients relapsing within a year after abstinence [12].

In addition to all these detrimental impacts PUDs have on individuals, it is also important to note that the families of patients and the whole society are significantly affected by these disorders. However, despite all these significant harms associated with PUDs, no FDA-approved therapy is available on the market, which makes it important to explore novel ways of developing an effective therapy toward PUD.

2 Critical Processes Involved in the Development of Psychostimulant Use Disorders

2.1 Dopaminergic Pathways

DA is synthesized from its precursor, that is, tyrosine and is converted into L-DOPA by the rate-limiting enzyme tyrosine hydroxylase (TH). L-DOPA is then converted into DA by aromatic amino acid decarboxylase (AADC). DA is then packed into secretory vesicles via the vesicular monoamine transporter 2 (VMAT2) through direct synthesis-packaging coupling and released through exocytosis [13, 14]. Released DA subsequently binds dopamine receptors on the postsynaptic neurons, leading to downstream signaling. Following release, DA will bind postsynaptic dopamine receptor to trigger DA signaling. Excessive DA is removed from the synaptic cleft and taken back into the presynaptic DA neuron by the dopamine transporter (DAT). As a result, DAT is important for recycling DA and terminating DA signaling. Cocaine and the other two most misused psychostimulants—METH and AMPH—all target DAT but by different mechanisms. Cocaine is an inhibitor of DAT, whereas both METH and AMPH are substrates of DAT, triggering both release of DA and inhibition of DAT. Consequently, psychostimulant exposure leads to elevated DA levels in the synaptic cleft, which results in an amplified dopaminergic signal that produces the rewarding experience associated with psychostimulant use.

Dopaminergic neurons project their signals to a wide range of anatomical regions and affect executive function, learning, reward, motivation, and neuroendocrine control [15]. In addition, DA signaling has been implicated in many diseases such as psychostimulant use disorder (PUD), Parkinson's disease (PD), and attention deficit hyperactivity disorder (ADHD) [15]. Dopaminergic pathways can be divided into three pathways. Two of them are known to be disrupted in PUDs: the mesocortical and mesolimbic pathway [16]. The third one is the nigrostriatal pathway, with dopaminergic neurons projecting from substantia nigra (SN) to dorsal striatum and is mainly involved in mediating voluntary movement [17]. Different from the nigrostriatal pathway, both mesolimbic and mesocortical pathways are involved in reward-related behaviors and contribute to the rewarding properties of the psychostimulants [18]. Major dopaminergic neurons in the mesolimbic and mesocortical pathways originate from the ventral tegmental area (VTA). However, dopaminergic neurons in the mesolimbic pathway mainly project to the nucleus accumbens (NAc), while dopaminergic neurons in the mesocortical pathways mainly project to the

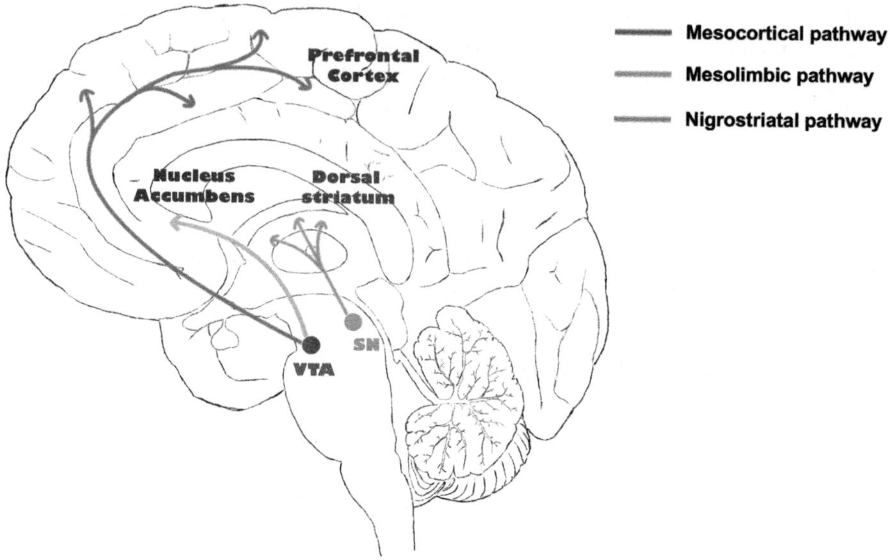

Fig. 1 Dopaminergic pathways. The orange arrow indicates the projections through the nigrostriatal pathway that connects substantia nigra (SN) to dorsal striatum. The small blue arrows indicate the mesolimbic and mesocortical pathways (also called reward pathway), which connects the VTA in the midbrain to either the nucleus accumbens (NAc) or prefrontal cortex (PFC)

prefrontal cortex (PFC) (Fig. 1). The release of dopamine within the mesolimbic pathway regulates and reinforces reward learning and may also produce the sense of euphoria associated with psychostimulants. Amplified dopaminergic signals caused by drugs of abuse as described above are usually observed in the NAc, further highlighting the importance of NAc in the dopaminergic pathway as well as its characteristic of having high density of DAT compared to other regions in the brain [19, 20].

2.2 *DAT*

DAT, together with serotonin transporter (SERT) and norepinephrine transporter (NET), are all members of the monoamine transporter (MAT) family. MATs are 12 transmembrane domain surface membrane proteins that belong to the solute carrier 6 (SLC6) family. They are Na^+/Cl^--dependent sodium symporters that utilize the naturally occurring Na^+ gradient to drive the transport of substrates against their concentration gradient [21]. Many studies characterizing MAT structure and mechanism have provided valuable knowledge about MAT structures and conformations. Based on information provided by crystallography and cryo-electron microscopy (cryo-EM) structures of human MATs, both a high-affinity primary substrate-binding site (orthosteric site) and allosteric sites were identified. The primary orthosteric substrate-binding site is named the S1 site, which lies at the core of the translocation pathway located between TM1 and TM6. The S1 site binds substrates together with one or two Na^+ ions [22]. Crystal and cryo-EM structures have found substrates including DA [23–25], NET-specific reuptake inhibitors (NRIs) [24, 26], norepinephrine and dopamine reuptake inhibitor (NDRI) [23], selective serotonin-reuptake inhibitors (SSRIs) [27–30], tricyclic antidepressants (TCAs) [31, 32], serotonin-norepinephrine-reuptake inhibitors (SNRIs) [26], cocaine and its analogs [26, 33], and AMPH [24] in the orthosteric S1 site.

It has been proposed that ions and most substrates like DA enter DAT through a process of binding and conformational changes known as the "alternating access mechanism" of transport [34, 35]. It is initiated by the binding of a Na^+ ion on the extracellular side and is followed by substrate binding within the central S1 site (orthosteric site) with the transporter forming an outward-facing conformation exposed to the extracellular side. As substrate binds to the S1 site, a conformational change is triggered to close the extracellular gate of the transporter to form an occluded state. In this state, both the ion and the substrate are inside a sealed pocket in DAT with no access to the intracellular and extracellular spaces. Finally, the intracellular gate opens, leading to an inward-facing conformation, releasing the ions and substrate into the cytoplasm through diffusion facilitated by hydration of the substrate-binding site (Fig. 2).

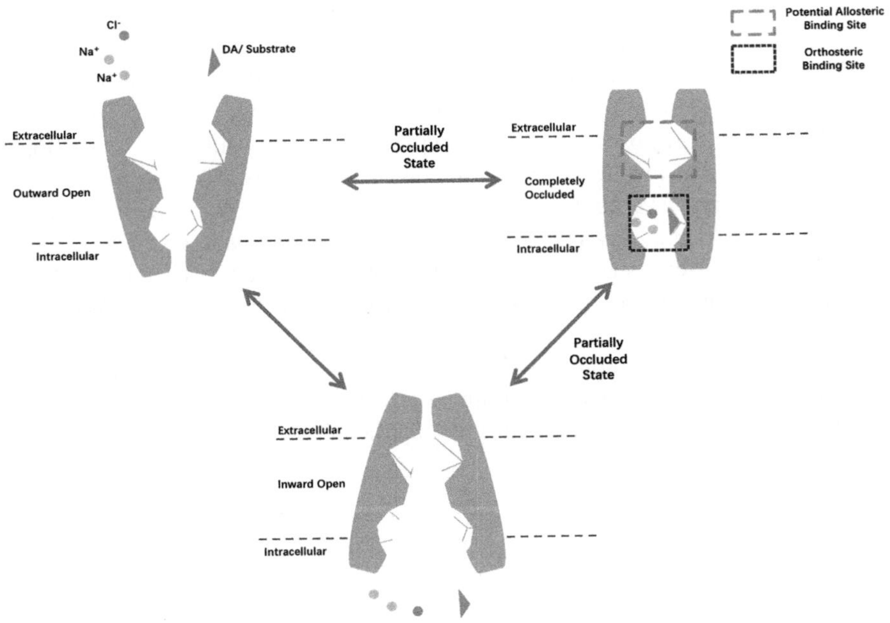

Fig. 2 Proposed alternating access mechanism of translocation of substrates and ions through the dopamine transporter

In addition to the orthosteric S1 site, it has been suggested that secondary substrate/ligand binding sites exist in the extracellular domain of the MATs and might have functional relevance. Computational and biochemical evidence indicates that ligands binding to these secondary sites can influence the binding at the substrate-binding site. One molecular dynamic study revealed that a secondary site S2 exists in the bacterial MAT homologue LeuT in the extracellular vestibule. It also showed that the binding of another leucine molecule at S2 triggers the release of the bound leucine from the substrate-binding site S1 [36, 37]. Several crystal structures of LeuT have also been discovered in complex with different tricyclic antidepressants and SSRIs bound at the extracellular vestibule [38–40].

Recently, we have discovered allosteric sites in both human DAT (hDAT) and human SERT (hSERT) [41, 42]. We also identified ligands binding to this site through virtual screening and demonstrated that these ligands have modulatory effects on the transporters [42, 43]. In further support of the existence of this allosteric site, the crystal structure of hSERT, in complex with two (S)-citalopram molecules, found that one citalopram molecule bound at the S1 site and the other bound within an extracellular vestibular region similar to the allosteric site we described. Earlier evidence has shown that the binding of a second (S)-citalopram molecule to hSERT slows the dissociation rate of (S)-citalopram bound in the S1 pocket [44]. In other cryo-EM studies on hSERT, the substrate serotonin [25] and the atypical antidepressant vortioxetine [32] was also found bound to this domain of the transporter located away from the orthosteric site. We have also in a recent functional mode analysis unraveled structural conformational changes of hSERT and confirmed the existence of allosteric sites in hSERT [41].

Additional recent studies have further confirmed the existence of allosteric binding sites in MATs. For human NET (hNET), an extracellular allosteric site was identified at the C-terminal end of transmembrane helix 1b (TM1b) in the hNET cryo-EM structure. Substrates when bound to hNET in this allosteric site will keep the transporter in an inward-open position instead of an outward-open position [45]. Additionally, cryo-EM studies on hDAT show that MRS7292, a non-competitive inhibitor of hDAT that slows the unbinding of the cocaine analogue β-CFT, binds to hDAT in an allosteric site. Residues such as W84, D385, Y394, and M414 were identified as important residues that contribute to the binding affinity of MRS7292 [46].

Collectively, these studies provide a compelling argument for the presence of functionally relevant secondary binding sites in the extracellular domains of the MATs and provide opportunities for utilizing the allosteric sites for novel modulatory effects on the transporters.

2.3 Cocaine: A Competitive Inhibitor of DAT

Cocaine has an inhibitory effect on all three MATs. However, the study of knock-in mice with cocaine-insensitive DAT (DAT-CI mice) indicates that DAT inhibition is responsible for mediating the rewarding effect of cocaine [47]. Cocaine is well established as a competitive inhibitor of DAT, suggesting that cocaine-binding sites overlap with the dopamine-binding sites [33]. This has also been confirmed with biochemical and mutational studies as a high overlap between the cocaine- and dopamine-binding sites was determined by trapping the radiolabeled cocaine analog in the transporter, either by cross-linking engineered cysteines or with an engineered Zn^{2+}-binding site that was situated extracellularly to the predicted common binding pocket, showing that cocaine and DA share a competitive profile [48]. Utilizing cryo-EM, Srivastava and colleagues visualized the structure of hDAT in complex with the cocaine analog β-CFT. In their study, utilizing the DA uptake assay, they further confirmed that β-CFT binds to the central orthosteric S1 site of hDAT, with residues D79, A81, F76, and G323 as key residues in binding [46]. Another recent cryo-EM study with full-length hDAT also confirmed that cocaine binds in the central binding site while keeping hDAT in an outward-open position [33].

3 DAT-Targeting Compounds for Treating Psychostimulant Use Disorders

There are numerous known DAT-targeting compounds. Because those compounds display different behavioral effects as a result of binding DAT, they have been classified into two different types: typical and atypical DAT inhibitors [49, 50]. In general, DAT inhibitors are expected to inhibit DA uptake and compete with other DAT blockers. It has also been demonstrated that DAT inhibitors can inhibit the release of substrate resulting from reversed transport. Compared to typical DAT inhibitors such as cocaine, atypical DAT inhibitors also display these properties but do not share all the in vivo behavioral effects of the typical DAT inhibitors. While typical DAT inhibitors, as expected, show abuse liability and stimulate locomotor behaviors in rodents, the atypical DAT inhibitors have been found to display unique behavioral aspects including less stimulatory effects on locomotion and less cocaine-like rewarding effects in studies employing assays such as cocaine self-administration. The atypical DAT inhibitors displaying such different mechanisms of action compared to cocaine have therefore become promising candidates for cocaine-use disorder treatment.

3.1 Properties of Atypical DAT Inhibitors

All known typical DAT inhibitors like cocaine bind to DAT in the orthosteric site. It is proposed that the atypical DAT inhibitors bind to DAT in different ways such as in an allosteric site or bind the orthosteric site but stabilize DAT conformations differently [50]. More specifically, the atypical DAT inhibitors are proposed to be more likely to induce and/or prefer other DAT conformations such as a more "closed" DAT conformation. In this "closed" conformation, DAT opens more to the intracellular side, which makes it more likely to display a slower binding rate of compounds. This is different to how cocaine binds a more "open" conformation, where DAT opens more to the extracellular side [50]. This notion of binding different conformations was initially based on biochemical assays [51] and this has now also been supported by structural cryo-EM studies [23]. It has also become clear that there is not a single mechanism that determines whether a compound will act as an atypical DAT inhibitor. For example, there are atypical DAT inhibitors who exert their functions through binding an allosteric site, through the inhibition of the sigma receptor, and/or displaying slower pharmacokinetics compared to cocaine, whose onset of behavioral effects is often much faster than most atypical DAT inhibitors. In addition, the atypical DAT inhibitors also display a slower association rate with DAT [50].

Because of the promise of these compounds, several studies have pursued these compounds and their analogs. The most studied atypical DAT inhibitors are the benztropine (BZT) and modafinil analogs. There are also drugs developed to non-competitively inhibit cocaine-DAT binding, including GBR12909. Finally, there are other atypical DAT inhibitors, including allosteric modulators, that could have therapeutic potential in treating PUD.

3.1.1 Benztropine and Its Analogs

Among the drugs that can induce and/or prefer the "closed" conformation, only BZT and modafinil have been tested in clinical trials [50]. In 2001, a clinical trial on BZT on cocaine craving started, but no result was posted at NIDA [50]. Although not approved to treat cocaine-related disorders as an atypical DAT inhibitor, BZT is also an anticholinergic drug and is approved as an adjunctive therapy for Parkinson disease [50, 52]. A direct effect of BZT on DA reuptake through DAT in the striatum was confirmed by in vivo microdialysis on anesthetized rats, which demonstrated a dose-dependent inhibition of reuptake of DA [53]. Because BZT and its analogs are representative drugs of atypical DAT inhibitors that bind the orthosteric pocket in an inward conformation [23], they have been studied in detail to reveal mechanisms of action and also for the consideration as candidates for treatments for PUD [54].

The study of the binding site of a BZT analog, GA II 34, on hDAT using ^{125}I revealed that the DAT structure stabilized by GA II 34 is different from the structure stabilized by cocaine [55]. The interaction of GA II 34 at Asp79 in TM1 showed that it binds in the orthosteric substrate-binding S1 site at a similar binding site of cocaine and its analogs. Although further assays confirmed this finding by substituting the residue with cysteine, Tomlinson and colleagues found that there are differences in the DAT structure that GA II 34 stabilizes compared with cocaine. Utilizing the substituted cysteine accessibility method, GA II 34 was found to keep DAT in a more open binding position compared to the cocaine-bound structure, which is more outwardly occluded, and this was suggested to explain the difference between the neurochemical outcomes between GA II 34 and cocaine [55].

Other studies have shown that JHW007, an N-butyl analog of BZT, could have therapeutic potential in preventing PUDs. Association binding assays in WT and DAT-KO mice showed that JHW007 has two binding sites: one high-affinity binding site in DAT and another low-affinity non-DAT-binding site [56]. Computational modeling and site-directed mutagenesis studies found that amino acids Asn157, Ala 479, and Ala480 in transmembrane segment (TM) 10 are important for JHW007 binding. Moreover, these studies found that the JHW007-binding site overlaps with the substrate-binding site and may compete with those substances in DAT binding [57]. When used in combination with cocaine, JHW007 inhibits cocaine-induced DA increase in the NAc shell area [58]. JHW007 did not produce sensitized locomotor activity when given alone but could prevent the sensitized motor behavior induced by chronic AMPH administration, and even prevent the long-term structural changes induced by chronic AMPH in NAc [59]. JHW007, as well as AHN1-055, another N-methyl-BZT analog, was shown to have an inhibitory effect on cocaine-induced hyperlocomotion, conditioned place preference, and self-administration [58, 60, 61]. In another study, the binding affinities of BZT analogs could be adjusted by changes in area and intensity of the negative electrostatic potential areas of diphenylmethoxyl group [62]. This indicates that BZT analogs of different effectiveness can be obtained and provide more approaches to generate new candidate molecules. Even though many molecular and preclinical behavioral studies on BZT analogs show great promise, some issues remain. AHN1-055 was observed to increase response rate during extinction of cocaine self-administration [63]. Moreover, more studies are needed to examine the role of the low-affinity non-DAT-binding site for N-butyl analogs of BZT like JHW007. As a result, to develop BZT analogs as a therapy, more preclinical studies will be needed.

3.1.2 GBR12909

GBR12909 was developed as a noncompetitive inhibitor to mitigate the effects of cocaine. GBR12909 is a potent inhibitor for DAT and is more selective for DAT compared to the other MATs [64]. It was designed to have high affinity toward DAT so that it can dissociate slower and inhibit the binding of other abused stimulants while changing the DA level in a more moderate and prolonged way [65, 66]. A site-directed mutagenesis study revealed that Tyr88 and His547 played an important role in DAT binding for both cocaine and GBR12909, indicating an overlap between their binding sites [67]. A recent cryo-EM study has shown that GBR12909, similar to BZT, binds and stabilizes an inward-facing conformation of DAT [23]. When GBR12909 is systemically administered, it does not affect extracellular DA. However, the same dose of GBR12909 can largely diminish the effect of a modest dose of cocaine [66]. A Rhesus monkeys behavior study showed that GBR12909 could decrease cocaine-maintained response without affecting food-maintained response [66]. In a behavior study in rats, GBR12909 also produced a dose-dependent decrease in cocaine self-administration [68]. Although GBR12909 does not produce any psychostimulant effect in the therapeutic range and successfully completed phase I clinical trials, its study was discontinued due to its detrimental cardiac effects [69, 70]. However, studies on GBR12909 provided important experimental data for the development of atypical DAT inhibitors as a potential therapy for PUDs.

3.1.3 Modafinil and Its Analogs

Modafinil is another drug besides benztropine in the class of atypical DAT inhibitors that has been pursued in human clinical studies with inconsistent results [1, 50]. Modafinil is a drug that is currently used to treat sleepiness due to narcolepsy, shift work sleep disorder, and obstructive sleep apnea. It is also used as a purported cognitive enhancer [71] and to overcome fatigue in the air force [72]. It was also demonstrated that the cognition-enhancing effects of modafinil might be associated with its ability to increase motivation [73]. Nevertheless, this wide range of activities raises some concerns, especially for the use of modafinil as a nootropic drug for students [74]. Regarding PUD, in a double-blind, placebo-controlled clinical trial on modafinil, researchers found that a single morning dose of 400 mg modafinil can reduce the time needed to achieve cocaine-protracted abstinence [75]. Promising preliminary evidence also demonstrated that modafinil improves some clinical outcomes for cocaine dependence when being used with cognitive-behavioral therapy [75]. Modafinil also has other benefits, as studies showed that after modafinil treatment, slow-wave sleep and memory is improved in cocaine users [76, 77].

A post hoc analysis showed a significant effect of modafinil increasing the weekly percentage of nonuse days in the subgroup of those cocaine patients without a history of alcohol dependence [78]. In another report with a total sample of 210 patients of cocaine-use disorders, no significant difference between modafinil and placebo group was observed in the change in average weekly percent of cocaine nonuse days over the 12-week treatment period. A reduction in craving was observed in both groups [78]. In a study testing the combination therapy of modafinil and escitalopram, it was found that modafinil significantly weakened the subject-rated effect produced by cocaine, and escitalopram did not enhance the efficacy of modafinil to reduce any symptoms [79]. To examine how modafinil affects cocaine seeking under different conditions of high or low level of effort for each cocaine administration and valuable alternative, researchers recruited 15 current cocaine users that were not on any therapy for CUD. In this test, individuals were either assigned to the low response effort (500 responses/dose) low value alternative (two game plays for money), or the high response effort (2500 responses/dose) high value alternative (four game plays for money). In each group, individuals were given either placebo or modafinil in a double-blind manner. It was found that modafinil only decreases cocaine choice in the high effort high value alternative group, indicating the limitation of efficacy of modafinil [80].

Some preclinical results indicate that modafinil may not be a perfect candidate to treat CUD. Male mice when given moderate and high levels of modafinil showed pronounced locomotor sensitization that cross-sensitized in a bidirectional way with cocaine. Furthermore, additive behavioral sensitization was observed when cocaine and modafinil were repeatedly given to mice sequentially [81]. This result demonstrated a potential risk of modafinil to trigger the relapse of CUD. In fact, modafinil successfully reinstates cocaine-induced conditioned place preference following extinction in rats [82]. However, in human clinical trials, no evidence was found of modafinil to promote cocaine use in patients, which suggest that modafinil does not have abuse liability in human [83].

Modafinil has two enantiomers: R- and S-modafinil. Molecular docking studies showed that both enantiomers bind in a site that overlaps with the DAT orthosteric site. Compared to the S-modafinil, R-modafinil has around threefold higher affinity to DAT. Moreover, a unique interaction was found between R-modafinil and Tyr156 in DAT but not with the S-enantiomer [84]. Even though modafinil enantiomers bind to DAT in a similar site compared to cocaine, radioactive dopamine uptake assay revealed that modafinil enantiomers stabilize DAT in a different conformation compared to cocaine. More specifically, when using Cys159 and 2-(trimethylammonium)ethyl-methanethiosulfonate (MTSET) to assess the conformation of DAT, both enantiomers, differently from cocaine, prevented Cys159 from reacting with MTSET, which suggested that they bind to DAT in a more inward-facing conformation compared to cocaine [84]. To improve the pharmacological profile while keeping modafinil's effect on CUD, JJC8-016, a novel analog of modafinil, was designed and tested in Long Evans rats. Structural studies on JJC8-016 demonstrated that it stabilizes DAT in a more occluded conformation, which is also preferred by other atypical DAT inhibitors [85–87]. JJC8-016 was also shown to inhibit the effect of cocaine by interfering with cocaine-DAT binding. Behavioral tests on JJC8-016 in Long Evans rats showed that JJC8-016 significantly reduced cocaine-induced hyperlocomotion as well as cocaine self-administration. Moreover, JJC8-016 was also found to inhibit cocaine-primed reinstatement in a dose-dependent manner [88]. More recently, another compound in the same series has been pursued that shows less cardiotoxic effects compared to JJC8-016. This compound, JJC8-091, reduces METH and cocaine self-administration and inhibits cocaine-primed reinstatement of cocaine seeking [86, 89]. Of translational significance, this compound also shows some promise in nonhuman primate models of PUD [90, 91].

In conclusion, modafinil can ameliorate CUD to some extent; however, the risk of modafinil triggering relapse and its limited efficacy limits its usage as treatment for PUDs. Compared to modafinil, the novel analogs of modafinil, JJC8-016 and JJC8-091, are more effective in inhibiting cocaine's effect and have a lower abuse potential, which could be a more favorable and safer choice for PUDs' treatment. However, only preclinical studies of these compounds in rodents and nonhuman primates have been performed.

3.1.4 Sydnocarb

Sydnocarb is a drug that was originally developed in the USSR in the 1970s for a variety of psychological indications. Earlier studies have been conducted of this drug in Russia, but recently, some studies have also been performed outside of Russia and sydnocarb has gone through Phase II clinical trials for Parkinson's disease [92]. Extensive in vivo evidence shows that sydnocarb acts on dopaminergic neurotransmission [93]. Even though it was once believed that sydnocarb may act as a prodrug and D-amphetamine (D-AMPH) could be its metabolite, recent studies show differences between sydnocarb and D-AMPH. Sydnocarb and D-AMPH differ in the way they increase DA, their selectivity for DAT, and half-life, indicating a different mechanism of action from D-AMPH [94]. Compared to D-AMPH, sydnocarb also has less impact on locomotor activity, self-administration, and stereotyped behaviors [94].

Our lab has established that sydnocarb is a selective noncompetitive DAT inhibitor that binds to an allosteric site in DAT [95]. Molecular docking and dynamics simulations of sydnocarb showed that sydnocarb binds to an allosteric site and stabilizes DAT in an outward-facing conformation. This finding was further confirmed by biotinylation assays and DA uptake kinetic assay demonstrating that sydnocarb is a noncompetitive inhibitor of DA uptake and interacts with DAT in a different way from cocaine. Sydnocarb also dose-dependently decreased the potency of cocaine toward DAT [95]. Taken together, this suggests sydnocarb is an atypical DAT inhibitor with a potential for treating PUDs, but additional preclinical studies are required to further establish the therapeutic potential of sydnocarb for treating PUDs.

3.1.5 DAT Allosteric Modulator KM822

As mentioned above, our group has previously identified an allosteric site in DAT and characterized small-molecule allosteric modulators that bind to DAT in this allosteric site. Our studies have shown that they can modulate the function of DAT in unique ways. One of these allosteric DAT modulators, KM822, was identified utilizing computational modeling and virtual screening [43]. Using site-directed mutagenesis and biotinylation assays, KM822 was found to bind to a more extracellular site compared to both cocaine and sydnocarb [95]. Interestingly, a recent cryo-EM study has confirmed the binding of an analog of KM822 to DAT [96]. In COS-7 cells transfected with hDAT, KM822 decreases the Vmax of DA uptake in a dose-dependent manner, indicating KM822 is a noncompetitive inhibitor of DAT. When given as a pretreatment before cocaine, KM822, like sydnocarb, decreases the potency of cocaine to DAT in a dose-dependent manner. This result implies that DAT allosteric modulators like KM822 can interfere with cocaine-DAT binding [43]. When KM822 is given to planarians, it inhibits cocaine-induced hyperlocomotion while not changing locomotor activity by itself [43]. When used as a pretreatment in wild-type *Caenorhabditis elegans*, which has been found to become paralyzed when extracellular level of DA is increased by AMPH treatment, KM822 successfully decreased the ratio of paralysis in AMPH-treated *Caenorhabditis elegans* [97]. Taken together, these findings make KM822 a promising therapy for PUDs. Unfortunately, the water solubility of KM822 is low and this compound can therefore only act as a strong proof of principle for this strategy. In summary, although the DAT allosteric modulator KM822 showed promising effects on cocaine-DAT binding and psychostimulant-elicited behaviors, the solubility of this compound needs to be resolved. Moreover, more behavioral preclinical testing demonstrating effects on behavioral models of PUDs is required before it can be concluded that DAT allosteric modulators like KM822 have the potential of being used as an effective therapy for PUDs.

3.2 *Atypical DA Releasers*

Because amphetamines trigger the release of DA from intracellular vesicles through their interaction with DAT and the vesicular monoamine transporter VMAT2, they are known as DA releasers and have strong abuse liability through a different mechanism compared to DAT inhibitors like cocaine [13]. Different from DAT inhibitors, DA releasers are transported from the synaptic cleft to the cytosol by DAT similar to the endogenous substrate DA [98].

Low doses of AMPH are used clinically and found to improve the severity of ADHD symptoms [99]. It is also used as an athletic performance enhancer, because its stimulating effect on dopaminergic system can lead to reduced tiredness and increased alertness [100, 101]. Due to the similar effect of elevating extracellular DA generated by AMPH and cocaine, according to studies on experienced cocaine and amphetamine users, most of them cannot sense the difference between the two [102]. When administered together with cocaine, 0.25 mg/kg AMPH is shown to have a moderately mild additive effect; however, not all doses of AMPH were observed to have additive effect [103, 104]. Also, due to the damage by AMPH and METH on dopaminergic and serotonin neurons, long-term use of these drugs can lead to neuroinflammation and neurotoxicity and can have harmful effect on mental health [105, 106]. Other studies showed that AMPH use at early stage of life can lead to early onset of dementia and a higher chance of developing Alzheimer's disease in human [107, 108]. Studies also suggest that AMPH has the potential for resulting in persecutory delusions, visual hallucinations, and schizophrenia-like symptoms [109].

Interestingly, there are other DA releasers that show lower efficacy to release dopamine and have some features different from the classic ones [50]. Those DA releasers are known as atypical DA releasers and are suggested to be less likely to cause addiction. Atypical DA releasers would have different properties that make them potential candidates for PUD therapy. The atypical features of those compounds can include bigger molecule size, hybrid receptor profile, lower maximum release rate (partial releaser), and lack of effects on VMAT2. It is observed that atypical DA releasers will result in increased locomotion but at a smaller magnitude compared to typical DA releasers such as AMPH and METH [50]. As a result, atypical DA releasers could be a potential substitution treatment for PUD. But obstacles remain as demonstrated by methyldiethanolamine (MDEA), a partial DA releaser known to cause pathological behavior changes in both animal models and humans [110]. When rats were injected with MDEA, significant retention deficits were observed [111]. Also, only a few studies on the impact of atypical DA releasers on addictive stimulants have been conducted and more research is needed to demonstrate the potential of these types of molecules as a therapy for PUD.

3.2.1 Methylphenidate

Methylphenidate (MPH) is a stimulant that is used as a first-line treatment of ADHD, with a lower risk of abuse [112]. MPH binds to DAT with a similar affinity as cocaine, but it has a lower affinity for the SERT compared to cocaine. MPH interacts differently with DAT compared to cocaine. For example, while cocaine showed no significant change in affinity for DAT with double mutations in serine 356 and 359 compared to WT DAT, MPH showed a reduced affinity for these mutants. This result suggests that the binding site for MPH and cocaine does not fully overlap [113]. A recent cryo-EM study has shown that MPH binds DAT in an outward-facing conformation similar to the one cocaine binds [23]. When given to rats, MPH decreases DAT in rostral caudate putamen, with no change in NAc, while none of the treatments with cocaine affected DAT density [114].

Some reports showed that adolescent rats previously treated with MPH were not as likely to develop abuse-like behaviors with cocaine [115]. This result implies that MPH treatment taken during developmental stages can diminish cocaine reward in adulthood [116]. MPH treatment could also improve ADHD symptoms and inhibitory control abilities for cocaine-dependent individuals with adult ADHD, and these improvements were thought to be associated with the reduction in cocaine use [117, 118]. Moreover, a reduction in impulsive behaviors is observed in ADHD children treated with MPH [119]. It is known that both ADHD and PUD patients have poor performance in executive functions and cognitive control. As a result, MPH, as the most widely used treatment for ADHD, could have a potential as a treatment for PUD [120]. However, even though MPH has therapeutic potential due to the similarity between ADHD and PUD in a theoretical view, short-term pretreatment of adults with a high dose of MPH seems to sensitize the reinforcing effects of cocaine [121], which brought up a concern of cross-sensitization. Moreover, results from randomized controlled trials suggest that MPH is not effective for PUD, particularly in patients without ADHD [122].

3.2.2 Ibogaine

Ibogaine use originated in West Africa. It is a hallucinogenic compound extracted from the root bark of the Iboga shrub [123]. It has been used in religious ceremonies in local cultures [124]. Ibogaine has a complex mechanism of action acting on multiple targets. It is known as a noncompetitive SERT inhibitor and a weak opioid agonist, and has been studied as an addiction treatment [125]. Other than the two mechanisms mentioned above, Ibogaine can also lower the DA level and enhance the metabolism of DA [126]. It was demonstrated in a microdialysis study that a 40 mg/kg systemic injection of ibogaine successfully inhibited DA release [127]. Moreover, due to the similar structure of SERT and DAT, ibogaine is also reported to be a noncompetitive inhibitor of DAT [128, 129]. An augmentation in DA in the NAc is observed when ibogaine is given before cocaine; however, 18-MC, an *iboga* alkaloid congener, does not lead to the same effect [130, 131]. Other than influencing the DA level in NAc, acute ibogaine is also observed to increase serotonin in the NAc [132]. Different from other compounds that have therapeutic potential toward PUD, ibogaine stabilizes the transporters in both outward- and inward-open conformations [30, 133]. Also, the affinity of ibogaine is greatly affected by the conformation of SERT and ionic conditions as the affinity of ibogaine is attenuated when SERT is in an outward-open conformation or when there is no potassium ion present [30]. A recent study using cryo-EM visualized the structure of SERT in complex with ibogaine. The cryo-EM study revealed that Asn177, Tyr176, Asp98, and Phe341 are important residues in ibogaine binding with SERT [30]. Another recent study looked deeper into the unique binding characteristic of ibogaine using large library docking techniques. It found that cocaine and ibogaine affect SERT confirmations differently, especially for their impact on residue Cys277. Cocaine decreases the accessibility of Cys277 and keeps SERT in an outward-open position while ibogaine has an opposing effect on Cys277 and keeps SERT in an inward-open conformation [133]. It was observed that a single 40 mg/kg injection of ibogaine leads to a long-lasting significant decrease of cocaine self-administration in rats, which indicates that either ibogaine or its metabolites has a prolonged inhibitory effect on cocaine seeking [134]. Since the known half-life of ibogaine is 1 h in rodents, it implies that the prolonged attenuating effect on cocaine self-administration is because of active and long-lasting ibogaine metabolites. It was also found that ibogaine injection could abolish the expression of cocaine-induced dopamine sensitization and reverse neuroadaptations produced by chronic cocaine administration that contribute to its putative antiaddictive property [135]. Even though ibogaine is effective for suppression of cocaine seeking, its toxicity and severe side effects could be a severe concern if used as a treatment. The median lethal dose (LD50) of ibogaine is very low. As a result, numerous cardiotoxicity and GI toxicity reactions and sudden deaths caused by ibogaine have been reported [125]. In conclusion, ibogaine and its analogs might have therapeutic potential in treating PUDs; however, proper dosage should be established through animal and human study before it is used as a therapy.

4 Conclusion

PUDs have been a serious issue to individuals and society for a long period, and substantial efforts have been spent on the discovery of molecules that could be used to treat PUDs. Even though different categories of candidate molecules have been tested in animals, only a very small proportion of drugs have been explored in clinical trials. In addition, and unfortunately, none of them met the safety criteria and/or had sufficient efficacy. Most of the candidate molecules pursued in this research are known drugs with other implications for mental diseases. Only a few compounds with new mechanisms have been discovered. Due to this lack of effective treatment options, PUDs remain a severe issue. As a result, the future of PUD treatment should focus on combined therapies with both medical and behavioral therapies. For the medical therapy part, targeting DAT in novel ways including through atypical inhibition and/or allosteric modulation as described in this chapter remains a promising field to pursue.

Acknowledgments This work was funded by National Institutes of Health grants MH106912, MH121453, and DA057982.

References

1. Hersey M et al (2021) Psychostimulant use disorder, an unmet therapeutic goal: can modafinil narrow the gap? Front Neurosci 15:656475–656475
2. NIDA (2018) Understanding drug use and addiction DrugFacts. Available from: https://nida.nih.gov/publications/drugfacts/understanding-drug-use-addiction
3. Sanz Cortés A (2022) Diagnostic and Statistical Manual of Mental Disorders, Fifth Edition, Text Revision (DSM-5-TR). American Psychiatric Association, 2022. https://doi.org/10.1176/appi.books.9780890425787. Psicooncología 19(2):339–340
4. SAMHSA (2020) Treatment of stimulant use disorders. Substance Abuse and Mental Health Services Administration, Rockville
5. Mottarlini F et al (2020) Repeated cocaine exposure during adolescence impairs recognition memory in early adulthood: a role for BDNF signaling in the perirhinal cortex. Dev Cogn Neurosci 43:100789
6. Substance Abuse and Mental Health Services Administration (US); Office of the Surgeon General (US) (2016) Chapter 2 – The neurobiology of substance use, misuse, and addiction. In: Facing addiction in America: the surgeon general's report on alcohol, drugs, and health. US Department of Health and Human Services, Washington, DC
7. Zhao W (2008) Mechanisms mediating sex differences in the effects of cocaine. PhD thesis, University of Michigan
8. Brady KT et al (1991) Cocaine-induced psychosis. J Clin Psychiatry 52(12):509–512
9. Stankovic M et al (2009) P01-72 Cocaine dependence produce psychiatric symptoms – a case report. Eur Psychiatry 24:S460
10. Roncero C et al (2017) Higher severity of cocaine addiction is associated with tactile and somatic hallucinations. Eur Psychiatry 42:63–69
11. CDC (2024) Stimulant overdose. [cited 2024 Feb 11]. Available from: https://www.cdc.gov/drugoverdose/deaths/stimulant-overdose.html

12. FAS (2022) 10 Most common reasons for addiction relapse: family addiction specialist: addiction counselor. [cited 2022 Jan 21]. Available from: https://www.familyaddictionspecialist.com/blog/10-most-common-reasons-for-addiction-relapse
13. Eiden LE et al (2004) The vesicular amine transporter family (SLC18): amine/proton antiporters required for vesicular accumulation and regulated exocytotic secretion of monoamines and acetylcholine. Pflugers Arch 447(5):636–640
14. Chen R et al (2003) Demonstration of functional coupling between dopamine synthesis and its packaging into synaptic vesicles. J Biomed Sci 10(6):774–781
15. Alcaro A, Huber R, Panksepp J (2007) Behavioral functions of the mesolimbic dopaminergic system: an affective neuroethological perspective. Brain Res Rev 56(2):283–321
16. Blum K et al (2012) Sex, drugs, and rock "N" roll: hypothesizing common mesolimbic activation as a function of reward gene polymorphisms. J Psychoactive Drugs 44(1):38–55
17. Boot N et al (2017) Creative cognition and dopaminergic modulation of fronto-striatal networks: integrative review and research agenda. Neurosci Biobehav Rev 78:13–23
18. Luo SX, Huang EJ (2016) Dopaminergic neurons and brain reward pathways: from neurogenesis to circuit assembly. Am J Pathol 186(3):478–488
19. Gardner EL (2011) Addiction and brain reward and antireward pathways. Adv Psychosom Med 30:22–60
20. Chavez C et al (2010) The effect of estrogen on dopamine and serotonin receptor and transporter levels in the brain: an autoradiography study. Brain Res 1321:51–59
21. Gedeon PC et al (2010) Molecular dynamics of leucine and dopamine transporter proteins in a model cell membrane lipid bilayer. Proteins 78(4):797–811
22. Aggarwal S, Mortensen OV (2017) Overview of monoamine transporters. Curr Protoc Pharmacol 79(1):12.16.1–12.16.17
23. Li Y et al (2024) Dopamine reuptake and inhibitory mechanisms in human dopamine transporter. Nature (London) 632(8025):686–694
24. Penmatsa A, Wang KH, Gouaux E (2015) X-ray structures of Drosophila dopamine transporter in complex with nisoxetine and reboxetine. Nat Struct Mol Biol 22(6):506–508
25. Yang D, Gouaux E (2021) Illumination of serotonin transporter mechanism and role of the allosteric site. Sci Adv 7(49):eabl3857
26. Pidathala S et al (2021) Structural basis of norepinephrine recognition and transport inhibition in neurotransmitter transporters. Nat Commun 12(1):2199–2199
27. Coleman JA et al (2020) Chemical and structural investigation of the paroxetine-human serotonin transporter complex. elife 2020(9):e56427–e56438
28. Coleman JA, Gouaux E (2018) Structural basis for recognition of diverse antidepressants by the human serotonin transporter. Nat Struct Mol Biol 25(2):170–175
29. Coleman JA, Green EM, Gouaux E (2016) X-ray structures and mechanism of the human serotonin transporter. Nature (London) 532(7599):334–339
30. Coleman JA et al (2019) Serotonin transporter–ibogaine complexes illuminate mechanisms of inhibition and transport. Nature 569(7754):141–145
31. Penmatsa A, Wang KH, Gouaux E (2013) X-ray structure of dopamine transporter elucidates antidepressant mechanism. Nature 503(7474):85–90
32. Plenge P et al (2021) The antidepressant drug vilazodone is an allosteric inhibitor of the serotonin transporter. Nat Commun 12(1):5063–5012
33. Nielsen JC et al (2024) Structure of the human dopamine transporter in complex with cocaine. Nature (London) 632(8025):678–685
34. Forrest LR, Rudnick G (2009) The rocking bundle: a mechanism for ion-coupled solute flux by symmetrical transporters. Physiology 24(6):377–386
35. Jardetzky O (1966) Simple allosteric model for membrane pumps. Nature (London) 211(5052):969–970
36. Reyes N, Tavoulari S (2011) To be, or not to be two sites: that is the question about LeuT substrate binding. J Gen Physiol 138(4):467–471

37. Shi L et al (2008) The mechanism of a neurotransmitter: sodium symporter – inward release of Na^+ and substrate is triggered by substrate in a second binding site. Mol Cell 30(6):667
38. Loland CJ (2015) The use of LeuT as a model in elucidating binding sites for substrates and inhibitors in neurotransmitter transporters. Biochim Biophys Acta 1850(3):500–510
39. Yamashita A et al (2005) Crystal structure of a bacterial homologue of Na^+/Cl^--dependent neurotransmitter transporters. Nature 437(7056):215–223
40. Zhou Z et al (2007) LeuT-desipramine structure reveals how antidepressants block neurotransmitter reuptake. Science 317(5843):1390–1393
41. Nguyen H et al (2024) Allosteric modulation of serotonin and dopamine transporters: new insights from computations and experiments. Curr Res Physiol 7:100125
42. Kortagere S et al (2013) Identification of an allosteric modulator of the serotonin transporter with novel mechanism of action. Neuropharmacology 72:282–290
43. Aggarwal S et al (2019) Identification of a novel allosteric modulator of the human dopamine transporter. ACS Chem Neurosci 10(8):3718–3730
44. Coleman JA, Green EM, Gouaux E (2016) X-ray structures and mechanism of the human serotonin transporter. Nature 532(7599):334–339
45. Tan J et al (2024) Molecular basis of human noradrenaline transporter reuptake and inhibition. Nature (London) 632(8026):921–929
46. Srivastava DK et al (2024) Structure of the human dopamine transporter and mechanisms of inhibition. Nature (London) 632(8025):672–677
47. O'Neill B et al (2014) Behavior of knock-in mice with a cocaine-insensitive dopamine transporter after virogenetic restoration of cocaine sensitivity in the striatum. Neuropharmacology 79:626–633
48. Javitch JA et al (2008) The binding sites for cocaine and dopamine in the dopamine transporter overlap. Nat Neurosci 11(7):780–789
49. Aggarwal S et al (2023) Discovery and development of monoamine transporter ligands. Adv Neurobiol 30:101–129
50. Reith MEA et al (2015) Behavioral, biological, and chemical perspectives on atypical agents targeting the dopamine transporter. Drug Alcohol Depend 147:1–19
51. Loland CJ et al (2008) Relationship between conformational changes in the dopamine transporter and cocaine-like subjective effects of uptake inhibitors. Mol Pharmacol 73(3):813–823
52. Ahuja A, Patel P, Abdijadid S (2024) Benztropine. In: StatPearls. StatPearls Publishing, Treasure Island. Available from: https://www.ncbi.nlm.nih.gov/books/NBK560633/
53. Itoh K et al (1991) Intrastriatal dialysis evidence for a direct inhibitory effect of benztropine on dopamine re-uptake. Neurosci Lett 125(1):22–24
54. Tanda G, Newman AH, Katz JL (2009) Discovery of drugs to treat cocaine dependence: behavioral and neurochemical effects of atypical dopamine transport inhibitors. Adv Pharmacol 57:253–289
55. Tomlinson MJ et al (2019) Identification of the benztropine analog [125I]GA II 34 binding site on the human dopamine transporter. Neurochem Int 123:34–45
56. Kopajtic TA et al (2010) Dopamine transporter-dependent and -independent striatal binding of the benztropine analog JHW 007, a cocaine antagonist with low abuse liability. J Pharmacol Exp Ther 335(3):703–714
57. Bisgaard H et al (2011) The binding sites for benztropines and dopamine in the dopamine transporter overlap. Neuropharmacology 60(1):182–190
58. Zanettini C, Wilkinson DS, Katz JL (2018) Behavioral economic analysis of the effects of N-substituted benztropine analogs on cocaine self-administration in rats. Psychopharmacology 235(1):47–58
59. Velázquez-Sánchez C et al (2013) The atypical dopamine transport inhibitor, JHW 007, prevents amphetamine-induced sensitization and synaptic reorganization within the nucleus accumbens. Prog Neuropsychopharmacol Biol Psychiatry 44:73–80

60. Velázquez-Sánchez C et al (2010) The high affinity dopamine uptake inhibitor, JHW 007, blocks cocaine-induced reward, locomotor stimulation and sensitization. Eur Neuropsychopharmacol 20(7):501–508
61. Hiranita T et al (2014) Preclinical efficacy of N-substituted benztropine analogs as antagonists of methamphetamine self-administration in rats. J Pharmacol Exp Ther 348(1):174–191
62. Fengyi L et al (2004) Conformational and electrostatic potential study of the benztropine analogs at the dopamine transporter. J Mol Struct 712(1):207–214
63. Hiranita T et al (2017) σ Receptor effects of N-substituted benztropine analogs: implications for antagonism of cocaine self-administration. J Pharmacol Exp Ther 362(1):2–13
64. Sharif NA et al (1989) Comparative properties of the dopamine transport complex in dog and rodent brain: striatal [(3)H]GBR12935 binding and [(3)H]dopamine uptake. Neurochem Int 15(3):325–332
65. Rothman RB et al (1989) Tight binding dopamine reuptake inhibitors as cocaine antagonists: a strategy for drug development. FEBS Lett 257(2):341–344
66. Rothman RB et al (2008) Dopamine transport inhibitors based on GBR12909 and benztropine as potential medications to treat cocaine addiction. Biochem Pharmacol 75(1):2–16
67. Sun W-L et al (2019) Mutational effects of human dopamine transporter at tyrosine88, lysine92, and histidine547 on basal and HIV-1 Tat-inhibited dopamine transport. Sci Rep 9(1):3843–3843
68. Tella SR (1995) Effects of monoamine reuptake inhibitors on cocaine self-administration in rats. Pharmacol Biochem Behav 51(4):687–692
69. Westerink BHC et al (1987) Dopamine re-uptake inhibitors show inconsistent effects on the in vivo release of dopamine as measured by intracerebral dialysis in the rat. Eur J Pharmacol 135(2):123–128
70. Preti A (2000) Vanoxerine National Institute on Drug Abuse. Curr Opin Investig Drugs 1(2):241
71. Modafinil Monograph for Professionals (2024) [cited 2024 Feb 14]. Available from: https://www.drugs.com/monograph/modafinil.html#:~:text=Produces%20psychoactive%20and%20euphoric%20effects,schedule%20II%20CNS%20stimulants%20(e.g.%2C
72. Blocker DE et al (2008) In-flight fatigue management and modafinil trial in usaf tanker aircrew. Aviat Space Environ Med 79(3):327–327
73. Young JW, Geyer MA (2010) Action of modafinil—increased motivation via the dopamine transporter inhibition and D1 receptors? Biol Psychiatry 67(8):784–787
74. Wilms W et al (2019) Nootropic drugs: methylphenidate, modafinil and piracetam – population use trends, occurrence in the environment, ecotoxicity and removal methods – a review. Chemosphere 233:771–785
75. Dackis CAMD et al (2012) A double-blind, placebo-controlled trial of modafinil for cocaine dependence. J Subst Abus Treat 43(3):303–312
76. Morgan PT et al (2016) Modafinil and sleep architecture in an inpatient–outpatient treatment study of cocaine dependence. Drug Alcohol Depend 160:49–56
77. Kalechstein AD et al (2013) Modafinil, but not escitalopram, improves working memory and sustained attention in long-term, high-dose cocaine users. Neuropharmacology 64:472–478
78. Anderson AL et al (2009) Modafinil for the treatment of cocaine dependence. Drug Alcohol Depend 104(1):133–139
79. Verrico CD et al (2014) Treatment with modafinil and escitalopram, alone and in combination, on cocaine-induced effects: a randomized, double blind, placebo-controlled human laboratory study. Drug Alcohol Depend 141:72–78
80. Foltin RW et al (2016) Modafinil decreases cocaine choice in human cocaine smokers only when the response requirement and the alternative reinforcer magnitude are large. Pharmacol Biochem Behav 150–151:8–13
81. Wuo-Silva R et al (2011) Addictive potential of modafinil and cross-sensitization with cocaine: a pre-clinical study. Addict Biol 16(4):565–579

82. Bernardi RE et al (2009) Modafinil reinstates a cocaine conditioned place preference following extinction in rats. Behav Brain Res 204(1):250–253
83. Vosburg SK et al (2010) Modafinil does not serve as a reinforcer in cocaine abusers. Drug Alcohol Depend 106(2):233–236
84. Loland CJ et al (2012) R-modafinil (armodafinil): a unique dopamine uptake inhibitor and potential medication for psychostimulant abuse. Biol Psychiatry 72(5):405–413
85. Keighron JD et al (2019) Effects of (R)-modafinil and modafinil analogues on dopamine dynamics assessed by voltammetry and microdialysis in the mouse nucleus accumbens shell. ACS Chem Neurosci 10(4):2012–2021
86. Newman AH et al (2019) Translating the atypical dopamine uptake inhibitor hypothesis toward therapeutics for treatment of psychostimulant use disorders. Neuropsychopharmacology 44(8):1435–1444
87. Cao J et al (2016) Novel and high affinity 2-[(diphenylmethyl)sulfinyl]acetamide (Modafinil) analogues as atypical dopamine transporter inhibitors. J Med Chem 59(23):10676–10691
88. Zhang HY et al (2017) The novel modafinil analog, JJC8-016, as a potential cocaine abuse pharmacotherapeutic. Neuropsychopharmacology 42(9):1871–1883
89. Tunstall BJ et al (2018) Atypical dopamine transporter inhibitors attenuate compulsive-like methamphetamine self-administration in rats. Neuropharmacology 131:96–103
90. Allen MI et al (2024) Contrasting the reinforcing effects of the novel dopamine transport inhibitors JJC8-088 and JJC8-091 in monkeys: potential translation to medication assisted treatment. J Pharmacol Exp Ther 392:100033
91. Rahimi O et al (2023) The effects of the dopamine transporter ligands JJC8-088 and JJC8-091 on cocaine versus food choice in rhesus monkeys. J Pharmacol Exp Ther 384(3):372–381
92. Business Wire (2024) Melior Discovery announces spinout of Melior Pharmaceuticals II, LLC. Available from: https://www.businesswire.com/news/home/20160510005485/en/Melior-Discovery-Announces-Spinout-Melior-Pharmaceuticals-II
93. Gainetdinov RR et al (1997) Effects of a psychostimulant drug sydnocarb on rat brain dopaminergic transmission in vivo. Eur J Pharmacol 340(1):53–58
94. Gruner JA et al (2011) Characterization of pharmacological and wake-promoting properties of the dopaminergic stimulant sydnocarb in rats. J Pharmacol Exp Ther 337(2):380–390
95. Aggarwal S et al (2021) Functional characterization of the dopaminergic psychostimulant sydnocarb as an allosteric modulator of the human dopamine transporter. Biomedicines 9(6):634
96. Pedersen CN et al (2024) Cryo-EM structure of the dopamine transporter with a novel atypical non-competitive inhibitor bound to the orthosteric site. J Neurochem 168:2043
97. Refai O et al (2022) Allosteric modulator KM822 attenuates behavioral actions of amphetamine in Caenorhabditis elegans through interactions with the dopamine transporter DAT-1. Mol Pharmacol 101(3):123–131
98. Sulzer D et al (2005) Mechanisms of neurotransmitter release by amphetamines: a review. Prog Neurobiol 75(6):406–433
99. Castells X, Blanco-Silvente L, Cunill R (2018) Amphetamines for attention deficit hyperactivity disorder (ADHD) in adults. Cochrane Database Syst Rev 2018(8):CD007813
100. Namdari B, Kollins SH (2015) Stimulants: abuse and performance enhancement (or lack thereof). Focus 13(3):283–289
101. Avois L et al (2006) Central nervous system stimulants and sport practice. Br J Sports Med 40(suppl 1):i16–i20
102. Fischman MW et al (1976) Cardiovascular and subjective effects of intravenous cocaine administration in humans. Arch Gen Psychiatry 33(8):983–989
103. Scheckel CL, Boff E (1964) Behavioral effects of interacting imipramine and other drugs with d-amphetamine, cocaine, and tetrabenazine. Psychopharmacologia 5(3):198–208
104. Wenger GR, Wright DW (1990) Behavioral effects of cocaine and its interaction with d-amphetamine and morphine in rats. Pharmacol Biochem Behav 35(3):595–600

105. Rusyniak DEMD (2013) Neurologic manifestations of chronic methamphetamine abuse. Psychiatr Clin N Am 36(2):261–275
106. Fu Y-P, Wang Y (2023) Amphetamines abuse and depression: focus on TRPC channels. Exp Neurol 364:114391–114391
107. Alamri SH (2024) Young-onset dementia following chronic abuse of amphetamine-type stimulants. Pak J Med Sci 40(4):779–781
108. Tzeng NS et al (2020) Association between amphetamine-related disorders and dementia-a nationwide cohort study in Taiwan. Ann Clin Transl Neurol 7(8):1284–1295
109. Mullen JM, Richards JR, Crawford AT (2024) Amphetamine-related psychiatric disorders. In: StatPearls. StatPearls Publishing, Treasure Island
110. Freudenmann RW, Spitzer M (2004) The neuropsychopharmacology and toxicology of 3,4-methylenedioxy-N-ethyl-amphetamine (MDEA). CNS Drug Rev 10(2):89–116
111. Barrionuevo M et al (2000) Serotonergic deficits and impaired passive-avoidance learning in rats by MDEA: a comparison with MDMA. Pharmacol Biochem Behav 65(2):233–240
112. Methylphenidate Monograph for Professionals (2024) [cited 2024 Feb 14]. Available from: https://www.drugs.com/monograph/methylphenidate.html
113. Dar DE, Mayo C, Uhl GR (2005) The interaction of methylphenidate and benztropine with the dopamine transporter is different than other substrates and ligands. Biochem Pharmacol 70(3):461–469
114. Izenwasser S et al (1999) Chronic methylphenidate alters locomotor activity and dopamine transporters differently from cocaine. Eur J Pharmacol 373(2):187–193
115. Torres-Reveron A, Dow-Edwards DL (2005) Repeated administration of methylphenidate in young, adolescent, and mature rats affects the response to cocaine later in adulthood. Psychopharmacologia 181(1):38–47
116. Argento JK, Arvanitogiannis A, Flores C (2012) Juvenile exposure to methylphenidate reduces cocaine reward and alters netrin-1 receptor expression in adulthood. Behav Brain Res 229(1):202–207
117. Levin FR et al (2007) Treatment of cocaine dependent treatment seekers with adult ADHD: double-blind comparison of methylphenidate and placebo. Drug Alcohol Depend 87(1):20–29
118. Goldstein RZ et al (2010) Oral methylphenidate normalizes cingulate activity in cocaine addiction during a salient cognitive task. Proc Natl Acad Sci 107(38):16667–16672
119. Wang GJ et al (2013) Long-term stimulant treatment affects brain dopamine transporter level in patients with attention deficit hyperactive disorder. PLoS One 8(5):e63023
120. van Ruitenbeek P et al (2023) Methylphenidate as a treatment option for substance use disorder: a transdiagnostic perspective. Front Psych 14:1208120
121. Delavenne HMD et al (2013) Psychosis in a cocaine-dependent patient with ADHD during treatment with methylphenidate. Gen Hosp Psychiatry 35(4):451.e7–451.e9
122. Dürsteler KM et al (2015) Clinical potential of methylphenidate in the treatment of cocaine addiction: a review of the current evidence. Subst Abus Rehabil 6:61–74
123. Richer EJ (2009) Chapter 19 – Ibogaine and the treatment of opiate addiction. In: Complementary and alternative therapies and the aging population. Elsevier Inc, pp 393–401
124. Davis AK, Barsuglia JP, Windham-Herman AM, Lynch M, Polanco M (2017) Subjective effectiveness of ibogaine treatment for problematic opioid consumption: short- and long-term outcomes and current psychological functioning. J Psychedelic Stud 1(2):65–73
125. Calvey T (2018) Ibogaine as a treatment for substance misuse: potential benefits and practical dangers. Prog Brain Res 242:217–257
126. Glick SD, Maisonneuve IM, Szumlinski KK (2001) Chapter 2 – Mechanisms of action of ibogaine: relevance to putative therapeutic effects and development of a safer iboga alkaloid congener. In: The alkaloids: chemistry and biology. Elsevier Science & Technology, pp 39–53
127. Maisonneuve IM, Keller RW, Glick SD (1991) Interactions between ibogaine, a potential anti-addictive agent, and morphine: an in vivo microdialysis study. Eur J Pharmacol 199(1):35–42

128. Sandtner W et al (2011) Effects by ibogaine on the dopamine transporter. Biophys J 100(3):135a
129. Jacobs MT et al (2007) Ibogaine, a noncompetitive inhibitor of serotonin transport, acts by stabilizing the cytoplasm-facing state of the transporter. J Biol Chem 282(40):29441–29447
130. Maisonneuve IM, Glick SD (1992) Interactions between ibogaine and cocaine in rats: in vivo microdialysis and motor behavior. Eur J Pharmacol 212(2):263–266
131. Glick SD et al (1999) (±)-18-Methoxycoronaridine: a novel iboga alkaloid congener having potential anti-addictive efficacy. CNS Drug Rev 5(1):27–42
132. Wei D et al (1998) Acute iboga alkaloid effects on extracellular serotonin (5-HT) levels in nucleus accumbens and striatum in rats. Brain Res 800(2):260–268
133. Singh I et al (2023) Structure-based discovery of conformationally selective inhibitors of the serotonin transporter. Cell 186(10):2160–2175.e17
134. Cappendijk SLT, Dzoljic MR (1993) Inhibitory effects of ibogaine on cocaine self-administration in rats. Eur J Pharmacol 241(2):261–265
135. Szumlinski KK, Maisonneuve IM, Glick SD (2000) Differential effects of ibogaine on behavioural and dopamine sensitization to cocaine. Eur J Pharmacol 398(2):259–262

Monoamine-Uptake Inhibitors and Pain: Antidepressant Drugs Among First-Line Treatments Against Neuropathic Pain and Fibromyalgia

Michel Barrot, Henrico-Pio Basile, and Mélanie Kremer

Contents

Abstract Some monoamine-uptake inhibitors are among first-line treatments for specific chronic pain conditions. It concerns tricyclic antidepressant drugs, such as amitriptyline or nortriptyline, and the more selective serotonergic and noradrenergic reuptake inhibitors, such as duloxetine. They are recommended for treating neuropathic pain, which is pain caused by a lesion or disease of the somatosensory nervous system, and fibromyalgia, which is a chronic widespread pain. Clinically, their action on pain was proposed to be independent from the one on depression. Research in animal models provided some understanding of the pain-relieving mechanism. The noradrenergic component of monoamine-uptake inhibitors appears to be critical, with the therapeutic effect likely involving targets at peripheral, spinal and supraspinal levels. At least two independent mechanisms would contribute to pain relief. One is spinal, relying on the recruitment of aminergic descending controls of pain, with a downstream key role of the α_2 adrenergic receptors; the other may require more sustained treatment and relies on the noradrenergic recruitment of β_2 adrenergic receptors and a downstream anti-neuroimmune action. Both mechanisms require a functional endogenous opioid system. These insights, however, focused on the sensory component of pain, and the contribution of the supraspinal action of antidepressant drugs needs to be explored in detail.

M. Barrot (✉) · H.-P. Basile · M. Kremer (✉)
Institut des Neurosciences Cellulaires et Intégratives, Centre National de la Recherche Scientifique, Université de Strasbourg, Strasbourg, France
e-mail: mbarrot@unistra.fr; melaniekremer@unistra.fr

O. V. Mortensen (ed.), *Monoamine Transporters*, Advances in Neurobiology 46,
https://doi.org/10.1007/978-3-031-96364-3_12

Keywords Neuropathic pain · Antidepressants · Monoaminergic system · Opioidergic system · Neuroimmune

Abbreviations

CGRP	Calcitonin-gene-related peptide
CREB	cAMP response element–binding protein
DSP-4	N-(2-chloroethyl)-N-ethyl-2-bromobenzylamine
FRET	Fluorescence resonance energy transfer
IASP	International Association for the Study of Pain
ICAM-1	Intercellular adhesion molecule 1
IL	Interleukin
MCP-1	Monocyte chemoattractant protein 1
MIP-2	Macrophage inflammatory protein 2
NET	Noradrenaline transporter
NeuPSIG	IASP Neuropathic Pain Special Interest Group
NNH	Number needed to harm
NNT	Number needed to treat
PARP	ADP-ribose polymerase cleavage
RGS	Regulators of G protein signaling
SERT	Serotonin transporter
SSNRI	Selective serotonin- and noradrenaline-reuptake inhibitors
TLR	Toll-like receptor
TNF-α	Tumor necrosis factor α
TRPV1	Transient receptor potential vanilloid 1

Chronic pain, which deeply affects the quality of personal and professional life, is now considered as an illness per se. It can be defined as "nociceptive," that is, arising from actual or threatened damage to nonneural tissue and due to the activation of nociceptors, as in inflammatory pain; as "neuropathic," that is, caused by a lesion or disease of the somatosensory nervous system [118]; or as nociplastic, that is, arising from altered nociception despite no clear evidence of actual or threatened tissue damage or of a disease or lesion of the somatosensory system, as in fibromyalgia. Clinically, some monoamine-uptake inhibitors are presently first-line treatments for both neuropathic pain and fibromyalgia.

Neuropathic pain can be associated with pain, abnormal sensations, and neurological deficits in a given nervous territory [4, 18, 24]. It develops following alterations of the somatosensory pathways. A lesion or a disease affecting any of the structures and/or circuits allowing the perception of the sensory information coming from the skin and the musculoskeletal apparatus may thus lead to neuropathic pain [118]. Neuropathic pain can, for example, be a consequence of diabetic polyneuropathy, of a compression, lesion or section of a nerve, of a central lesion that affects the spinal cord or a given brain region, of viral infections impacting the nervous

Table 1 IASP definition and assessment of sensory symptoms or signs in neuropathic pain

Terms	Definitions
Pain	An unpleasant sensory and emotional experience associated with, or resembling that associated with, actual or potential tissue damage
Allodynia	Pain due to a stimulus that does not normally provoke pain
Analgesia	Absence of pain in response to stimulation which would normally be painful
Dysesthesia	An unpleasant abnormal sensation, whether spontaneous or evoked
Hyperalgesia	Increased pain from a stimulus that normally provokes pain
Hypoalgesia	Diminished pain in response to a normally painful stimulus
Neuropathic pain	Pain caused by a lesion or disease of the somatosensory nervous system
Nociception	The neural process of encoding noxious stimuli
Nociceptive pain	Pain that arises from actual or threatened damage to non-neural tissue and is due to the activation of nociceptors
Nociceptive stimulus	An actually or potentially tissue-damaging event transduced and encoded by nociceptors
Nociceptor	A high-threshold sensory receptor of the peripheral somatosensory nervous system that is capable of transducing and encoding noxious stimuli
Pain threshold	The minimum intensity of a stimulus that is perceived as painful
Paraesthesia	An abnormal sensation; this may be spontaneous or evoked
Sensitization	Increased responsiveness of nociceptive neurons to their normal input, and/or recruitment of a response to normally subthreshold inputs

system, or be of iatrogenic origin as with chemotherapies. The etiology and the topography of neuropathic pain can thus be diverse, and the diagnosis is mostly based on the combination of expressed symptoms, signs, and etiology. Spontaneous paroxysmal pain, similar to electrical shocks or stabbing, permanent spontaneous pain, felt as compression, burning, or coldness or as needle sensation or itching, evoked pain such as allodynia, hyperalgesia, or hyperpathia, abnormal sensations such as paresthesia and dysesthesia, and sensory deficits in a given nervous territory, can be present in neuropathic pain (see Table 1).

Fibromyalgia is characterized by chronic widespread pain, with no identified cause [122]. With a worldwide prevalence of 2–4%, it is more common in females than in males [142]. Its symptoms also include diffuse hyperalgesia and/or allodynia, coupled with a number of other comorbidities such as fatigue, memory difficulties, sleep, and mood disorders [122].

1 Monoamine-Uptake Inhibitors as First-Line Treatments in Given Pain Conditions

The pharmacological treatment of either neuropathic pain or fibromyalgia is challenging. Indeed, classical analgesics have a limited action on these types of pain, and while tramadol and opioids such as morphine or oxycodone can show some benefit, they are not recommended as first-line treatments. The evidence-based guidelines from the Neuropathic Pain Special Interest Group (NeuPSIG) of the International Association for the Study of Pain (IASP) and from various national and international medical and scientific societies converge [5, 36, 41, 96] to recommend as first-line drugs against neuropathic pain either monoamine-reuptake inhibitors, more specifically tricyclic antidepressants and selective serotonin-noradrenaline-reuptake inhibitors (SSNRIs), or gabapentinoids. Similar recommendations are present for fibromyalgia [117, 131]. Contrary to tricyclic antidepressants and SSNRIs, which are dual serotonergic-noradrenergic-reuptake inhibitors, the selective serotonin-reuptake inhibitors are clinically not or poorly effective against either neuropathic pain or fibromyalgia. This chapter will more specifically address present mechanistic knowledge concerning tricyclic antidepressants and SSNRIs in the context of pain.

1.1 Tricyclic Antidepressants

The first report of neuropathic pain relief by an antidepressant drug was done in 1960, in depressed painful patients treated with imipramine for their depression [105]. Twenty years later, clinical trials confirmed that amitriptyline can relieve postherpetic neuropathic pain [138] and painful diabetic polyneuropathy [86]. This action on neuropathic pain was shown to be independent of the neuropathic pain etiology [115], from the presence or absence of depression [91, 106] and from the action of these drugs on depression itself when copresent in pain patients [139]. While the first historical report concerned intravenous treatments, oral administration of these drugs is now preferred. Concerning tricyclic antidepressants, amitriptyline has been the most studied and prescribed clinically, at doses ranging from 25 to 150 mg/day. Its efficacy has been demonstrated in various neuropathic pain conditions, including diabetic painful polyneuropathy, postherpetic neuralgia, and central pain [115], with relief of different components of neuropathic pain, such as constant pain, shooting pain, and mechanical allodynia [139]. Such treatment is however not effective in all patients, as reflected by the NNT (number-needed-to-treat), which is the number of patients needed to treat to obtain at least moderate pain relief in one patient. The NNT is 3.6 (95% CI 3–4.5) for tricyclic antidepressants, with no clear evidence of a dose-response effect [115]. On the other hand, the NNH (number-needed-to-harm) for tricyclic antidepressants is 6 (95% CI 4.2–10.7). It is calculated based on adverse effects such as drowsiness, dizziness, dry mouth, constipation, nausea, urinary retention, sweating, headache, blurred vision, palpitations, irritability, and ataxia, which can result in interruption of treatment in around 20% of the patients [115].

1.2 Local Administration of Tricyclic Antidepressants

At high concentration, tricyclic antidepressants can also target ion channels [9], which are secondary targets shared with various other aminergic reuptake inhibitors. It is responsible for potential arrhythmia properties [98], contributing to cardiotoxic adverse effects with high doses. The action of tricyclic antidepressants on ion channels can provide anesthetic-like action when the drug is directly delivered to the sciatic nerve [45] or intrathecally [46] in rodents. The therapeutic interest is however limited by potential neurotoxicity at such high local concentrations, as illustrated by dose-dependent peripheral axonal damage and Wallerian degeneration on rat sciatic nerve [38, 42]. While local amitriptyline produces cutaneous analgesia in healthy subjects, it also induces local redness, itching, and burning sensations [47, 53], and does not appear to be more effective than placebo in neuropathic pain patients [53, 79]. Topical and intrathecal tricyclic antidepressants thus seem to have limited clinical interest.

1.3 SSNRIs

The SSNRIs duloxetine and venlafaxine have been demonstrated to relieve neuropathic pain [5, 10, 40]. The clinical evidence was first provided in diabetic painful polyneuropathy, leading to specific label for this condition by the Food and Drug Administration and by the European Medicines Agency, but the actual prescription usage is larger, both drugs being considered among main treatments for neuropathic pain conditions. These SSNRIs have an NNT around 6.4 (95% CI 5.2–8.4), but an NNH around 11.8 (95% CI 9.5–15.2) [41], with reported adverse effects such as nausea, dry mouth, headache, decreased libido, dizziness, somnolence or insomnia, reduced appetite, and hypertension. While SSNRIs are overall safer than tricyclic antidepressants, interruption of treatment is still observed in 15–20% of the patients [43, 48].

2 Insights into the Mechanism of Aminergic Uptake Inhibitors in a Neuropathic Pain Context

Preclinical studies provided mechanistic insights concerning monoamine-uptake inhibitors in a pain context (see Figs. 1, 2, and 3). These data were mostly obtained using rodent models of neuropathic pain.

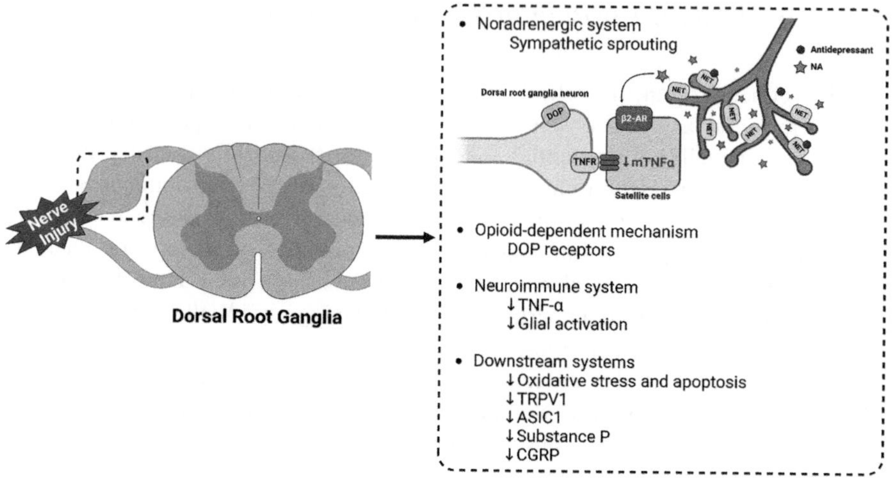

Fig. 1 Peripheral action of monoamine-uptake inhibitors on neuropathic pain. Tricyclic antidepressants and SSRNIs, first-line treatments of neuropathic pain, have a peripheral component that is relevant to their pain-relieving action. This peripheral action is preferentially seen after sustained treatment, and relies on noradrenaline recruitment of nonneuronal β_2 adrenergic receptors with a downstream antineuroimmune effect. This mechanism requires a functional opioid system. Abbreviations: β2-AR, β_2 adrenergic receptors; ASIC1, acid-sensing ion channel 1; CGRP, calcitonin gene–related peptide; DOP, delta opioid receptor; NA, noradrenaline; NET, noradrenaline transporter; TNF-α, tumor necrosis factor α; TNFR, TNF-α receptor; TRPV1, transient receptor potential vanilloid 1. (Created with BioRender.com)

2.1 *Long-Term Effects of Antidepressant Drugs in Rodents*

As observed in depression itself, the clinical effect of antidepressant drugs on neuropathic pain usually requires a prolonged treatment, which indicates that cellular plasticity may underlie the therapeutic action. A notable part of the knowledge concerning the pain-relieving mechanism of antidepressants was however obtained using acute pain procedures in animals [23, 30, 73, 91, 102, 127], which provided important information but may not fully model the clinical treatments. This chapter will thus report some of the mechanistic insights obtained with acute administration, but will mostly focus on studies based on sustained treatments in animal models of neuropathic pain.

To address antidepressant drugs' action on neuropathic pain, a large variety of animal models have been used, relying on nerve lesion, diabetes, or chemotherapies. Chronic nerve compression is, for example, modeled through a tight ligation of part of the sciatic nerve (partial sciatic nerve ligation model) [120], a tight ligation of L5 and L6 spinal nerves (spinal nerve ligation model) [66], loose ligatures of the sciatic nerve (chronic constriction injury model) [17], or insertion of a polyethylene tubing around the sciatic nerve (cuff model) [16, 146]. Axotomy of two of the three branches of the sciatic nerve (spared nerve injury model) [31] has also been used, as well as allograft of nucleus pulposus in the epidural space [64]. Obese

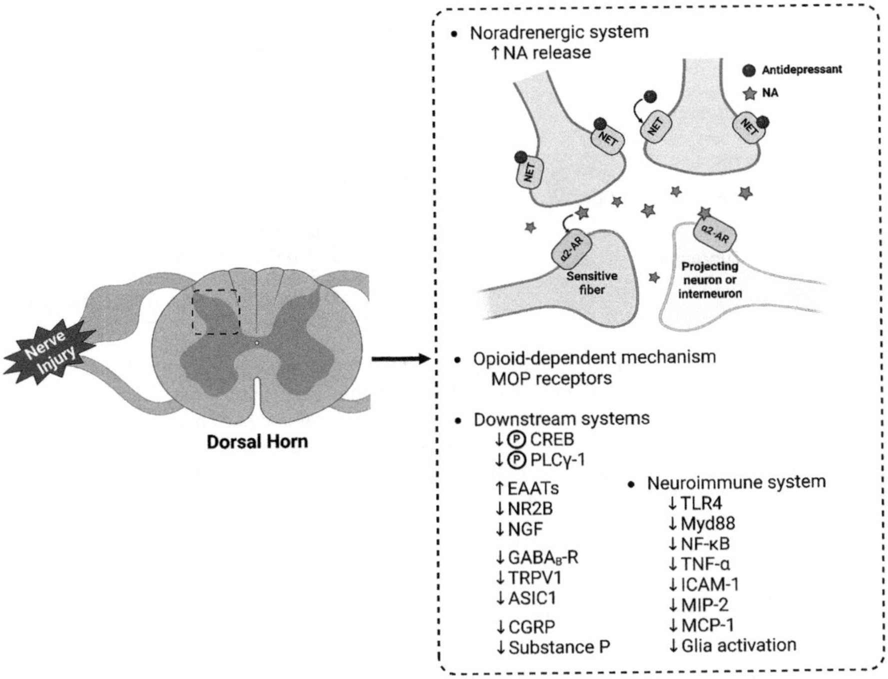

Fig. 2 Spinal action of monoamine-uptake inhibitors on neuropathic pain. Tricyclic antidepressants and SSRNIs, first-line treatments of neuropathic pain, have a spinal component that is relevant to their pain-relieving action. This mechanism requires a functional opioid system. An antineuroimmune action and other downstream modifications have also been reported in the spinal cord. Abbreviations: α2-AR, α_2 adrenergic receptors; ASIC1, acid-sensing ion channel 1; CGRP, calcitonin gene–related peptide; CREB, cAMP response element–binding protein; EAAT, excitatory amino-acid transporters; ERK1/2, extracellular signal-regulated kinase 1/2; $GABA_B$-R, gamma-aminobutyric acid B receptor; ICAM-1, intercellular adhesion molecule 1; MOP, mu opioid receptor; NA, noradrenaline; NET, noradrenaline transporter; NFκB, nuclear factor kappa-light-chain-enhancer of activated B cells; NR2B, N-methyl D-aspartate receptor subtype 2B; MCP-1, monocyte chemoattractant protein-1; MIP-2, macrophage inflammatory protein-2; Myd88, myeloid differentiation primary response 88; PLCγ-1, phosphoinositide phospholipase c gamma 1; TLR4, Toll-like receptor 4; TNF-α, tumor necrosis factor α; TRPV1, transient receptor potential vanilloid 1. (Created with BioRender.com)

leptin-deficient mice (*ob/ob*) as genetic model of type 2 diabetes [1, 28], streptozotocin to chemically induce diabetes [74], and oxaliplatin [25] to model chemotherapy-induced neuropathic pain have also been used to study the action of antidepressant drugs.

In neuropathic pain models, long-term treatment with various tricyclic antidepressants, such as amitriptyline, nortriptyline, imipramine, or desipramine, or with various SSNRIs, such as duloxetine, venlafaxine, or milnacipran, leads to an antiallodynic or antihyperalgesic effect [69]. This effect can be observed for doses that are ineffective acutely [14, 63, 70, 71, 73, 101, 113]. While some mechanistic aspects may be shared between the acute and long-term treatments, treatment

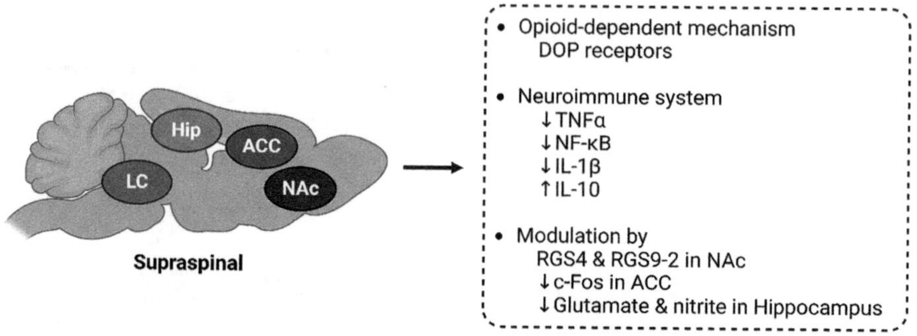

Fig. 3 Supraspinal action of monoamine-uptake inhibitors on neuropathic pain. Tricyclic antidepressants and SSRNIs, first-line treatments of neuropathic pain, have a supraspinal component whose action in pain relief has been less studied than the spinal and peripheral ones. Involvement of the opioid system, involvement of an antineuroimmune action, and a decrease in excitatory transmission have all been reported. Abbreviations: ACC, anterior cingulate cortex; DOP, delta opioid receptor; Hip, hippocampus; IL, interleukin; LC, locus coeruleus; NAc, nucleus accumbens; NFκB, nuclear factor kappa-light-chain-enhancer of activated B cells; RGS, regulators of G protein signaling; TNF-α, tumor necrosis factor α. (Created with BioRender.com)

duration is critical to consider, particularly regarding plasticity-related changes that may underlie the therapeutic action. Such molecular and neural plasticities have been well studied in the field of depression-related research, with evidence for chromatin regulation and changes in gene expression, in neurogenesis, and in dendritic and spine structure [35, 111, 134]. Regarding the pain context, the onset of the therapeutic action of antidepressant drugs on neuropathic pain has been proposed to be faster than on depression itself, but it still requires several days of treatment. Based on the therapeutic delay, it could thus be hypothesized that treatment-induced molecular and neural changes may take place and interfere with the molecular and neural changes responsible for neuropathic pain [29, 97, 141].

2.2 *Monoaminergic Component: Noradrenaline Versus Serotonin*

Monoamine uptake is the primary target of tricyclic antidepressant drugs and of SSNRIs. While the inhibitory action of antidepressant drugs on these uptakes is considered as underlying their effect on both depression and specific pain conditions, the various antidepressant drugs do not have the same efficacy on the serotonergic and the noradrenergic systems and both systems do not have the same importance in the therapeutic action of these drugs.

Among antidepressant drugs, tricyclic ones display the best therapeutic efficacy [5, 41, 96], even though their adverse effects are limiting their use. The tricyclic antidepressant amitriptyline, which is prescribed both in the psychiatry field and in the pain field, is sevenfold more potent on the serotonin transporter (SERT) than on the noradrenaline one (NET). However, amitriptyline can also be considered as a prodrug as it is metabolized by demethylation into nortriptyline that is 40-fold more potent on NET than SERT and this active metabolite may play a critical role against neuropathic pain. Similar to tricyclic antidepressant drugs, SSNRIs are also effective both clinically and in animal models of neuropathic pain [41, 49, 70, 81, 83, 140, 143]. Preclinically and using the SSNRI venlafaxine, it has been proposed that antidepressant drug action in a neuropathic pain model depends on both the serotonergic and the noradrenergic systems [81]. However, other reports highlighted a more critical role of the noradrenergic system in the antiallodynic or antihyperalgesic action of sustained treatment with either tricyclic antidepressants or SSNRIs [22, 70, 127, 143, 144]. It can thus be hypothesized that inhibition of noradrenaline uptake could be a necessary component of antidepressant drug action on neuropathic pain, and that the recruitment of the serotonergic system modulates this effect.

The critical role of noradrenaline is also supported by some data based on selective noradrenaline-reuptake inhibitors such as reboxetine [56, 68, 144], even though the adverse effects of these selective uptake inhibitors limit their clinical use. Conversely, the selective serotonin-reuptake inhibitors may have more limited adverse effects and be widely used to treat depression, but they are not or poorly effective against neuropathic pain [87, 103, 114, 121]. While clinical data do not support the use of selective serotonin-reuptake inhibitors against neuropathic pain [3, 5, 33, 115], preclinical findings are inconclusive, with some reporting either the presence [8, 63, 85] or lack [14, 59] of effect in neuropathic pain models. This limited action may partly be a consequence of the neuropathic pain pathology itself. Indeed, postsynaptic PDZ-domain-containing proteins sequester spinal 5-HT_{2A} receptors to the membrane in the context of neuropathic pain, and if spinal 5-HT_{2A} receptors are experimentally disconnected from their associated PDZ proteins, the selective serotonin reuptake inhibitors can become more effective [107].

Altogether, data suggest that a chronic inhibition of serotonin uptake may not be sufficient per se to relieve neuropathic pain symptoms. On the other hand, the noradrenergic component seems to play an important role (see follow-up sections). Before detailing this noradrenergic component, it should be mentioned that some antidepressant drugs may also affect, directly or indirectly, the dopaminergic systems. For example, thanks to the use of in vivo microdialysis, an increase in extracellular dopamine levels has been observed in the dorsal horn of the spinal cord following systemic injections of antidepressant drugs, and spinal blockade of D_2 receptors decreases the acute action of antidepressant drugs at high dose in a neuropathic pain model [27].

2.3 Monoaminergic Component: Adrenergic Receptors

Noradrenaline acts via three families of adrenergic receptors, namely, α_1, α_2, and β, that are G-protein-coupled receptors, either to Gq (α_1 receptors), Gi (α_2 receptors), or Gs (β receptors) [93]. The α_2 adrenergic receptors contribute to the action of antidepressant drugs, particularly at high doses and/or acutely. Indeed, the analgesic effect of a high dose of amitriptyline is impaired in mice without α_{2A} adrenergic receptors [104]; and antagonists of α_2 adrenergic receptors, such as yohimbine or idazoxan, can decrease or block the effect of high doses of the antidepressants amitriptyline, duloxetine, venlafaxine, or milnacipran in models of neuropathic pain [49, 55, 58, 70, 94]. It has more specifically been shown that duloxetine at a high dose acts transiently on neuropathic pain thanks to the secondary recruitment of spinal α_{2A} adrenergic receptors [70]. Some recent studies also suggested that α_1 adrenergic receptors, and in particular the α_{1B} subtype, can contribute to the therapeutic effect of duloxetine on neuropathic pain [55, 57]. On the other hand, the effect of a long-term treatment with tricyclic antidepressant drugs or with SSNRIs has been shown to more specifically require β_2 adrenergic receptors. Indeed, antagonists or knock-out of this receptor block the antiallodynic effect of chronic treatments with the antidepressants amitriptyline, nortriptyline, desipramine, duloxetine, or venlafaxine [22, 28, 70, 143, 144].

2.4 Monoaminergic Component: Neuroanatomical Site of Action

Aminergic uptake inhibitors can directly act within supraspinal structures [37, 77], but also in the spinal cord on terminals from the descending noradrenergic inhibitory controls of pain [77, 147], and even in the periphery. Indeed, nerve lesion leads to a sprouting of sympathetic noradrenergic fibers in the dorsal root ganglia [89, 110], providing a local target for noradrenergic uptake inhibitors (see Fig. 1).

Studies addressing the acute site of action of antidepressant drugs have highlighted the role of the spinal cord [7, 54, 70, 84, 149], through local increase in noradrenaline and serotonin levels after inhibition of NET and SERT. These findings strongly support the hypothesis that antidepressant drugs can present an analgesic effect through the recruitment of noradrenergic descending pathways [100] and the secondary action of noradrenaline on α_{2A} adrenergic receptors [70] (see Fig. 2).

Interestingly, the effect of long-term treatment with antidepressant drugs in animal models of peripheral neuropathic pain can be blocked by a lumbar intrathecal injection of the neurotoxin 6-hydroxydopamine [2], a lumbar intrathecal injection of an antagonist of β_2 adrenergic receptors [144], or an intraperitoneal administration of N-(2-chloroethyl)-N-ethyl-2-bromobenzylamine (DSP-4) [52] that deplete noradrenaline in the spinal cord. These observations have been used to support the potential role of the spinal cord. However, technically, lumbar intrathecal injections in animals can also affect the nearby dorsal root ganglia, which is also the case of peripheral injections. Thus, they do not always allow differentiating spinal from peripheral mechanisms.

Studies from the past decade have highlighted that a peripheral target can also contribute to antidepressant drugs' action on neuropathic pain. Indeed, a comparison of different noradrenergic lesions [22, 70] showed that noradrenergic fibers in the dorsal root ganglia are critical to the effect of sustained nortriptyline or duloxetine treatments on allodynia in a mouse neuropathic pain model.

While both aminergic descending pathways and peripheral mechanisms have been the most studied aspects of antidepressant drug action on neuropathic pain, some experimental evidence also suggests a supraspinal contribution (see Fig. 3). Indeed, after L5 spinal nerve ligation in rats and the experimental suppression of noradrenergic descending inhibitory systems, the antihyperalgesic action of repeated amitriptyline treatment remains partly present [52]. Moreover, with chronic desipramine treatment in mice with spared nerve injury, the onset of action is faster in *Rgs9-2*-deficient mice, which can be blunted by local overexpression of RGS9-2 in the nucleus accumbens [95]. The RGS proteins are regulators of G protein signaling [95], and RGS4 also influences desipramine action on neuropathic allodynia, as evidenced by moderate doses of desipramine being less effective in *Rgs4*-deficient mice [125].

2.5 Contribution of the Opioid System

Various studies have shown that the opioid system, a well-known modulator of pain [34, 44], contributes to the effect of antidepressant drugs on neuropathic pain. For example, naloxone, an opioid receptor antagonist, can block the action of antidepressant drugs in neuropathic pain models [14, 32, 82, 132, 140]. However, data concerning the implicated types of opioid receptors and their TextObjectanatomical sites of action are more diverse. Thus, clomipramine's action on allodynia has been proposed to require the recruitment of spinal mu opioid receptors and supraspinal delta opioid receptors [82] (see Figs. 2 and 3). On the other hand, using diverse opioid receptor antagonists and transgenic mice deficient for opioid receptors, the peripheral delta opioid receptors expressed in Nav1.8+ neurons have been shown to be critical to the antiallodynic action of long-term duloxetine and tricyclic antidepressant treatments [15, 26, 28, 70]. While the antiallodynic action of chronic amitriptyline, nortriptyline, or duloxetine treatments is lost in delta opioid-receptor-deficient mice, this action is still present in mu- or kappa-opioid-receptor-deficient mice [21, 70, 90], thus highlighting the specific influence of delta opioid receptors. On the other hand, both central mu and delta opioid receptors were shown to contribute to the acute action of duloxetine [70]. While an intact opioid system is required for the action of antidepressant drugs on neuropathic pain, the precise mechanism linking the aminergic action of these drugs and the opioid system has not been detailed yet.

It could be proposed that antidepressant drugs favor the production of opioid peptides. Indeed, chronic amitriptyline treatment can increase spinal and supraspinal enkephalin levels in nonpainful animals [50]. This is in agreement with a genomic analysis in naïve mice, showing that amitriptyline increases proenkephalin and prodynorphin mRNA levels in the nucleus accumbens [20]. It has also been shown that noradrenaline can favor the production of β-endorphin by nonneuronal immune cells, through a mechanism mediated by β adrenoceptors [19]. Besides a potential action on opioid peptide production, sustained amitriptyline treatment may also increase the spinal density of mu and delta opioid receptor–binding sites [50]. These mechanistic hypotheses have however not yet been directly linked to the effect of antidepressant drugs on neuropathic pain.

Concerning these mechanistic aspects, it is important to determine whether the adrenergic and opioid receptors that contribute to the action of antidepressant drugs are coexpressed in the same cells or not. Indeed, a coexpression would, for example, support potential direct molecular interactions. The experimental coexpression of delta or kappa opioid receptors and β_2 adrenoceptors in cell culture can indeed lead to their heteromerization [61]. Experimentally coexpressed mu opioid receptors and α_{2A} adrenoceptors can also interact physically, with functional cross-talk concerning intracellular signaling [62, 135]. However, if both types of receptors are on different cells, it would be more supportive of a sequential mechanism, with, for example, synthesis of opioid peptides or regulation of opioid receptors.

2.6 An Anti-neuroimmune Action of Antidepressant Drugs

The pathological mechanism leading to neuropathic pain implies neurogenic inflammation [6], with immune and neuroimmune components such as the proinflammatory cytokines tumor necrosis factor α (TNF-α), interleukin (IL)-6, and IL-1β playing significant roles [75, 92, 123, 148]. After spinal nerve ligation, sustained treatment with mirtazapine, a tetracyclic antidepressant drug that acts as an antagonist of α_2 adrenergic receptors and of some serotonergic receptors, attenuates the increase in brain TNF-α and IL-1β levels [151]. A similar influence of more classical antidepressant drugs acting as aminergic uptake inhibitors has also been observed. Acute amitriptyline can, for example, decrease the spinal mRNA expression of TNF-α, intercellular adhesion molecule 1 (ICAM-1), macrophage inflammatory protein 2 (MIP-2), and monocyte chemoattractant protein 1(MCP-1) [67] (see Fig. 2). These changes, observed in rats with L5 and L6 spinal nerve ligation, depend on the A3 adenosine receptor. Amitriptyline also favors the production of IL-10, an anti-inflammatory cytokine, and decreases the hyperactivity of the transcription factor NF-κB [151], known to control TNF-α expression. Similarly, amitriptyline in rodents with sciatic nerve constriction decreases TNF-α immunoreactivity, not only in the injured peripheral nerves, but also in the hippocampus [126] (see Figs. 1 and 3), and α_2 adrenoceptors have been proposed to contribute to this link with immune factors. In a rat model of neuropathic pain induced by autologous nucleus pulposus applied on the left L5 dorsal root ganglion, duloxetine downregulated TNF-α in the dorsal root ganglion, nerve growth factor in the spinal cord, and microglia activation [51] (see Figs. 1 and 2). In mice with sciatic nerve constriction, chronic treatment with amitriptyline, nortriptyline, duloxetine, or venlafaxine also suppresses neuropathic pain–induced overexpression of TNF-α in dorsal root ganglia [22, 70], and the use of knock-out mice demonstrated that this mechanism is in fact mediated by β_2 adrenergic receptors [22]. The therapeutic relevance of this anti-neuroimmune component (see Fig. 1) is further supported by the fact that etanercept, infliximab, adalimumab, and certolizumab, which are anti-TNF-α biotherapies, as well as an inhibitor of NF-κB, were all effective in alleviating neuropathic allodynia in an animal model [22, 70]. Studies more largely highlighted the contribution of other members of the TNF-α/NF-κB signaling pathway. Toll-like receptors (TLRs) are transmembrane receptors expressed in immune cells and these have been proposed to contribute to neuropathic pain [129]. The antidepressant drug duloxetine indirectly displays an anti-TLR2 action by regulating its overexpression in a neuropathic pain context, through a β_2 adrenoceptor-mediated mechanism [70]. Similarly, prolonged treatment with duloxetine in a rat model of diabetic neuropathic pain rescued the overexpression of TLR4, Myd88, and NF-κB in the spinal cord [150].

Overall, these studies converge to support the idea that antidepressant drugs can attenuate the neuroimmune component contributing to neuropathic pain.

2.7 Excitatory and Inhibitory Transmission

Antidepressant drugs could also affect the central sensitization of nociceptive pathways, which accompanies neuropathic pain [72]. In vitro studies showed that tricyclic antidepressant drugs have some affinity for NMDA receptors [12, 112, 124], and amitriptyline may act as a noncompetitive NMDA antagonist [65]. However, this direct action may require high doses that are not relevant to the clinical use and usual therapeutic doses of these drugs. On the other hand, an indirect action on glutamatergic transmission has also been shown. Thus, amitriptyline and mirtazapine attenuate the spinal upregulation of NR2B induced by oxaliplatin [76, 116] (see Fig. 2). Amitriptyline could also favor the expression of excitatory amino-acid transporters in the spinal cord after spared nerve injury [80] (see Fig. 2), and reduce glutamate and nitrite levels supraspinally in the hippocampus [39] (see Fig. 3). Neuropathic pain is also accompanied by a deficit in spinal inhibitory transmission [72, 108, 133], and early amitriptyline treatment after sciatic nerve ligation can prevent changes in dorsal horn $GABA_B$ receptor function [88] (see Fig. 2). However, studies are still needed to understand whether the action of aminergic antidepressant drugs on the excitatory or inhibitory transmissions is directly involved in their pain-relieving effect.

2.8 *Other Mechanistic Insights*

In models of chemo-induced neuropathy, duloxetine inhibits poly ADP-ribose polymerase cleavage (PARP) and the recruitment of tumor suppressor gene p53, recruitment of members of the Bax/Bcl2 family of apoptosis regulation and recruitment of apoptopic markers, which would contribute to counteract paclitaxel and oxaliplatin-induced oxidative stress and apoptosis in the dorsal root ganglia [78, 137] (see Fig. 1). In the dorsal root ganglia and spinal cord of rats with diabetic neuropathic pain, the tetracyclic mirtazapine corrects transient receptor potential vanilloid 1 (TRPV1) and acid-sensing ion channel 1 (ASIC1) levels [13] (see Figs. 1 and 2). Similarly, the SSNRI duloxetine suppresses TRPV1 upregulation in the lumbar dorsal root ganglia and spinal cord, reduces calcitonin gene–related peptide (CGRP) and substance P release, as well as spinal glia activation and TNF-α expression in a model of paclitaxel-induced neuropathic pain [136] (see Fig. 2). Centrally, sustained systemic imipramine treatment decreases phosphorylated levels of the transcription factor cAMP-responsive element-binding protein (CREB) and of PLCγ-1 in the spinal cord of mice with partial sciatic nerve ligation [71] (see Fig. 2), which differs from the known effect of the antidepressant drugs in supraspinal limbic structures [99, 130]. At supraspinal level, chronic milnacipran in rats decreases c-Fos recruitment in the anterior cingulate cortex of rats with sciatic nerve constriction [128] (see Fig. 3), a cortical structure important to the aversiveness of ongoing pain [11, 60, 109] and to the anxiodepressive consequences of neuropathic pain [11, 119, 145].

While many converging actors are now evidenced to explain how antidepressant drugs can act on neuropathic pain, studies are still needed to provide more detailed mechanistic insights linking these various actors between them and to the actual pain-relieving effect.

3 Conclusion

Some antidepressant drugs are first-line treatments for specific pain conditions, namely, neuropathic pain and fibromyalgia. The mechanisms underlying such action have been partly unraveled thanks to preclinical research on neuropathic pain models that highlighted peripheral and spinal actors. These findings suggest the copresence of at least two main mechanistic components: one directly acting on descending controls of pain to exert a spinal antinociceptive action, and one acting through cellular plasticity to interfere with neuroimmune and other molecular alterations accompanying neuropathic pain. Many aspects of these treatments however remain to be detailed. For example, the precise cascade of events linking the different actors of the cellular plasticity is far from being described. Moreover, all data presented in this chapter were obtained in models of neuropathic conditions, and it is unclear whether the same mechanism(s) generalize to all neuropathic pain etiologies or would also apply to fibromyalgia. These mechanistic insights were also obtained by studying the therapeutic action on nociceptive responses rather than pain itself. The supraspinal component of antidepressant drugs, which could theoretically also interfere with the emotional perception of pain, has been less explored even though it may constitute a third mechanism by which monoamine-uptake inhibitors may act on some pain pathologies.

Acknowledgments This work was supported by the Centre National de la Recherche Scientifique and the Université de Strasbourg (contract UPR3212), by the Agence Nationale de la Recherche (Euridol ANR-17-EURE-0022), and by the Fondation Université de Strasbourg. The authors declare no conflict of interest.

References

1. Allen TJ, Cooper ME, Lan HY (2004) Use of genetic mouse models in the study of diabetic nephropathy. Curr Diab Rep 4:435–440
2. Arsenault A, Sawynok J (2009) Perisurgical amitriptyline produces a preventive effect on afferent hypersensitivity following spared nerve injury. Pain 146:308–314
3. Attal N, Cruccu G, Haanpaa M, Hansson P, Jensen TS, Nurmikko T, Sampaio C, Sindrup S, Wiffen P (2006) EFNS guidelines on pharmacological treatment of neuropathic pain. Eur J Neurol 13:1153–1169
4. Attal N, Fermanian C, Fermanian J, Lanteri-Minet M, Alchaar H, Bouhassira D (2008) Neuropathic pain: are there distinct subtypes depending on the aetiology or anatomical lesion? Pain 138:343–353
5. Attal N, Cruccu G, Baron R, Haanpaa M, Hansson P, Jensen TS, Nurmikko T (2010) EFNS guidelines on the pharmacological treatment of neuropathic pain: 2010 revision. Eur J Neurol 17:1113–1e88
6. Austin PJ, Moalem-Taylor G (2010) The neuro-immune balance in neuropathic pain: involvement of inflammatory immune cells, immune-like glial cells and cytokines. J Neuroimmunol 229:26–50
7. Bannister K, Dickenson AH (2016) What do monoamines do in pain modulation? Curr Opin Support Palliat Care 10:143–148
8. Baptista-De-Souza D, Di Cesare Mannelli L, Zanardelli M, Micheli L, Nunes-De-Souza RL, Canto-De-Souza A, Ghelardini C (2014) Serotonergic modulation in neuropathy induced by oxaliplatin: effect on the 5HT2C receptor. Eur J Pharmacol 735:141–149
9. Barber MJ, Starmer CF, Grant AO (1991) Blockade of cardiac sodium channels by amitriptyline and diphenylhydantoin. Evidence for two use-dependent binding sites. Circ Res 69:677–696
10. Baron R, Binder A, Wasner G (2010) Neuropathic pain: diagnosis, pathophysiological mechanisms, and treatment. Lancet Neurol 9:807–819
11. Barthas F, Sellmeijer J, Hugel S, Waltisperger E, Barrot M, Yalcin I (2015) The anterior cingulate cortex is a critical hub for pain-induced depression. Biol Psychiatry 77:236–245

12. Barygin OI, Nagaeva EI, Tikhonov DB, Belinskaya DA, Vanchakova NP, Shestakova NN (2017) Inhibition of the NMDA and AMPA receptor channels by antidepressants and antipsychotics. Brain Res 1660:58–66
13. Bektur E, Sahin E, Ceyhan E, Donmez DB, Canbek M, Baycu C, Can OD (2019) Beneficial effect of mirtazapine on diabetes-induced hyperalgesia: involvement of TRPV1 and ASIC1 channels in the spinal cord and dorsal root ganglion. Neurol Res 41:544–553
14. Benbouzid M, Choucair-Jaafar N, Yalcin I, Waltisperger E, Muller A, Freund-Mercier MJ, Barrot M (2008) Chronic, but not acute, tricyclic antidepressant treatment alleviates neuropathic allodynia after sciatic nerve cuffing in mice. Eur J Pain 12:1008–1017
15. Benbouzid M, Gaveriaux-Ruff C, Yalcin I, Waltisperger E, Tessier LH, Muller A, Kieffer BL, Freund-Mercier MJ, Barrot M (2008) Delta-opioid receptors are critical for tricyclic antidepressant treatment of neuropathic allodynia. Biol Psychiatry 63:633–636
16. Benbouzid M, Pallage V, Rajalu M, Waltisperger E, Doridot S, Poisbeau P, Freund-Mercier MJ, Barrot M (2008) Sciatic nerve cuffing in mice: a model of sustained neuropathic pain. Eur J Pain 12:591–599
17. Bennett GJ, Xie YK (1988) A peripheral mononeuropathy in rat that produces disorders of pain sensation like those seen in man. Pain 33:87–107
18. Bennett MI, Attal N, Backonja MM, Baron R, Bouhassira D, Freynhagen R, Scholz J, Tolle TR, Wittchen HU, Jensen TS (2007) Using screening tools to identify neuropathic pain. Pain 127:199–203
19. Binder W, Mousa SA, Sitte N, Kaiser M, Stein C, Schafer M (2004) Sympathetic activation triggers endogenous opioid release and analgesia within peripheral inflamed tissue. Eur J Neurosci 20:92–100
20. Bohm C, Newrzella D, Herberger S, Schramm N, Eisenhardt G, Schenk V, Sonntag-Buck V, Sorgenfrei O (2006) Effects of antidepressant treatment on gene expression profile in mouse brain: cell type-specific transcription profiling using laser microdissection and microarray analysis. J Neurochem 97(Suppl 1):44–49
21. Bohren Y, Karavelic D, Tessier LH, Yalcin I, Gaveriaux-Ruff C, Kieffer BL, Freund-Mercier MJ, Barrot M (2010) Mu-opioid receptors are not necessary for nortriptyline treatment of neuropathic allodynia. Eur J Pain 14:700–704
22. Bohren Y, Tessier LH, Megat S, Petitjean H, Hugel S, Daniel D, Kremer M, Fournel S, Hein L, Schlichter R, Freund-Mercier MJ, Yalcin I, Barrot M (2013) Antidepressants suppress neuropathic pain by a peripheral beta2-adrenoceptor mediated anti-TNFalpha mechanism. Neurobiol Dis 60:39–50
23. Bomholt SF, Mikkelsen JD, Blackburn-Munro G (2005) Antinociceptive effects of the antidepressants amitriptyline, duloxetine, mirtazapine and citalopram in animal models of acute, persistent and neuropathic pain. Neuropharmacology 48:252–263
24. Bouhassira D, Attal N, Alchaar H, Boureau F, Brochet B, Bruxelle J, Cunin G, Fermanian J, Ginies P, Grun-Overdyking A, Jafari-Schluep H, Lanteri-Minet M, Laurent B, Mick G, Serrie A, Valade D, Vicaut E (2005) Comparison of pain syndromes associated with nervous or somatic lesions and development of a new neuropathic pain diagnostic questionnaire (DN4). Pain 114:29–36
25. Cavaletti G, Tredici G, Petruccioli MG, Donde E, Tredici P, Marmiroli P, Minoia C, Ronchi A, Bayssas M, Etienne GG (2001) Effects of different schedules of oxaliplatin treatment on the peripheral nervous system of the rat. Eur J Cancer 37:2457–2463
26. Ceredig RA, Pierre F, Doridot S, Alduntzin U, Salvat E, Yalcin I, Gaveriaux-Ruff C, Barrot M, Massotte D (2018) Peripheral delta opioid receptors mediate duloxetine antiallodynic effect in a mouse model of neuropathic pain. Eur J Neurosci 48:2231–2246
27. Chen M, Hoshino H, Saito S, Yang Y, Obata H (2017) Spinal dopaminergic involvement in the antihyperalgesic effect of antidepressants in a rat model of neuropathic pain. Neurosci Lett 649:116–123

28. Choucair-Jaafar N, Salvat E, Freund-Mercier MJ, Barrot M (2014) The antiallodynic action of nortriptyline and terbutaline is mediated by beta(2) adrenoceptors and delta opioid receptors in the ob/ob model of diabetic polyneuropathy. Brain Res 1546:18–26
29. Costigan M, Scholz J, Woolf CJ (2009) Neuropathic pain: a maladaptive response of the nervous system to damage. Annu Rev Neurosci 32:1–32
30. De Vry J, Kuhl E, Franken-Kunkel P, Eckel G (2004) Pharmacological characterization of the chronic constriction injury model of neuropathic pain. Eur J Pharmacol 491:137–148
31. Decosterd I, Woolf CJ (2000) Spared nerve injury: an animal model of persistent peripheral neuropathic pain. Pain 87:149–158
32. Deng L, Lee WH, Xu Z, Makriyannis A, Hohmann AG (2016) Prophylactic treatment with the tricyclic antidepressant desipramine prevents development of paclitaxel-induced neuropathic pain through activation of endogenous analgesic systems. Pharmacol Res 114:75–89
33. Dharmshaktu P, Tayal V, Kalra BS (2012) Efficacy of antidepressants as analgesics: a review. J Clin Pharmacol 52:6–17
34. Dierich A, Kieffer BL (2004) Knockout mouse models in pain research. Methods Mol Med 99:269–299
35. Duman CH, Duman RS (2015) Spine synapse remodeling in the pathophysiology and treatment of depression. Neurosci Lett 601:20–29
36. Dworkin RH, O'Connor AB, Backonja M, Farrar JT, Finnerup NB, Jensen TS, Kalso EA, Loeser JD, Miaskowski C, Nurmikko TJ, Portenoy RK, Rice AS, Stacey BR, Treede RD, Turk DC, Wallace MS (2007) Pharmacologic management of neuropathic pain: evidence-based recommendations. Pain 132:237–251
37. El Mansari M, Guiard BP, Chernoloz O, Ghanbari R, Katz N, Blier P (2010) Relevance of norepinephrine-dopamine interactions in the treatment of major depressive disorder. CNS Neurosci Ther 16:e1–e17
38. Estebe JP, Myers RR (2004) Amitriptyline neurotoxicity: dose-related pathology after topical application to rat sciatic nerve. Anesthesiology 100:1519–1525
39. Farghaly HSM, Elbadr MM, Ahmed MA, Abdelhaffez AS (2019) Effect of single and repeated administration of amitriptyline on neuropathic pain model in rats: focus on glutamatergic and upstream nitrergic systems. Life Sci 233:116752
40. Finnerup NB, Sindrup SH, Jensen TS (2010) The evidence for pharmacological treatment of neuropathic pain. Pain 150:573–581
41. Finnerup NB, Attal N, Haroutounian S, Mcnicol E, Baron R, Dworkin RH, Gilron I, Haanpaa M, Hansson P, Jensen TS, Kamerman PR, Lund K, Moore A, Raja SN, Rice AS, Rowbotham M, Sena E, Siddall P, Smith BH, Wallace M (2015) Pharmacotherapy for neuropathic pain in adults: a systematic review and meta-analysis. Lancet Neurol 14:162–173
42. Fukushima FB, Barros GA, Marques ME, Vidal EI, Ganem EM (2009) The neuraxial effects of intraspinal amitriptyline at low concentrations. Anesth Analg 109:965–971
43. Gahimer J, Wernicke J, Yalcin I, Ossanna MJ, Wulster-Radcliffe M, Viktrup L (2007) A retrospective pooled analysis of duloxetine safety in 23,983 subjects. Curr Med Res Opin 23:175–184
44. Gaveriaux-Ruff C, Kieffer BL (2002) Opioid receptor genes inactivated in mice: the highlights. Neuropeptides 36:62–71
45. Gerner P, Mujtaba M, Sinnott CJ, Wang GK (2001) Amitriptyline versus bupivacaine in rat sciatic nerve blockade. Anesthesiology 94:661–667
46. Gerner P, Mujtaba M, Khan M, Sudoh Y, Vlassakov K, Anthony DC, Wang GK (2002) N-phenylethyl amitriptyline in rat sciatic nerve blockade. Anesthesiology 96:1435–1442
47. Gerner P, Kao G, Srinivasa V, Narang S, Wang GK (2003) Topical amitriptyline in healthy volunteers. Reg Anesth Pain Med 28:289–293
48. Goldstein DJ, Lu Y, Detke MJ, Lee TC, Iyengar S (2005) Duloxetine vs. placebo in patients with painful diabetic neuropathy. Pain 116:109–118

49. Hajhashemi V, Banafshe HR, Minaiyan M, Mesdaghinia A, Abed A (2014) Antinociceptive effects of venlafaxine in a rat model of peripheral neuropathy: role of alpha2-adrenergic receptors. Eur J Pharmacol 738:230–236
50. Hamon M, Gozlan H, Bourgoin S, Benoliel JJ, Mauborgne A, Taquet H, Cesselin F, Mico JA (1987) Opioid receptors and neuropeptides in the CNS in rats treated chronically with amoxapine or amitriptyline. Neuropharmacology 26:531–539
51. Handa J, Sekiguchi M, Krupkova O, Konno S (2016) The effect of serotonin-noradrenaline reuptake inhibitor duloxetine on the intervertebral disk-related radiculopathy in rats. Eur Spine J 25:877–887
52. Hiroki T, Suto T, Saito S, Obata H (2017) Repeated administration of amitriptyline in neuropathic pain: modulation of the noradrenergic descending inhibitory system. Anesth Analg 125:1281–1288
53. Ho KY, Huh BK, White WD, Yeh CC, Miller EJ (2008) Topical amitriptyline versus lidocaine in the treatment of neuropathic pain. Clin J Pain 24:51–55
54. Hoshino H, Obata H, Saito S (2015) Antihyperalgesic effect of duloxetine and amitriptyline in rats after peripheral nerve injury: influence of descending noradrenergic plasticity. Neurosci Lett 602:62–67
55. Huang Y, Chen H, Chen SR, Pan HL (2023) Duloxetine and amitriptyline reduce neuropathic pain by inhibiting primary sensory input to spinal dorsal horn neurons via alpha1- and alpha2-adrenergic receptors. ACS Chem Neurosci 14:1261–1277
56. Hughes S, Hickey L, Donaldson LF, Lumb BM, Pickering AE (2015) Intrathecal reboxetine suppresses evoked and ongoing neuropathic pain behaviours by restoring spinal noradrenergic inhibitory tone. Pain 156:328–334
57. Ishibashi T, Sueto D, Yoshikawa Y, Koga K, Yamaura K, Tsuda M (2022) Identification of spinal inhibitory interneurons required for attenuating effect of duloxetine on neuropathic allodynia-like signs in rats. Cells 11:4051
58. Ito S, Suto T, Saito S, Obata H (2018) Repeated administration of duloxetine suppresses neuropathic pain by accumulating effects of noradrenaline in the spinal cord. Anesth Analg 126:298–307
59. Jett MF, Mcguirk J, Waligora D, Hunter JC (1997) The effects of mexiletine, desipramine and fluoxetine in rat models involving central sensitization. Pain 69:161–169
60. Johansen JP, Fields HL, Manning BH (2001) The affective component of pain in rodents: direct evidence for a contribution of the anterior cingulate cortex. Proc Natl Acad Sci USA 98:8077–8082
61. Jordan BA, Trapaidze N, Gomes I, Nivarthi R, Devi LA (2001) Oligomerization of opioid receptors with beta 2-adrenergic receptors: a role in trafficking and mitogen-activated protein kinase activation. Proc Natl Acad Sci USA 98:343–348
62. Jordan BA, Gomes I, Rios C, Filipovska J, Devi LA (2003) Functional interactions between mu opioid and alpha 2A-adrenergic receptors. Mol Pharmacol 64:1317–1324
63. Katsuyama S, Sato K, Yagi T, Kishikawa Y, Nakamura H (2013) Effects of repeated milnacipran and fluvoxamine treatment on mechanical allodynia in a mouse paclitaxel-induced neuropathic pain model. Biomed Res 34:105–111
64. Kawakami M, Tamaki T, Weinstein JN, Hashizume H, Nishi H, Meller ST (1996) Pathomechanism of pain-related behavior produced by allografts of intervertebral disc in the rat. Spine (Phila Pa 1976) 21:2101–2107
65. Kiefer G, Fischer W, Feuerstein TJ (1999) Effects of amitriptyline, amitriptylinoxide, doxepine and clozapine on N-methyl-D-aspartate-evoked release of [H-3]-acetylcholine in rat caudatoputamen. Arzneimittelforschung-Drug Res 49:820–823
66. Kim SH, Chung JM (1992) An experimental model for peripheral neuropathy produced by segmental spinal nerve ligation in the rat. Pain 50:355–363
67. Kim Y, Kwon SY, Jung HS, Park YJ, Kim YS, In JH, Choi JW, Kim JA, Joo JD (2019) Amitriptyline inhibits the MAPK/ERK and CREB pathways and proinflammatory cytokines through A3AR activation in rat neuropathic pain models. Korean J Anesthesiol 72:60–67

68. Krell HV, Leuchter AF, Cook IA, Abrams M (2005) Evaluation of reboxetine, a noradrenergic antidepressant, for the treatment of fibromyalgia and chronic low back pain. Psychosomatics 46:379–384
69. Kremer M, Salvat E, Muller A, Yalcin I, Barrot M (2016) Antidepressants and gabapentinoids in neuropathic pain: mechanistic insights. Neuroscience 338:183–206
70. Kremer M, Yalcin I, Goumon Y, Wurtz X, Nexon L, Daniel D, Megat S, Ceredig RA, Ernst C, Turecki G, Chavant V, Theroux JF, Lacaud A, Joganah LE, Lelievre V, Massotte D, Lutz PE, Gilsbach R, Salvat E, Barrot M (2018) A dual noradrenergic mechanism for the relief of neuropathic allodynia by the antidepressant drugs duloxetine and amitriptyline. J Neurosci 38:9934–9954
71. Kusuda R, Ravanelli MI, Cadetti F, Franciosi A, Previdelli K, Zanon S, Lucas G (2013) Long-term antidepressant treatment inhibits neuropathic pain-induced CREB and PLCgamma-1 phosphorylation in the mouse spinal cord dorsal horn. J Pain 14:1162–1172
72. Latremoliere A, Woolf CJ (2009) Central sensitization: a generator of pain hypersensitivity by central neural plasticity. J Pain 10:895–926
73. Le Cudennec C, Castagne V (2014) Face-to-face comparison of the predictive validity of two models of neuropathic pain in the rat: analgesic activity of pregabalin, tramadol and duloxetine. Eur J Pharmacol 735:17–25
74. Lenzen S (2008) The mechanisms of alloxan- and streptozotocin-induced diabetes. Diabetologia 51:216–226
75. Leung L, Cahill CM (2010) TNF-alpha and neuropathic pain--a review. J Neuroinflammation 7:27
76. Liu X, Zhang G, Dong L, Wang X, Sun H, Shen J, Li W, Xu J (2013) Repeated administration of mirtazapine attenuates oxaliplatin-induced mechanical allodynia and spinal NR2B up-regulation in rats. Neurochem Res 38:1973–1979
77. Llorca-Torralba M, Borges G, Neto F, Mico JA, Berrocoso E (2016) Noradrenergic locus coeruleus pathways in pain modulation. Neuroscience 338:93–113
78. Lu Y, Zhang P, Zhang Q, Yang C, Qian Y, Suo J, Tao X, Zhu J (2020) Duloxetine attenuates paclitaxel-induced peripheral nerve injury by inhibiting p53-related pathways. J Pharmacol Exp Ther 373:453–462
79. Lynch ME, Clark AJ, Sawynok J, Sullivan MJ (2005) Topical 2% amitriptyline and 1% ketamine in neuropathic pain syndromes: a randomized, double-blind, placebo-controlled trial. Anesthesiology 103:140–146
80. Mao QX, Yang TD (2010) Amitriptyline upregulates EAAT1 and EAAT2 in neuropathic pain rats. Brain Res Bull 81:424–427
81. Marchand F, Alloui A, Chapuy E, Jourdan D, Pelissier T, Ardid D, Hernandez A, Eschalier A (2003) Evidence for a monoamine mediated, opioid-independent, antihyperalgesic effect of venlafaxine, a non-tricyclic antidepressant, in a neurogenic pain model in rats. Pain 103:229–235
82. Marchand F, Ardid D, Chapuy E, Alloui A, Jourdan D, Eschalier A (2003) Evidence for an involvement of supraspinal delta- and spinal mu-opioid receptors in the antihyperalgesic effect of chronically administered clomipramine in mononeuropathic rats. J Pharmacol Exp Ther 307:268–274
83. Marchand F, Pelissier T, Eschalier A, Ardid D, Alloui A, Soto-Moyano R, Mondaca M, Laurido C, Constandil L, Hernandez A (2004) Blockade of supraspinal 5-HT1A receptors potentiates the inhibitory effect of venlafaxine on wind-up activity in mononeuropathic rats. Brain Res 1008:288–292
84. Matsuoka H, Suto T, Saito S, Obata H (2016) Amitriptyline, but not pregabalin, reverses the attenuation of noxious stimulus-induced analgesia after nerve injury in rats. Anesth Analg 123:504–510
85. Matsuzawa-Yanagida K, Narita M, Nakajima M, Kuzumaki N, Niikura K, Nozaki H, Takagi T, Tamai E, Hareyama N, Terada M, Yamazaki M, Suzuki T (2008) Usefulness of

antidepressants for improving the neuropathic pain-like state and pain-induced anxiety through actions at different brain sites. Neuropsychopharmacology 33:1952–1965
86. Max MB, Culnane M, Schafer SC, Gracely RH, Walther DJ, Smoller B, Dubner R (1987) Amitriptyline relieves diabetic neuropathy pain in patients with normal or depressed mood. Neurology 37:589–596
87. Max MB, Lynch SA, Muir J, Shoaf SE, Smoller B, Dubner R (1992) Effects of desipramine, amitriptyline, and fluoxetine on pain in diabetic neuropathy. N Engl J Med 326:1250–1256
88. Mccarson KE, Ralya A, Reisman SA, Enna SJ (2005) Amitriptyline prevents thermal hyperalgesia and modifications in rat spinal cord GABA(B) receptor expression and function in an animal model of neuropathic pain. Biochem Pharmacol 71:196–202
89. Mclachlan EM, Janig W, Devor M, Michaelis M (1993) Peripheral nerve injury triggers noradrenergic sprouting within dorsal root ganglia. Nature 363:543–546
90. Megat S, Bohren Y, Doridot S, Gaveriaux-Ruff C, Kieffer BL, Freund-Mercier MJ, Yalcin I, Barrot M (2015) kappa-Opioid receptors are not necessary for the antidepressant treatment of neuropathic pain. Br J Pharmacol 172:1034–1044
91. Mico JA, Ardid D, Berrocoso E, Eschalier A (2006) Antidepressants and pain. Trends Pharmacol Sci 27:348–354
92. Mika J, Jurga AM, Starnowska J, Wasylewski M, Rojewska E, Makuch W, Kwiatkowski K, Malek N, Przewlocka B (2015) Effects of chronic doxepin and amitriptyline administration in naive mice and in neuropathic pain mice model. Neuroscience 294:38–50
93. Millan MJ (2002) Descending control of pain. Prog Neurobiol 66:355–474
94. Minami K, Tamano R, Kasai E, Oyama H, Hasegawa M, Shinohara S, Asaki T (2018) Effects of duloxetine on pain and walking distance in neuropathic pain models via modulation of the spinal monoamine system. Eur J Pain 22:355–369
95. Mitsi V, Terzi D, Purushothaman I, Manouras L, Gaspari S, Neve RL, Stratinaki M, Feng J, Shen L, Zachariou V (2015) RGS9-2-controlled adaptations in the striatum determine the onset of action and efficacy of antidepressants in neuropathic pain states. Proc Natl Acad Sci USA 112:E5088
96. Moisset X, Bouhassira D, Attal N (2021) French guidelines for neuropathic pain: an update and commentary. Rev Neurol (Paris) 177:834–837
97. Nardone R, Holler Y, Brigo F, Seidl M, Christova M, Bergmann J, Golaszewski S, Trinka E (2013) Functional brain reorganization after spinal cord injury: systematic review of animal and human studies. Brain Res 1504:58–73
98. Nattel S (1985) Frequency-dependent effects of amitriptyline on ventricular conduction and cardiac rhythm in dogs. Circulation 72:898–906
99. Nestler EJ, Barrot M, Dileone RJ, Eisch AJ, Gold SJ, Monteggia LM (2002) Neurobiology of depression. Neuron 34:13–25
100. Obata H (2017) Analgesic mechanisms of antidepressants for neuropathic pain. Int J Mol Sci 18:2483
101. Obata H, Saito S, Koizuka S, Nishikawa K, Goto F (2005) The monoamine-mediated antiallodynic effects of intrathecally administered milnacipran, a serotonin noradrenaline reuptake inhibitor, in a rat model of neuropathic pain. Anesth Analg 100:1406–1410. table of contents
102. Onal A, Parlar A, Ulker S (2007) Milnacipran attenuates hyperalgesia and potentiates antihyperalgesic effect of tramadol in rats with mononeuropathic pain. Pharmacol Biochem Behav 88:171–178
103. Otto M, Bach FW, Jensen TS, Brosen K, Sindrup SH (2008) Escitalopram in painful polyneuropathy: a randomized, placebo-controlled, cross-over trial. Pain 139:275–283
104. Ozdogan UK, Lahdesmaki J, Mansikka H, Scheinin M (2004) Loss of amitriptyline analgesia in alpha 2A-adrenoceptor deficient mice. Eur J Pharmacol 485:193–196
105. Paoli F, Darcourt G, Cossa P (1960) Preliminary note on the action of imipramine in painful states. Rev Neurol (Paris) 102:503–504
106. Perahia DG, Pritchett YL, Desaiah D, Raskin J (2006) Efficacy of duloxetine in painful symptoms: an analgesic or antidepressant effect? Int Clin Psychopharmacol 21:311–317

107. Pichon X, Wattiez AS, Becamel C, Ehrlich I, Bockaert J, Eschalier A, Marin P, Courteix C (2010) Disrupting 5-HT(2A) receptor/PDZ protein interactions reduces hyperalgesia and enhances SSRI efficacy in neuropathic pain. Mol Ther 18:1462–1470
108. Prescott SA (2015) Synaptic inhibition and disinhibition in the spinal dorsal horn. Prog Mol Biol Transl Sci 131:359–383
109. Qu C, King T, Okun A, Lai J, Fields HL, Porreca F (2011) Lesion of the rostral anterior cingulate cortex eliminates the aversiveness of spontaneous neuropathic pain following partial or complete axotomy. Pain 152:1641–1648
110. Ramer MS, Bisby MA (1998) Normal and injury-induced sympathetic innervation of rat dorsal root ganglia increases with age. J Comp Neurol 394:38–47
111. Rantamaki T, Yalcin I (2016) Antidepressant drug action--from rapid changes on network function to network rewiring. Prog Neuro-Psychopharmacol Biol Psychiatry 64:285–292
112. Reynolds IJ, Miller RJ (1988) Tricyclic antidepressants block N-methyl-D-aspartate receptors: similarities to the action of zinc. Br J Pharmacol 95:95–102
113. Rode F, Brolos T, Blackburn-Munro G, Bjerrum OJ (2006) Venlafaxine compromises the antinociceptive actions of gabapentin in rat models of neuropathic and persistent pain. Psychopharmacology 187:364–375
114. Rowbotham MC, Reisner LA, Davies PS, Fields HL (2005) Treatment response in antidepressant-naive postherpetic neuralgia patients: double-blind, randomized trial. J Pain 6:741–746
115. Saarto T, Wiffen PJ (2007) Antidepressants for neuropathic pain. Cochrane Database Syst Rev 2007:CD005454
116. Sada H, Egashira N, Ushio S, Kawashiri T, Shirahama M, Oishi R (2012) Repeated administration of amitriptyline reduces oxaliplatin-induced mechanical allodynia in rats. J Pharmacol Sci 118:547–551
117. Schmidt-Wilcke T, Clauw DJ (2011) Fibromyalgia: from pathophysiology to therapy. Nat Rev Rheumatol 7:518–527
118. Scholz J, Finnerup NB, Attal N, Aziz Q, Baron R, Bennett MI, Benoliel R, Cohen M, Cruccu G, Davis KD, Evers S, First M, Giamberardino MA, Hansson P, Kaasa S, Korwisi B, Kosek E, Lavand'homme P, Nicholas M, Nurmikko T, Perrot S, Raja SN, Rice ASC, Rowbotham MC, Schug S, Simpson DM, Smith BH, Svensson P, Vlaeyen JWS, Wang SJ, Barke A, Rief W, Treede RD, Classification Committee of the Neuropathic Pain Special Interest Group (2019) The IASP classification of chronic pain for ICD-11: chronic neuropathic pain. Pain 160:53–59
119. Sellmeijer J, Mathis V, Hugel S, Li XH, Song Q, Chen QY, Barthas F, Lutz PE, Karatas M, Luthi A, Veinante P, Aertsen A, Barrot M, Zhuo M, Yalcin I (2018) Hyperactivity of anterior cingulate cortex areas 24a/24b drives chronic pain-induced anxiodepressive-like consequences. J Neurosci 38:3102–3115
120. Seltzer Z, Dubner R, Shir Y (1990) A novel behavioral model of neuropathic pain disorders produced in rats by partial sciatic nerve injury. Pain 43:205–218
121. Sindrup SH, Otto M, Finnerup NB, Jensen TS (2005) Antidepressants in the treatment of neuropathic pain. Basic Clin Pharmacol Toxicol 96:399–409
122. Sluka KA, Clauw DJ (2016) Neurobiology of fibromyalgia and chronic widespread pain. Neuroscience 338:114–129
123. Sorkin LS, Xiao WH, Wagner R, Myers RR (1997) Tumour necrosis factor-alpha induces ectopic activity in nociceptive primary afferent fibres. Neuroscience 81:255–262
124. Stepanenko YD, Boikov SI, Sibarov DA, Abushik PA, Vanchakova NP, Belinskaia D, Shestakova NN, Antonov SM (2019) Dual action of amitriptyline on NMDA receptors: enhancement of Ca-dependent desensitization and trapping channel block. Sci Rep 9:19454
125. Stratinaki M, Varidaki A, Mitsi V, Ghose S, Magida J, Dias C, Russo SJ, Vialou V, Caldarone BJ, Tamminga CA, Nestler EJ, Zachariou V (2013) Regulator of G protein signaling 4 [corrected] is a crucial modulator of antidepressant drug action in depression and neuropathic pain models. Proc Natl Acad Sci USA 110:8254–8259

126. Sud R, Spengler RN, Nader ND, Ignatowski TA (2008) Antinociception occurs with a reversal in alpha 2-adrenoceptor regulation of TNF production by peripheral monocytes/macrophages from pro- to anti-inflammatory. Eur J Pharmacol 588:217–231
127. Suzuki T, Ueta K, Tamagaki S, Mashimo T (2008) Antiallodynic and antihyperalgesic effect of milnacipran in mice with spinal nerve ligation. Anesth Analg 106:1309–1315. table of contents
128. Takeda R, Watanabe Y, Ikeda T, Abe H, Ebihara K, Matsuo H, Nonaka H, Hashiguchi H, Nishimori T, Ishida Y (2009) Analgesic effect of milnacipran is associated with c-Fos expression in the anterior cingulate cortex in the rat neuropathic pain model. Neurosci Res 64:380–384
129. Thakur KK, Saini J, Mahajan K, Singh D, Jayswal DP, Mishra S, Bishayee A, Sethi G, Kunnumakkara AB (2017) Therapeutic implications of toll-like receptors in peripheral neuropathic pain. Pharmacol Res 115:224–232
130. Thome J, Sakai N, Shin K, Steffen C, Zhang YJ, Impey S, Storm D, Duman RS (2000) cAMP response element-mediated gene transcription is upregulated by chronic antidepressant treatment. J Neurosci 20:4030–4036
131. Tzadok R, Ablin JN (2020) Current and emerging pharmacotherapy for fibromyalgia. Pain Res Manag 2020:6541798
132. Ucel UI, Can OD, Demir Ozkay U, Ozturk Y (2015) Antihyperalgesic and antiallodynic effects of mianserin on diabetic neuropathic pain: a study on mechanism of action. Eur J Pharmacol 756:92–106
133. Ueda H (2006) Molecular mechanisms of neuropathic pain-phenotypic switch and initiation mechanisms. Pharmacol Ther 109:57–77
134. Vialou V, Feng J, Robison AJ, Nestler EJ (2013) Epigenetic mechanisms of depression and antidepressant action. Annu Rev Pharmacol Toxicol 53:59–87
135. Vilardaga JP, Nikolaev VO, Lorenz K, Ferrandon S, Zhuang Z, Lohse MJ (2008) Conformational cross-talk between alpha2A-adrenergic and mu-opioid receptors controls cell signaling. Nat Chem Biol 4:126–131
136. Wang J, Zhou F, Zhang S, Mao M, Feng S, Wang X (2022) Participation of transient receptor potential vanilloid 1 in the analgesic effect of duloxetine for paclitaxel induced peripheral neuropathic pain. Neurosci Lett 773:136512
137. Wang M, Zhang L, Liu X, Qiu S, Xu R, Yang C, Lu Y, Zhang P, Yan M, Zhu J (2022) Duloxetine alleviates oxaliplatin-induced peripheral neuropathy by regulating p53-mediated apoptosis. Neuroreport 33:437–444
138. Watson CP, Evans RJ, Reed K, Merskey H, Goldsmith L, Warsh J (1982) Amitriptyline versus placebo in postherpetic neuralgia. Neurology 32:671–673
139. Watson CP, Chipman M, Reed K, Evans RJ, Birkett N (1992) Amitriptyline versus maprotiline in postherpetic neuralgia: a randomized, double-blind, crossover trial. Pain 48:29–36
140. Wattiez AS, Libert F, Privat AM, Loiodice S, Fialip J, Eschalier A, Courteix C (2011) Evidence for a differential opioidergic involvement in the analgesic effect of antidepressants: prediction for efficacy in animal models of neuropathic pain? Br J Pharmacol 163:792–803
141. West SJ, Bannister K, Dickenson AH, Bennett DL (2015) Circuitry and plasticity of the dorsal horn – toward a better understanding of neuropathic pain. Neuroscience 300:254–275
142. Wolfe F, Ross K, Anderson J, Russell IJ, Hebert L (1995) The prevalence and characteristics of fibromyalgia in the general population. Arthritis Rheum 38:19–28
143. Yalcin I, Choucair-Jaafar N, Benbouzid M, Tessier LH, Muller A, Hein L, Freund-Mercier MJ, Barrot M (2009) Beta(2)-adrenoceptors are critical for antidepressant treatment of neuropathic pain. Ann Neurol 65:218–225
144. Yalcin I, Tessier LH, Petit-Demouliere N, Doridot S, Hein L, Freund-Mercier MJ, Barrot M (2009) Beta2-adrenoceptors are essential for desipramine, venlafaxine or reboxetine action in neuropathic pain. Neurobiol Dis 33:386–394
145. Yalcin I, Barthas F, Barrot M (2014) Emotional consequences of neuropathic pain: insight from preclinical studies. Neurosci Biobehav Rev 47:154–164

146. Yalcin I, Megat S, Barthas F, Waltisperger E, Kremer M, Salvat E, Barrot M (2014) The sciatic nerve cuffing model of neuropathic pain in mice. J Vis Exp 89:51608
147. Yoshimura M, Furue H (2006) Mechanisms for the anti-nociceptive actions of the descending noradrenergic and serotonergic systems in the spinal cord. J Pharmacol Sci 101:107–117
148. Zhang JM, Li H, Liu B, Brull SJ (2002) Acute topical application of tumor necrosis factor alpha evokes protein kinase A-dependent responses in rat sensory neurons. J Neurophysiol 88:1387–1392
149. Zhang TT, Xue R, Zhu L, Li J, Fan QY, Zhong BH, Li YF, Ye CY, Zhang YZ (2016) Evaluation of the analgesic effects of ammoxetine, a novel potent serotonin and norepinephrine reuptake inhibitor. Acta Pharmacol Sin 37:1154–1165
150. Zhou DM, Zhuang Y, Chen WJ, Li W, Miao B (2018) Effects of duloxetine on the toll-like receptor 4 signaling pathway in spinal dorsal horn in a rat model of diabetic neuropathic pain. Pain Med 19:580–588
151. Zhu J, Wei X, Feng X, Song J, Hu Y, Xu J (2008) Repeated administration of mirtazapine inhibits development of hyperalgesia/allodynia and activation of NF-kappaB in a rat model of neuropathic pain. Neurosci Lett 433:33–37

Index

O. V. Mortensen (ed.), *Monoamine Transporters*, Advances in Neurobiology 46,
https://doi.org/10.1007/978-3-031-96364-3

9783031963636